REDESIGNING LIFE

John Parrington is an Associate Professor in Molecular and Cellular Pharmacology at the University of Oxford, and a Tutorial Fellow in Medicine at Worcester College, Oxford. He has published over 80 peer-reviewed articles in science journals including *Nature, Current Biology, Journal of Cell Biology, Journal of Clinical Investigation, The EMBO Journal, Development, Developmental Biology*, and *Human Reproduction*. He has extensive experience writing popular science, having published articles in *The Times, The Guardian, New Scientist, Chemistry World*, and *The Biologist*. He has also written science reports for the Wellcome Trust, British Council, and Royal Society. He is the author of *The Deeper Genome* (OUP, 2015)

Praise for *Redesigning Life*

'There is a revolution going on in the life sciences, one that has already transformed scientific discovery and will soon change medicine. It could even see us altering the ecosystem in a precise, targeted way. This revolution has a name—CRISPR—and the key part of John Parrington's *Redesigning Life* is a good summary of the gene-editing technique that lies behind the acronym.'
New Scientist

'a thorough and comprehensive account of the methodologies for altering life that have been or are being developed, and the directions that they may take in future.'
Nature

'painstakingly researched... the examples themselves are clear, concise and often arresting... an engaging and comprehensive introduction.'
LSE Business Review

'a comprehensive history of the research and discoveries underlying genome editing, as well as a broad coverage of research in the present day.'
Bionews

'John Parrington's engaging and thoughtful book explains the science behind recent rapid advances in genetic engineering that mean it is increasingly possible to enact precise changes at a molecular level... Parrington's clear descriptions and diagrams combined with interesting snippets from the narrative and biography of scientific discovery make the science behind these developments readily accessible.'
Socialist Review

JOHN PARRINGTON

redesigning life

how genome editing will transform the world

OXFORD
UNIVERSITY PRESS

Great Clarendon Street, Oxford, OX2 6DP,
United Kingdom

Oxford University Press is a department of the University of Oxford.
It furthers the University's objective of excellence in research, scholarship,
and education by publishing worldwide. Oxford is a registered trade mark of
Oxford University Press in the UK and in certain other countries

First published 2016
Revised paperback edition published in 2020

Impression: 1

Published in the United States of America by Oxford University Press
198 Madison Avenue, New York, NY 10016, United States of America

British Library Cataloguing in Publication Data
Data available

Library of Congress Control Number: 2020934976

ISBN 978–0–19–876683–4

Printed and bound in Great Britain by
Clays Ltd, Elcograf S.p.A.

ACKNOWLEDGEMENTS

I would like to thank a number of people who have helped bring this, my second book, to fruition. I owe a particular debt to Latha Menon, my editor at Oxford University Press, who always expertly balances her critical comments with great encouragement about the work at hand. I would also like to thank Jenny Nugee of the OUP editorial team, for her help on a multitude of practical matters, and Elizabeth Stone at Bourchier Limited, Chandrakala Chandrasekaran at SPi GLobal and Christina Fleischer at OUP for their meticulous copy-editing of the book. I gained some extremely valuable insights and suggestions for changes to the text from Margarida Ruas, as well as some very helpful comments from three anonymous reviewers who read my original proposal and one who read a late draft of the book. Margarida Ruas also produced a superb set of line drawings. I also owe many thanks to Anthony Morgan for producing the author photo for the book cover. For their expert assistance with marketing and publicity, and answers to my many questions on this front, I would like to thank Phil Henderson and Kate Farquhar-Thomson of OUP. I owe thanks also to friends and colleagues who have indulged my many speculations about the new technologies described in the book, as well as providing helpful feedback and suggestions. My final set of thanks is to my family, for providing such a warm and lovely home environment that meant so much to me during the long hours spent on researching and writing. Finally, there is a debt that is hard to express adequately in words, and that is to my mother. As well as her encouragement at every stage of my life, through both ups and downs, she also communicated her great love of books and reading, her ability to look for the positive in every situation, and a questioning and enquiring attitude to life. It is to her memory that I dedicate this book.

CONTENTS

LIST OF PLATES

LIST OF FIGURES

All figures by Margarida Ruas

Introduction

The Gene Revolution

Imagine if living things were as easy to modify as a computer Word file. Think of the possibilities if the genetic code of organisms could be tweaked a little here, changed a bit there, to give organisms slightly different properties, or even radically different ones. Let's take things a step further. Picture life forms whose very genetic recipe was manufactured in a chemistry lab using new components never seen before on Earth. In such a world, microorganisms might be adapted to produce new types of fuel, and farm animals or plants engineered to produce leaner meat or juicier fruit, but also to withstand extremes of temperature or drought to meet the increasing demands of climate change. Medical research too would be transformed if we could easily modify the genomes of different species in order to generate mutant animals to model human disease.

If genomes really could be modified like computer text, medicine itself might look very different. Instead of people having to suffer terrible diseases like cystic fibrosis or muscular dystrophy,[1] the gene defects associated with these conditions could be corrected in the affected tissues. If tinkering with genomes were both precise and efficient, such conditions might become a thing of the past, with genetic defects corrected at the embryo stage or even earlier, in the eggs or sperm within the gonads. Of course, this could pose the question of what was considered a defect. For instance, would the availability of personalized genome information and the ability to manipulate this information lead to a situation where parents clamoured to have offspring engineered to kick a football like Cristiano Ronaldo, compose music like Mozart, or have Einstein's scientific genius? And if life forms in the future become truly synthetic, does that mean that one day we'll have synthetic people too?

But there are other, more troubling future scenarios that can be imagined if genetic modification were to become as easy as cutting and pasting a Word file. For what is to stop such technology being used to engineer new lethal types of viruses, or synthetic life forms escaping and taking over the world? And how could we ensure that new types of genetically modified (GM) food—whether animal or plant—were safe to eat? Could such plants pose a risk to the environment? And what about the welfare of the modified animals? This is also potentially an issue if scientists develop new types of mutant animals to model human disease. Will this lead to pain and suffering in a wider range of species, including our biological cousins—monkeys and other primates? And if researchers developed GM primates to study how the human brain works, could this lead to a *Planet of the Apes*-type scenario? For that matter, could the technology be used as a tool to recreate long-extinct organisms such as woolly mammoths or fearsome dinosaurs like *Tyrannosaurus rex*?

The prospect of being able to modify life in such a routine manner may either excite or horrify, depending on your viewpoint. Yet although these imagined future scenarios sound like science fiction rather than fact, it's time we discussed the new technologies that are transforming our ability to manipulate life. For while the scenarios above are fictional, thanks to the new technologies—particularly one called 'genome editing'[2] but also a new branch of science called 'synthetic biology'—many soon may not be.[3]

Of course, you could be forgiven for thinking genome manipulation is nothing new. After all, isn't that the scientific basis for all those debates about GM crops, gene therapy, or designer babies? And indeed, we've had the technology to cut and paste gene sequences in a test tube since the 1970s,[4] while in the 1980s it became possible to modify the genome of an organism as complex as a mouse.[5] But the difference between genome editing and past forms of genetic engineering is a bit like comparing letterpress printing with the first word processor, or modern motor cars with horse-drawn carriages, in terms of scope and potential. So, as Jennifer Doudna of the University of California, Berkeley, a pioneer of a type of genome editing called CRISPR/CAS9, recently put it: 'It's a technology that gives scientists a capability that has not been in our hands in the past. We're basically now able to have a molecular scalpel for genomes. All the technologies in the past were sort of like sledgehammers.'[6] Quite why this is so, is a topic that will be explored in detail in this book.

A Scientific Revolution

Perhaps most astonishing is the speed at which the new genetic engineering is taking place.[7] Despite genome editing being a recent development, it's been taken up and applied in various different ways at a pace that has surprised many scientists. And it is for this reason that *Science* magazine named CRISPR/CAS9 its 'Breakthrough of the Year' in 2015, instead of the Pluto flyby or the discovery of a new human ancestor.[8] 'We all kind of marvel at how fast this took off as a technology,' said Doudna. 'There's just a really tremendous feeling of excitement for the potential of CRISPR.'[9] One way the technology is having a major impact in biomedical science is through a new-found ability to modify genomes from species ranging from simple bacteria through to mammals—not only mice, but large animals like pigs and monkeys. At the same time, the capacity of genome editing to genetically modify animals and plants important for agriculture looks set to have a huge impact on food production.

Yet amidst the excitement, the new technology is also creating controversy, precisely because of its greatly enhanced accuracy and power compared to past genetic engineering approaches. That isn't only due to the potential impact in already controversial areas like GM crops and animal models of disease, but because genome editing is equally applicable to human cells. In November 2015, the technology was used to treat a baby with an aggressive form of child leukaemia, producing what doctors called a 'near miracle' recovery.[10] More controversially, genome editing has been used to modify the genome of human embryos for the first time in history. In the hands of some scientists, such an approach is being used to study the genes involved in human embryo development, which could help in the diagnosis and treatment of miscarriage and birth disorders, and there was never any intention to implant the edited embryos into women. However, in December 2018, Jiankui He, a scientist at Southern University of Science and Technology in Shenzhen, China, shocked the world when he revealed that he had edited the genomes of two human embryos that were later born as twin baby girls.[11] The twins have been engineered to be resistant to HIV for reasons that I will explain when I discuss this case in more detail later in this book. In fact there were all sorts of problems with his scientific approach and he also broke the law in creating the twins; as a consequence, he was recently jailed for three years. Yet his actions have effectively breached what was thought to be a 'no-go area' in medical ethics, with all sorts of implications for the future.

The range of potential applications of genome editing are such that Dustin Rubinstein, who's applying this technology at the University of Wisconsin-Madison, believes 'it's really going to just empower us to have more creativity...to get into the sandbox and have more control over what you build. You're only limited by your imagination.'[6] Yet, as Jennifer Doudna recently pointed out, 'Great things can be done with the power of technology—and there are things you would not want done. Most of the public does not appreciate what is coming.'[12] Surely, given the scale of the scientific revolution now underway, this is a deficit that needs correcting if the wider public are to influence the way genome editing is employed. But taking part in such a debate requires a proper understanding of the underlying science and what distinguishes the new technology from previous approaches to modifying life. Such was the stimulus for writing this book.

There are other major developments taking place in biotechnology too. For instance, take the new field of optogenetics.[13] This employs laser beams to stimulate—or inhibit—nerve cells in the brain of a mouse, allowing scientists to better understand how this organ works, but also to control behaviour. This approach is revolutionizing neuroscience by making it possible to uncover how particular nerve cells contribute to complex brain functions like learning, memory, pain, and pleasure. 'Optogenetics is not just a flash in the pan,' believes neuroscientist Robert Gereau of Washington University in Saint Louis. 'It allows us to do experiments that were not doable before. This is a true game changer like few other techniques in science.'[14] In addition, scientists are identifying other ways to manipulate nerve cell activity with electromagnetism and ultrasound. What's more, this technology has recently been used to manipulate other cell types, such as those of the heart or the pancreatic cells that secrete insulin.

Or consider the latest in stem cell technology. The development of stem cells with 'pluripotent' potential—the ability to give rise to any cell type in the body—is a huge growth area in biomedicine.[15] Pluripotent cells can be derived from human embryos, which has caused some controversy, but recently even ordinary human skin cells have been transformed into such pluripotent stem cells.[16] Perhaps most startling has been the demonstration that such cells can self-organize into structures similar to organs like the gut, pancreas, liver, eye, and even brain.[17] Currently such technologies are primarily being used to further understanding of brain function or organ development, but the potential direct clinical applications are huge. And the ability to grow human organs in culture—both for research and as a potential replacement for a diseased or aged heart, liver, or pancreas—is itself being boosted by the ability to modulate gene activity that is now made possible by genome editing.

Synthetic biology is an even more radical approach to redesigning life. This approach has led to the creation of the first artificial bacterial genome and yeast chromosome.[18] The long-term aim of such research is to use such artificial constructions as a starting point for more radical changes to genomes than can be achieved even with genome editing. Meanwhile, other synthetic biologists are seeking to change the very structure of DNA and of the proteins that it encodes.[19] In the future, this approach may make it possible to radically redesign bacteria so that they can perform important practical functions, like producing fuel or food, detecting the presence of toxins in the body, or acting as construction materials. And the application of synthetic biology to more complex organisms might one day create animals or plants that are totally resistant to viruses.

These exciting advances pose serious ethical issues that shouldn't be ducked. For instance, in agriculture, how can we ensure that genome editing is used to benefit the majority of the world's population and not just increase the profits of giant companies? In biomedicine, such technology might revolutionize the understanding and treatment of disease, but what are the risks? This question is particularly controversial when we consider the possibility of using genome editing to genetically modify a human embryo, for instance for the treatment of disease. For, as well as the potential risks, would such embryo modification eventually lead to other genetic changes being introduced into a new human life, such as those that alter looks, skills, or personality? Optogenetics is revealing new information about how the brain works but could it one day be used as a form of mind control? And while synthetic biology might lead to the production of novel life forms with a variety of practical uses, how can we be sure these new life forms will not overrun the planet and cause harm?

Such are the themes of the book. But it's now time to step back a little and, in the first chapter, consider whether, for all the novelty of genome editing and other technologies for transforming life, the human capacity for modifying life might not be quite so new after all.

1

Natural Born Mutants

Super-sized salmon. Glow-in-the-dark cats. Goats that produce spider silk in their milk.[20] It's not surprising many people view genetic engineering with suspicion, given the way we often encounter it in the media. We will look at some weird and wonderful uses of this technology, but in this book I'll be making the case for genetic engineering as a vital tool for understanding life and manipulating it for human benefit. And I hope to demonstrate that, far from being of interest only to scientists, these new powers are something everyone should know about. Very soon, their influence will affect us all. Yet they represent just the latest step in a unique human characteristic: our ability to consciously transform the world around us. This capacity is based on two key human traits: our ability to make and use tools, and our self-conscious awareness, which allows us to plan how to employ such tools.[21]

Now some may argue that a scientist manipulating genes in a test tube or creating a GM plant or animal, is quite different from a prehistoric cave dweller fashioning a spear from a stick and sharpened flint. But is it true that genome manipulation by humans is a totally new phenomenon? Certainly this is the case if we only consider the direct modification of genetic material, using tools first developed in the 1970s, a topic we'll shortly explore in Chapter 2. But indirect manipulation of genomes is something we humans have been engaged upon for many thousands of years, through our domestication of the various plants and animals that provide us with food, clothing, transport; through aesthetic pleasures like gardening or sport; and even through our yearning for company, in the form of pets.

We domesticated other organisms by taking wild species and transforming them in size, looks, behaviour, and other properties—ultimately by altering their genes. And while this was done without knowledge of the material basis of inheritance, recent advances in genome analysis mean we can now pinpoint in precise molecular detail the genetic changes that occurred during the agricultural revolution that transformed human society 12,000 years ago.[22] Such changes resulted from humans selecting certain wild variants over others and, in the process, creating rice or wheat from wild grasses or the domestic pig from a wild boar. Yet although the agricultural revolution was the main driver for such changes, it's not the first example of humans transforming the genome of a species. For that, we must look even further back, to a time when all humans lived as small groups of hunter gatherers. It was then that we adopted one particular wild species which not only transformed our hunting capacity but also evolved into a faithful companion—a status it has maintained to this day. By now you've probably guessed that I'm talking about the dog.

From Wolf to Hound

An image I spotted recently on the Internet shows a dog reclining on a sofa, above the words: 'We were wolves once, wild and wary, stealth and cunning, then we noticed you had couches.'[23] Now if we forgo some historical inaccuracy on the home furniture front, unless a cave alcove covered in furs counts as a couch, this image isn't so far off the truth. It correctly identifies the dog's ancestor as a wolf, and draws attention to the behavioural changes that accompanied the progression from wary wild animal to the lolling pet of today. That dogs evolved from wolves has long been known. What modern science can contribute is insight into how and when this evolution took place. It can also reveal the molecular changes that made our canine companions the way they are today.

We know the point in prehistory when wolves were first domesticated from archaeology—through the discovery of burials by humans of wolf skeletons that already show dog-like features like a smaller frame and shorter jaw[24]—and from comparing the genomes of different breeds of dogs with those of wolves. Since genomes acquire random mutations in their deoxyribonucleic acid (DNA) over time, such a comparison is used as a 'molecular clock' to estimate the time a species has been in existence. Genetic comparisons of existing human beings have helped us piece together the chronology and geography of our evolution, and has shown that modern humans first evolved in East Africa 200,000–150,000 years ago.[25] Similarly, genetic studies suggest dogs originated in South-East Asia and have been part of human society for at least 33,000 years.[26,27] What's more, DNA analysis by Love Dalén and colleagues at the Swedish Museum of Natural History of a wolf bone from the Taimyr Peninsula in northern Siberia that was radiocarbon dated to be 35,000 years old, indicated that the DNA already showed signs of genetic differences associated with modern dogs. Dalén believes this means that dogs were either domesticated at that time, or the population split into modern wolves and a wild ancestor of modern dogs that later became extinct. According to him, 'the simplest explanation is that dogs were domesticated at the time of the split.'[28]

Exactly how the wild wolf started on its journey to faithful human companion is a matter of some debate. One theory is that humans and wolves first came into contact while hunting. Both species tend to hunt in packs, and while they may have encountered each other pursuing a common prey, sometimes wolves must have hunted humans and vice versa. Perhaps during one hunt a wolf cub became separated from its pack, or the latter were killed and a human hunter took a cub home with him. Then, in a reverse situation to Rudyard Kipling's *The Jungle Book*, where the lost child Mowgli is raised by wolves,[29] the wolf cub would have grown up in human society. As the cub grew into an adult wolf, eventually it might have become too dangerous to remain in the settlement. Assuming this happened successively to wolves with slightly different temperaments, gradually, by natural selection, those animals that fitted best into human society would have been the ones that remained in the settlement and bred with similar tame wolves living there.

A second theory is probably not best to ponder if you're planning on eating soon. For this proposes that humans and wolves first came into close contact because of the latter's fondness for consuming the refuse that a settlement's occupants would throw out,[30] including the human faeces that accumulated on the edge of primitive settlements.[31] Based on this unsavoury starting point, the wolves that were more docile and least afraid of humans would thereby be most likely to consume such refuse: eventually this would lead to interactions with humans and the tame wolves being accepted into the life of the settlement.

For either theory to work, our human ancestors would have had to recognize some value in keeping such a canine companion, and here the hunting skills of wolves may have been a key factor. Subsequently, through natural selection, the human settlements that hunted using tame wolves would have been more likely to survive because of their greater efficiency at bringing home meat. In line with this possibility, even today some tribes in Nicaragua depend on dogs to detect prey, while traditional moose hunters in Arctic regions bring home 56 per cent more prey when accompanied by dogs.[30]

As well as ensuring that a bond developed between humans and their tame wolves, this joint activity between the two species may have helped select wolves most suited to this role, and possibly also humans that could work best with their wolf helpers. For instance, Pat Shipman of Pennsylvania State University believes working with wolves could have led to the evolution of a distinctive human characteristic—our eyes with their white sclera, coloured iris, and black pupil—which contrast with those of other primates that have a dark sclera. Since wolves' eyes also have a white sclera, Shipman thinks this feature evolved in humans to aid communication with our new canine companions. It is much easier to work out what humans/animals are gazing at if they have a white sclera. Shipman argues that this change probably aided communication between humans themselves, since 'it provides a very useful form of non-verbal communication and would have been of immense help to early hunters. They would have been able to communicate silently but very effectively.'[32]

Shipman believes the new-found alliance between humans and wolves was so important that it may have been a key factor in the disappearance of the Neanderthals, who became extinct between 30,000 and 40,000 years ago. Explanations for this extinction range from the effects of climate change, genocide by modern humans, or a battle for scarce resources between Neanderthals and humans—which our species eventually won thanks to our superior skills. Shipman favours the last scenario but with a twist, namely that tame wolves gave us the edge in this competition. 'Early wolf-dogs would have tracked and harassed animals like elk and bison and would have hounded them until they tired,' she says. 'Then humans would have killed them with spears or bows and arrows. This meant the dogs did not need to approach these large cornered animals to finish them off—often the most dangerous part of a hunt—while humans didn't have to expend energy in tracking and wearing down prey.'[32] While speculations like these are thought-provoking, they are difficult to confirm in the absence of a machine to replay prehistory. Yet what genome analysis is revealing with much higher certainty are the precise molecular changes that occurred as wolves evolved into dogs.

This analysis suggests that dogs are a product of a process called neoteny, which also seems to have been central to human evolution. Through neoteny, an evolving species retains juvenile features into adulthood. So human adults have many physical features—a flatter and broader face, hairless body, large head-to-body ratio—in common with young, but not older, apes.[33] This slowing of development was crucial in human evolution, allowing a greater capacity for learning. And recent studies indicate that, through genetic changes that promote neoteny, adult dogs have also retained an inclination to learn into adulthood that is absent in wolves. This helped them gain important skills to aid their human masters, like tracking and fetching prey, as well as a love of learning and performing tricks that may have enhanced their attractiveness as pets. Importantly, while dogs often compete over objects when playing with other dogs, they are usually more cooperative when the play partner is a human.[34] This may have helped the dog–owner relationship develop.

Another important characteristic in dogs is their capacity to eat a varied diet. Genetic analysis by Kerstin Lindblad-Toh of Uppsala University, Sweden, has revealed that dogs differ from wolves in several genes that aid the digestion of starches into sugars.[35] As a consequence, although our pet pooches love steak, they're also happy wolfing down rice or potatoes. This change would have been important in easing the adaptation of dogs to life around humans because it meant they could consume scraps that humans found unpalatable and left uneaten. Clearly, a number of complex changes—both behavioural and physical—occurred during the transition from wolf to dog, but what is clear to conservation biologist Peter Smith, chief executive of the British Wildwood Trust, is that 'the deep, deep connection has existed between man and wolves—now our dogs—for many tens of thousands of years and that is why we love dogs so much. They are part of our own evolution into a modern society.'[36]

A Feline Interloper

While dogs may be humankind's oldest friend, cats are the other main rivals for our affections, so much so that they are now the world's most popular pet, outnumbering dogs by around three to one.[37] This popularity is undoubtedly helped by the fact that cats are far more self-reliant than dogs—so they require virtually no training; they can groom themselves; and they may be left alone without pining for their owner, but nevertheless (usually) greet us affectionately when we return home.

Some behavioural differences between cats and dogs have a basis in the biology of the two species, namely wild cats and wolves, from which these pets evolved. However, they also reflect the distinct ways in which these species entered our lives and their subsequent evolution. For while humans have cohabited with dogs for at least 30,000 years, genome analysis indicates that cats only joined our households 10,000–12,000 years ago. And here there's a clear link with the period during the agricultural revolution when humans first began to grow enough cereal plants to accumulate a surplus, and thereby a need to store grain.[38]

The accumulation of grain in this way made it possible, for the first time, for humans to begin living together in large groups, and led to the birth of cities. But grain stores also attracted rats and mice. And wild cats that preyed on these rodents must have made a sufficient difference to the ability of people in the new cities to keep their grain safe for them to be welcomed into these places of human habitation. The Natufians, who inhabited the Levant region of the Middle East and are widely regarded to have invented agriculture, seem to have been the first people to attract the wild Arabian cat into their lives in this way.[37] Subsequently, cats in ancient Egypt became so important they were worshipped as gods.[39]

While the initial relationship of cats to people in these first cities was probably like today's urban foxes, which are adapted to a human environment while retaining an essential wildness, the usefulness and subsequent evolution of wild cats led to them being assimilated into human households. And recent analysis of domesticated cat genomes in comparison to those of wild cats has provided some fascinating insights into this evolution. Such analysis has revealed that domestic cats differ in genes that control aggressive behaviour, formation of memories, and the ability to learn from both fear- and reward-based stimuli.[40] According to 'anthrozoologist' John Bradshaw of Bristol University, such genetic changes 'give domestic kittens the ability to become sociable with people—but if they don't encounter humans until they're over 10 weeks old, they can remain as "wild" as any wildcat'.[41] In addition, like dogs, our feline companions tolerate a more varied diet than just meat, making it easier for them to fit into a domestic lifestyle by eating table scraps. So cats have a longer intestine than their wild cousins and enhanced activity of genes that aid digestion of fatty plant matter.[40]

In other respects though, cats are more similar to wild cats than dogs are to wolves, or as Bradshaw puts it, modern cats have 'three paws firmly planted in the wild'.[42] This may be due to their more recent evolution, but probably also reflects the fact that, while the first cats showed their usefulness to city-dwelling people because of their rodent-catching skills, there was probably no human selection for these skills analogous to the way dogs were selected for specific characteristics that were useful during a hunt. Maybe this explains why cats can be affectionate or aloof, serene or savage, since they are that much closer to wild cats than dogs are to wolves.

Taming the Earth

While cats protected our grain stores in ancient times, that grain had to come from plants, and a key feature of the agricultural revolution was the cultivation of wild plant species in a fixed location for the first time. Humans are not the only organisms that cultivate other species—certain ants, beetles, and termites 'farm' fungus[43]—but we are the only ones that do so on such a widespread scale and with so many different farmed varieties. Importantly, like our technologies in general, farming is constantly evolving. Thus, from the moment humans began to cultivate wild plants we also began to select for specific characteristics in these plants, and so the evolution of the different domesticated species began.

Now, through genome analysis, we're beginning to understand the molecular basis of the evolution of staple crops like rice, maize, and wheat.[44] In fact, all such cereal crops are varieties of grass, and at the genetic level they're much more similar than might appear from their superficial differences. Human selection led to the development of these distinct types of plants with very different attributes as foodstuffs. In some cases, different populations have also made selections based on distinct properties of the same plant: so in some parts of the world people traditionally eat long-grain rice, while others prefer short-grain, sticky varieties.[44]

Moreover, the characteristics for which a particular plant is bred have on occasion changed over time. So lettuce was originally grown in ancient Egypt for its seeds, which were used to produce oil, and the original plant formed no head of leaves; only subsequently was there selection for plants with large leaves.[45] Meanwhile, cabbage, which is part of the mustard family, was originally so toxic it was eaten in small quantities for medicinal properties. It was only through selective breeding that it evolved into the non-toxic plant we can eat today with impunity.[46] Such changes in what are viewed as important attributes even in the same plant pose a challenge for genomic studies of crop plants. Quite contradictory tendencies may have been selected at different points in a plant's history, and this complicates the genetic analysis.[44]

As well as identifying the genetic differences that underlie the specific characteristics of a particular domestic plant, genome analysis has also led to more general insights into the molecular mechanisms underlying domestication. So we now know that every domestic plant went through an evolutionary process called 'domestication syndrome' that made it easier to grow as a crop. For instance, in rice this syndrome included loss of shattering (the seeds don't break off from the central stalk before harvesting), an increase in seed size, and the seeds all germinating simultaneously so they can be harvested at once.[47] Similar features have been identified in the evolution of other cereal crops.

Of course, agriculture is not just about plants. The genomes of animals and birds kept for meat production, such as pigs, sheep, cows, goats, and chickens, have all now been sequenced and comparisons are underway, where possible, with the genomes of the wild species from which the domestic animal originated. Such a comparison between domestic pigs and wild boars by Merete Fredholm and colleagues at Copenhagen University has revealed the genetic differences that underlie one very noticeable domestic pig characteristic—its much greater length compared to a wild boar.[48] Longer pigs are an advantage to farmers because this means more pork on each individual animal. Fredholm explains how 'there are certain genetic variants that make the pig grow longer. Originally, it was a rare gene mutation, but today it occurs in just about all domesticated pigs.[48] One such mutation led to domestic pigs having an increased number of bones in their spine, compared to wild boars.

You might assume that the domestication of animals like pigs for meat production would only lead to changes in the genomes of the animals that are eaten. Yet new research has also identified some fascinating genetic changes in the humans who first began to farm pigs for their meat. Scientists, including Matthew Cobb of the University of Manchester, studied genetic differences across human populations in genes coding for so-called 'olfactory receptors' which detect different odours, and found that one gene, olfactory receptor 7D4 (OR7D4), which enables us to detect a smell called androstenone—a sex hormone produced by male pigs and found in pork—

differs between different populations.[49] Depending on their particular OR7D4 gene variant, people respond differently to androstenone—so some individuals find it foul, some sweet, and others cannot smell it at all. However, the concentration of these distinct types shows an intriguing distribution in different parts of the world. 'We found the variants that enable you to smell this odour are very highly represented in Africa which is where we all came from,' says Cobb. 'In Europe and Asia there is an increased tendency to not be able to smell this odour due to changes in their DNA.'[49] But these regions are precisely the places where pigs were first domesticated, and Cobb believes there is a good reason for this: 'Our hypothesis is that these mutations enabled people to eat pork without being disgusted by boar meat and that this helped those populations survive.'[49]

While production of meat is a primary reason for the domestication of animals, it's not the only one. So our ancestors bred sheep and goats for their wool, and oxen and horses as means of transport, and here too genomic analysis is providing new insights. For instance, analysis of the genomes of cashmere goats showed that the genes coding for keratins—the main proteins in skin, nails, and hair—and those that regulate hair growth are subtly different in this breed, accounting for their particularly fine wool.[50] Meanwhile, by studying the genomes of different horse breeds, Vera Warmuth and colleagues at Cambridge University have shown that wild horses were first domesticated in the western Eurasian Steppe—modern-day Ukraine and West Kazakhstan—around 6,000 years ago. Michael Hofreiter of York University believes that uncovering the timescale and geography of horse domestication in this way will reveal more than just the history of these animals. 'Horse domestication has changed human cultures a lot,' he says. 'It has changed warfare, it has changed transportation. Studying the past of horses can tell us a lot about our own past.'[51]

It's not only domestic farm animals whose genomes are now under scrutiny, but also the microorganisms that cause them disease. Such research is challenging long-held assumptions about the spread of certain diseases. One such assumption is that most diseases that plague humans originally came from farm animals. This is one reason why Jared Diamond of the

University of California, Los Angeles, believes the agricultural revolution was 'the worst mistake in the history of the human race'.[52] The idea that most human diseases are a by-product of agriculture comes from observations that smallpox is similar to cowpox and measles to rinderpest, a disease of cattle, and so on. And, certainly, comparisons of the genomes of life forms that infect humans and farm animals have tended to confirm the idea that agriculture, while generally a blessing, has also been a curse in this respect.[53]

Yet recent studies have also uncovered interesting exceptions in which it seems that humans infected animals first. So while it's long been assumed that tuberculosis, found in both cows and humans, was passed to us from these animals, new genomic evidence indicates that humans first became susceptible to this bacterium and subsequently passed it to cows.[53] Another common assumption is that we first acquired tapeworms from pigs, since we can be infected by these parasites after eating undercooked, contaminated meat. Yet genetic analysis indicates that here too it was humans who first infected these farm animals.[53]

Many New Breeds

Another reason for keeping domestic animals is for primarily aesthetic purposes. I say primarily because a prize-winning dog or cat can clearly be a pet as well as a show object. In fact, pedigree breeding is a relatively recent phenomenon; so the large number of dog breeds we now take for granted date back less than two hundred years, to the Victorian era.[54] The Industrial Revolution, which fuelled the rise of modern capitalism, also, for the first time, generated a significant middle class with both the time and money to indulge themselves.[55] And one form that such indulgence took was breeding specific types of dogs and cats, but also pigeons, and even mice, and presenting these at shows. Driven purely by aesthetic notions, in only a short period of time such competitive breeding created the 400 types of dogs with their fantastic range of size, shape, and looks, from the tiny Chihuahua to the huge Great Dane.[54]

The study of pedigree breeds has played an important role in the biological sciences, for instance stimulating Charles Darwin's theory of evolution by natural selection.[56] His voyage around the world in the HMS *Beagle* is rightly credited with helping him formulate this theory, since it exposed him to many natural varieties of animals and plants.[57] Yet an equally important stimulus was Darwin's observations of the many pigeon breeds being created in Britain by 'pigeon fanciers', for these showed how rapidly size, shape, and form could be influenced by artificial selection. A particular object of selection for such fanciers was the crest—feathers on the neck and head that grow upwards rather than downwards, as in wild pigeons.[58] Michael Shapiro of the University of Utah, who studies the genetics of crest formation, has described the different types. 'Some are small and pointed,' he says. 'Others look like a shell behind the head; some people think they look like mullets. They can be as extreme as an Elizabethan collar.'[58] Darwin's study of pigeons, which started in March 1855, was to be strictly a way of collecting facts about the variation in a domestic species: there was to be 'no amusement'.[59] However, by November of that year he wrote to his friend the geologist Charles Lyell, who was planning a visit: 'I will show you my pigeons! Which are the greatest treat, in my opinion, which can be offered to a human being.'[59] Darwin had been caught up in a craze so popular in mid-nineteenth-century Britain that it crossed the class divide, counting miners, weavers, and Queen Victoria herself amongst its many enthusiasts.

Whitwell Elwin, who reviewed Darwin's unpublished manuscript detailing his ideas about evolution, dismissed the work as a whole, calling it 'a wild & foolish piece of imagination…for an outline it is too much & for a thorough discussion of the question it is not near enough'.[56] However, he liked the section about pigeons, and recommended that Darwin scrap the main manuscript and write a short book about pigeons instead. 'Everybody is interested in pigeons,' he wrote, and a book like this would 'be reviewed in every journal in the kingdom and soon be on every table.'[56] Luckily Darwin and his publisher ignored this advice, and we ended up with *The Origin of Species*, not *The Little Book of Pigeons*. None the less, Darwin's observations of artificial selection and how it produced the many different pigeon breeds was a key step in the evolution of his theory. It led to a recognition that selection could also arise in nature, with competition for scarce resources resulting in the variants in the population which thrive in such circumstances being more likely to survive and produce offspring, which would inherit their traits.[60]

Recently, studies of pedigree breeds have become important in a different way, by providing new insights into the genetic basis of human disease. Since pet dogs and cats share our homes and our food (though less so now in societies in which people buy specialized pet foods), their environment is also more similar to ours than other mammalian species. Most importantly, the huge number of dog and cat varieties are not just different in size, shape, and behaviour, but also have different susceptibility to specific diseases.[61] And now, by analysing the genomes of these different breeds, it's becoming possible to pinpoint the molecular basis of such differences, with importance not only for veterinary, but also human, medicine.

So narcolepsy—the brain disorder that causes a person to suddenly fall asleep at inappropriate times and which can be potentially fatal if, say, driving—is particularly common in Dobermann Pinschers. Genome analysis of this breed identified a link between this condition and a gene that regulates the brain's uptake of a neurotransmitter called hypocretin.[61] Subsequent analysis of brain fluid from human narcoleptics showed that they lacked this chemical. Studies of how hypocretin can block sleep may identify ways to prevent narcolepsy, or conversely lead to treatments for insomnia. Meanwhile, a study led by Leslie Lyons of the University of Missouri found that an important cause of kidney failure in old age—polycystic kidney disease—is associated with mutations in the same gene in both people and cats.[62] And since some cat breeds are susceptible to type 2 diabetes, asthma, and other conditions found in humans, the search is now on for the feline genes associated with these conditions.

The study of naturally occurring mutations in different pedigree breeds is also leading to the identification of genes involved in body morphology. Dwarfism is a defining feature of dog breeds like Dachshunds, Pekingese, and Basset Hounds. In all such breeds, this is due to a change in the gene coding for fibroblast growth factor, which, as its name suggests, is involved in growth.[54] Changes in another gene, coding for bone morphogenesis protein (BMP), are responsible for the differently shaped skulls of sheepdogs, with their long snouts, compared to flat-faced bulldogs.[63] Understanding how BMP performs this role could lead to new insights into the molecular basis of human disorders of the skull and face.

Studies of the behavioural quirks of pets may even lead to new insights into human psychiatric disorders.[64] Dobermann Pinschers are particularly prone to a condition that causes them to chase their tails for hours on end, or suck on a toy or one of their paws so compulsively that it interferes with their sleeping or eating. Such canine compulsive disorder is thought to have similarities with obsessive-compulsive disorder (OCD) in humans. And Border Collies sometimes overreact to loud noises in a manner similar to people with anxiety disorders. But while behavioural quirks are often particular to specific breeds, in other cases a dog will display a behaviour that is unusual for its pedigree. A new project called Darwin's Dogs, run by scientists at the University of Massachusetts, is now trying to identify genetic links to such behavioural characteristics.[64] Miranda Workman of Buffalo, New York, is one pet owner who has enrolled her dogs in the scheme, because she wants to know why her Dutch Shepherd Athena has a jovial side not usually found in this guard dog breed, and why Sherlock, her Jack Russell, is more shy and sensitive than most terriers. 'I have some dogs that don't necessarily fit the stereotype,' she says. 'Is it their environment that's different or are they different? It will be fun to find out why they are that way.'[64]

Natural Born Mutants

Although the study of naturally occurring mutations in dogs and cats for biomedical purposes may be novel, as applied to other species in an experimental setting it's far from a new idea. Indeed, genetics has been intimately bound up with the investigation of mutants from its origins in the late nineteenth century. In the popular consciousness the word mutant tends to conjure up images of bug-eyed monsters, or beings with incredible powers like Spiderman or the Incredible Hulk.[65] However, in scientific terms a mutant is merely an organism with a new characteristic resulting from a change in its DNA.[66] Mutation is a natural consequence of the fact that DNA is subject to environmental insults, whether from radiation or chemicals; another important source of mutation is the DNA copying process.

The discovery of the famous double helix structure of DNA by Watson and Crick in 1953 was important because it immediately indicated how this 'molecule of life' replicates itself.[67] What happens is that the two strands of the helix split in two and a mirror image copy is formed from each, the DNA polymerase enzyme that does this assembling the new strand from the 'nucleotide' units that form the DNA molecule (see Figure 1).

While highly accurate, this replication process occasionally suffers from errors. To combat mutations—whether arising from radiation, chemical damage, or the replication process—organisms ranging from bacteria to humans employ various forms of DNA repair. Just how important such repair mechanisms are is demonstrated by the unfortunate individuals who lack them. People with a condition called xeroderma pigmentosum are highly prone to DNA damage caused by the Sun's UV rays.[68] Even brief

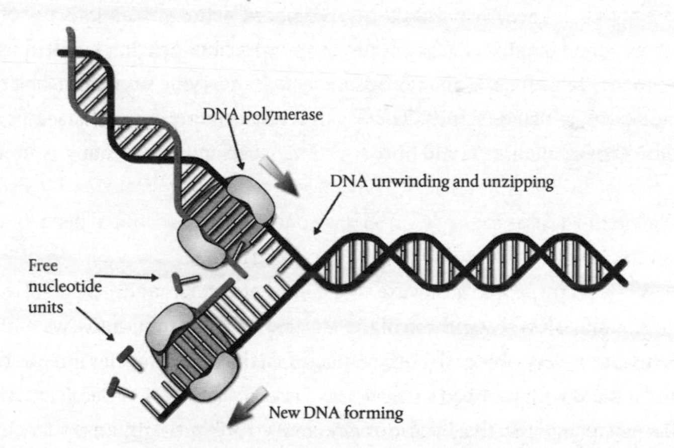

Fig. 1. DNA replication process

exposure to normal sunlight causes the skin of such people to blister, and they are prone to skin cancer. Another DNA repair defect, Cockayne syndrome, leads to premature ageing,[68] while defects in the breast cancer 1 and 2 (BRCA1 and BRCA2) genes, both involved in DNA repair, causes a strong tendency to develop breast and ovarian cancer. This latter defect made the headlines when actress Angelina Jolie made public her decision to have a double mastectomy and removal of her ovaries after her mother and aunt, with whom she shared the BRCA1 gene defect, died prematurely from ovarian and breast cancer, respectively.[69] That defects in the same cellular process can have such varied effects reflects the fact that different types of DNA repair are more important in some parts of the body than others.

Mutations can cause cancer because any change in DNA that results in abnormalities in the proteins regulating cell growth and division may lead to tumour growth. In fact multiple changes in the genome of a cell are generally required for this to happen, which is why our likelihood of succumbing to cancer increases with age, reflecting the gradual accumulation of mutations in our bodies over time.[70] However, if a mutation occurs in the sperm or egg DNA, a susceptibility to a particular type of cancer can be inherited. This was the case with Angelina Jolie, whose inherited BCRA1 gene defect meant that, before her mastectomy, her risk of getting breast cancer before the age of 70 was 65 to 87 per cent compared to a 12.5 per cent risk in most women.[71] In fact, each one of us is likely to be carrying numerous harmful mutations in our genomes. These can result in other serious diseases besides cancer, such as cystic fibrosis.[72] That such mutations don't generally result in disease is because we have every gene in duplicate, and it's sufficient in most cases for a person to have only a single functional gene for this not to cause a problem.

While many people are aware of the link between mutations and disease, there is much less recognition of the fact that, without mutations, we human beings, or indeed any of the other species on the planet, would not exist. To understand why, we need to return to Darwin's theory of natural selection. Darwin recognized that evolutionary change took place because some variants in a population are more suited to survive in a particular environment, and therefore to reproduce, than others.[73] And we now know that mutations in DNA underlie such differences between individuals, some of which enhance survival. This ensures the spread of such mutations through a population and ultimately the evolution of a species.

A Model of Life

Darwin never identified the material basis of inheritance. That was left to the monk Gregor Mendel, who, in the 1860s, studying the inheritance of pea plants, first proposed that inherited characteristics are passed down as discrete 'factors', which we now call genes.[74] Mendel showed that the inheritance of pea characteristics, like purple or white flowers, or short or long stems, followed precise mathematical patterns, which he divided into two types: dominant, in which a single defective gene can cause the characteristic, being passed down from an affected individual to half their offspring; and recessive, in which two defective genes are required, so that two unaffected carriers of the characteristic have a one in four chance of having an affected offspring (see Figure 2). Mendel's work complemented Darwin's ideas by providing a material basis for species variation and its transmission across generations. But Darwin died not knowing of Mendel's findings, which were also overlooked by other scientists. It was only when Mendel's findings were rediscovered in 1900 that Darwinism and Mendelism were united into a single theory of evolution and inheritance.[73]

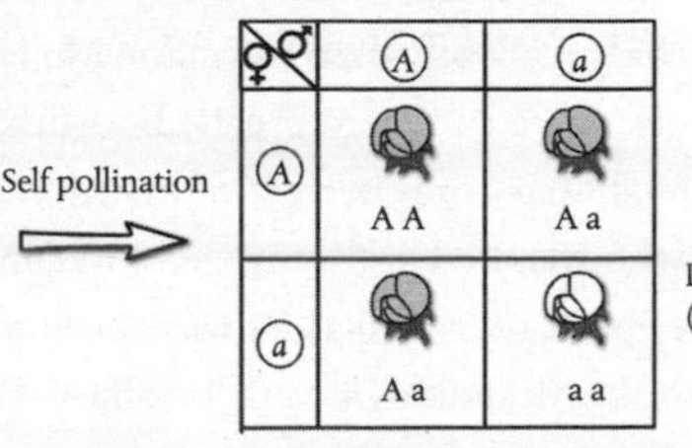

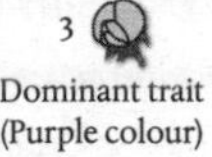

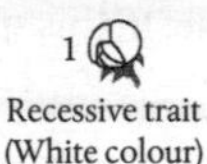

Fig. 2. Recessive and dominant characteristics in Mendel's pea flowers

The realization that genes are made out of DNA had to wait almost another half century, when Oswald Avery at the Rockefeller University in New York showed in 1944 that DNA was the molecule of inheritance,[75] while Watson and Crick identified DNA's double helix structure at Cambridge University in 1953.[76] Not only did this latter discovery reveal how the molecule replicated, subsequent studies showed how it acted as a code for the assembly of proteins. However, long before these discoveries, in the 1900s, Thomas Hunt Morgan, of Columbia University in New York, had recognized that mutant organisms offered a way to study the material basis of genes, since the pattern of inheritance of their abnormal characteristics could be studied, and the genes associated with the mutation identified.

Initially working with mice, Morgan quickly switched his studies to fruit flies (*Drosophila melanogaster*), because he realized that, with their rapid reproduction time and large numbers of offspring, there was far more chance of spotting rare mutants that did arise.[77] And through patient identification and characterization of such mutants, Morgan's team confirmed Mendel's findings in an animal species. They also identified new patterns of inheritance, such as a mutation that caused white eyes instead of the normal red ones, but typically only in males. The discovery that some characteristics are sex-linked led to the realization that the associated genes must be located on the X chromosome. This explained why human disorders like haemophilia generally only affect males: haemophilia is a recessive disorder and females with two X chromosomes are therefore usually protected, since it is unlikely that both copies are abnormal.[78]

Despite these promising initial findings with naturally occurring mutations, fruit fly genetics really only took off when Hermann Muller, a former member of Morgan's team, discovered a way to greatly boost the mutation rate in this species.[79] Muller might have only been 5′ 2″ in height, but he was a larger-than-life character who inspired and outraged in equal measure. For Muller was as passionate about socialism as he was about science, and he seemed to believe that a Bolshevik should be, well, bolshie—a stance that would get him into trouble throughout his life. In Morgan's group, Muller made some important contributions, such as showing that mutations in one gene could alter the expression of another gene, implying that genes interact. However, Muller didn't feel his ideas were given sufficient credit in Morgan's publications, and he moved to set up his own lab at the University of Texas. Here he showed that irradiating fruit flies with X-rays dramatically increased the number of mutants in subsequent offspring. According to James Crow, who was a graduate student at Texas and later became a professor at the University of Wisconsin-Madison: 'In a few months, Muller found more mutant genes than the total from all the Drosophila labs up to that time.'[80]

Unfortunately, Muller's socialist views led to trouble with the authorities. He helped publish a Communist newspaper at his university, and the FBI tracked his activities. In 1932, Muller moved to Russia, expecting to find himself amongst kindred spirits, only to discover that the country was in the grip of Stalin's clampdown on both personal and academic freedom. By the time he left in 1937, many of Muller's students and colleagues had 'disappeared' or been shipped to Siberia, and he was lucky not to meet a similar fate, as traditional genetics was increasingly seen as a 'bourgeois deviation' in the new totalitarian state.

Despite these troubles, Muller's greatest triumph was still to come. In 1945, he was awarded the Nobel Prize.[81] The award not only recognized the importance of Muller's findings for basic science, but also reflected increasing awareness of the dangerous effects of radiation on human genes.[79] This was shown in practice by the tragically early death of Marie Curie, who, with her husband Pierre, had isolated the naturally radioactive elements polonium and radium.[82] Describing these studies, Marie wrote how 'one of our joys was to go into our workroom at night; we then perceived on all sides the feebly luminous silhouettes of the bottles or capsules containing our products.... The glowing tubes looked like faint, fairy lights.'[83] Marie paid a terrible price for her lack of awareness of the health risks associated with radiation. She succumbed to aplastic anaemia, a cancer of the blood brought on by massive exposure to radiation during her work, which would result in her death in 1934.[82] The devastating effects of radiation on human beings was demonstrated on a far greater scale by the dropping of atomic bombs by the US Air Force on the Japanese cities of Hiroshima and Nagasaki in the same year that Muller was awarded his Nobel Prize.[84]

The foundations of experimental genetics were built upon Morgan, Muller, and their colleagues' studies of the fruit fly, and this organism remains highly important today in biomedical research. Studies of embryo development in this species have identified many genes involved in important processes in human beings. The most famous are the homeotic (HOX) genes that regulate the patterning of the body in both flies and humans,[85] but we have also learned a great deal about genes that regulate the development and function of the brain and nervous system from fly studies.[86] Indeed, the continuing importance of this species was shown by a recent study that used fruit fly larvae to capture the activity of an entire central nervous system in a complex animal for the first time.[87] The study, by Philipp Keller and colleagues at the Howard Hughes Medical Institute in Ashburn, Virginia, used a new technique called light-sheet microscopy, which illuminates a specimen with laser light from both sides, while twin cameras record images from front and back. The researchers genetically modified the larva's nerve cells—using techniques that we'll explore in Chapter 3—to make them fluoresce when they fire. 'By imaging different parts of the nervous system at the same time, we can see how behaviours are controlled and then build models of how it all works,' says Keller.[87] This approach makes it possible to study how the brain and nerves work simultaneously to generate behaviour, and may provide new insights into ways to treat spinal injuries in people.

Fancy Mice

While fruit fly studies have furthered our understanding of many basic biological processes, ultimately scientists need to study a mammalian 'model' of human health and disease, given that we are mammals ourselves. And the species that has been the favoured choice in this respect is the mouse. In addition to its rapid reproductive cycle and small size, the establishment of the mouse as a model organism was helped by the existence of naturally occurring mutant mice thanks to the nineteenth-century interest of breeding 'fancy' mice with differently coloured coats and other characteristics as a hobby.[88] Mendel himself first began studying inheritance of coat colour in mice, breeding these in his living quarters.[89] However, the local religious leader, Bishop Schaffgotsch, was outraged that a monk who had taken the vow of chastity was encouraging—and watching—rodent sex, and ordered Mendel to 'stop the work with the smelly creatures'.[90] In response, Mendel turned to cultivating peas instead, remarking that it was lucky that the bishop 'did not understand that plants also had sex!'[89] So it was that pea plants, not mice, became the first model organism of genetics.

When geneticists did begin studying mice in the early twentieth century, they were greatly helped by the activities of a woman called Abbie Lathrop.[91] She initially trained as a teacher but gave up this vocation because of chronic ill health. This didn't, however, prevent her beginning a new career, breeding fancy mice. These turned out to be of great interest not only to mice 'fanciers', but also to geneticists. The business was such a success that at one point Lathrop had more than 11,000 mice.[91] The animals were fed oats and crackers; Lathrop got through one and a half tons of oats and 12 barrels of crackers each month. She also paid local children seven cents an hour to clean the cages.[92] But, most importantly, Lathrop kept careful records of the different mouse breeds, and these would later prove vital for scientists interested in determining the inheritance pattern of interesting mouse characteristics.

At one point Lathrop noticed that some mouse breeds were particularly prone to developing lesions on their skin.[91] She sent samples to several scientists asking for advice about their origin, and one of these, Leo Loeb at Pennsylvania University, diagnosed the lesions as malignant. Lathrop and Loeb's joint interest in the genetic basis of this cancer susceptibility subsequently developed into a valuable collaboration. Amongst the important findings the pair made was the discovery that removing the ovaries from mice susceptible to mammary gland tumours reduced the incidence of such tumours. This finding had eventual relevance for treatment of breast cancer in humans, as one way of treating this cancer is to block the effects of the hormone oestrogen, which is secreted by the ovaries and inhibited by the anti-cancer drug tamoxifen, also known as Nolvadex®. Such was the importance of Lathrop's mice that, when she died in 1918, many were used to populate a new mouse breeding and research institute founded by Clarence Little at Bar Harbor, Maine,[93] despite him once patronizingly describing Lathrop as a 'talented pet-shop owner'.[90] The institute, now known as the Jackson Laboratory, continues to be the world's largest supplier of inbred mouse breeds to this day.

Enhancing Abnormality

Although scientists have learned a huge amount from studying naturally occurring mouse breeds, nature sometimes needs a helping hand, as Muller had shown in fruit flies. But although X-rays were shown to induce mutations in mice as early as 1923, the logistics of inducing large numbers of mutations by this method and subsequently trying to determine the genetic basis of such mutations seemed too daunting at this time.[94] More recently, the use of artificial mutagenesis to create and study mutations in mice has taken on a new lease of life, because the sequencing of the mouse genome makes it much easier to identify the genetic basis of mutants. Instead of employing X-rays, a mutagenic chemical ethylnitrosourea (ENU) is used.[94] By treating male mice with this chemical, then mating them with females, many mutant offspring can be generated. These then undergo a number of tests to detect abnormalities; mice are weighed and measured, X-rayed to detect skeletal defects, and their blood chemical composition analysed. Tests are also carried out to detect abnormalities in vision, hearing, and behaviour.

One research area that has particularly benefited from this approach is the study of deafness. Around one in six people in Britain has some kind of hearing loss, including most people over 70, and the number is rising,[95] with a similar situation being the case in the USA.[96] Karen Steel of King's College London has tracked down many of the genes involved by identifying mutant mice that fail to respond to a sound stimulus, or have problems in balance, which can be linked to deafness. Steel likens the quest to identify genes associated with each defect to solving a puzzle. 'You have no idea of what the mechanism might be before you start studying the genetics,' she says. 'So, it's a bit like putting a jigsaw together, or unwrapping a parcel, as you find out what's going on inside.'[97] According to her, 'characterising these mutants taught us many lessons. First, many of the genes that we found had never been linked to deafness before. That told us that there are many different genes that can cause deafness. Second, there are a wide variety of mechanisms that can cause hearing impairment.'[97] The studies showed that genetic defects can cause deafness at birth, but also create susceptibility to hearing loss in later years. Characterization of the genes associated with deafness may lead to greater understanding of the molecular mechanisms those genes regulate, and hopefully to new drugs for treating both congenital and progressive hearing loss.[95]

Although the study of mouse mutants is proving of great value for biomedical research it does raise important issues about animal welfare, since defects in a particular gene could potentially result in an abnormality that causes pain or distress. In fact, surprisingly, many mouse mutations have quite subtle effects on the body, perhaps because the developing embryo compensates for loss of a particular gene by enhancing or repressing the activity of other genes.[98] But this is not always the case. A naturally occurring deaf mutant, the whirler mouse, is so-named because of its rapid circling and head-tossing motions. This odd behaviour is due to a defect in a gene involved in forming hair-like projections in the cochlea, a component of the inner ear.[99] Because these projections play important roles in hearing and balance, study of this mutant has led to important insights into both these processes.

Another naturally occurring mouse mutant eats excessively, becomes severely obese, and develops diabetes, because of a defect in a gene coding for the hormone leptin, which regulates appetite by signalling to the brain that the animal is full.[100] Study of this mutant may reveal new ways to understand and treat obesity and diabetes, currently reaching epidemic proportions in many countries, despite regular calls for people to eat less and exercise more.[101] But it's not surprising some people find such obviously abnormal mouse mutants disturbing, and it's important that scientists can explain the benefits for human health to be gained from studying such mutants, in order to justify their use in biomedical research.

There will be always be people fundamentally opposed to any experimentation on animals no matter what the benefits for medicine. Others may see the potential value of such research, but still feel queasy about this cultivation of abnormality in another mammalian species. Such unease should certainly not be dismissed as irrelevant or irrational; it is based on a reasonable wish that other species be treated with respect and dignity. The deliberate generation of mutants by radiation or chemicals may also raise concerns that echo centuries-old fears about the dangers of humans 'playing God'. Such issues are also relevant for so-called 'transgenic' organisms in which, instead of relying on nature, radiation, or chemicals to mutate the genetic material, there is a conscious effort to directly manipulate the DNA that is the molecular basis of genes. For while the identification of the genes mutated in naturally occurring or ENU mutants is easier because of the availability of whole genome sequences, it still remains a laborious job to 'map' the defect to its position in the genome. In addition, generating mutations by either chemicals or radiation is very much a hit-and-miss affair. Because of this, scientists have sought to directly modify the genome of a living organism. We will now look at how they first achieved this and the potential of such technology for human society, but also some of the ethical issues it raises.

2

Supersize My Mouse

A tension has long existed in science between those who believe we should merely seek to understand the natural world and those who think we should actively manipulate and control it for human benefit. Contemplation of nature has long been regarded as a noble pursuit, but suspicion and even fear may be attached to scientists who seek to transform living things.[102] That such fears have ancient roots is shown by the existence of myths like that of Prometheus, the Titan who defied the Greek gods by giving fire to humankind, or the Jewish Golem, a powerful being created from clay. The same fears are reflected in literature, in characters such as Mary Shelley's Dr Frankenstein, with his monstrous creation, or Marlowe and Goethe's Faust, the scholar who sells his soul to the devil in order to gain powers that allow him to transform and manipulate the natural world.

A central theme of such myths and stories is the havoc and chaos such acts create, and the punishments awaiting those who disturb the natural order.[103] And such images certainly helped frame much popular discussion of the first attempts to directly modify the genetic material of living organisms. So GM crops have been referred to as 'Frankenfoods'[104] that pose grave risks to both human health and the environment, while media coverage of transgenic animals—so-called because foreign genes have been transferred into their genomes—often focuses on bizarre examples like animals that glow in the dark or plants that produce fish oils. A focus on such curiosities may help sell newspapers or advertising space on the Internet, but it can mislead people about the benefits of genetic engineering, as well as its real limitations and risks. Any serious discussion of this topic needs to examine the actual science of genetic modification, and to do this, we need to return to Saturday, 28 February 1953, when Jim Watson and Francis Crick first determined the double helix structure of DNA. At their celebratory drinks at the Eagle pub in Cambridge, Crick's boast that he and Watson had discovered the 'secret of life'[105] may have bemused anyone else listening, but wasn't so far wrong in terms of the impact of the discovery on our understanding of the natural world.

The discovery initiated the age of molecular biology by providing a unifying principle to genetics that had been lacking: the recognition that life can be viewed as a linear code. The genetic blueprint, DNA, can be seen as a long string of four different letters—defined by the chemical names adenine, cytosine, guanine, and thymine, generally abbreviated to A, C, G, and T. Importantly, the order of these four letters is not random, but occurs in a precise sequence of triplets, each coding for a particular amino acid, the units from which proteins are composed.[106] So a linear code based on the four DNA letters is 'transcribed' first into ribonucleic acid (RNA)—DNA's chemical cousin—and then 'translated' into proteins, themselves linear molecules, but composed of 20 different units—the amino acids (see Figure 3).

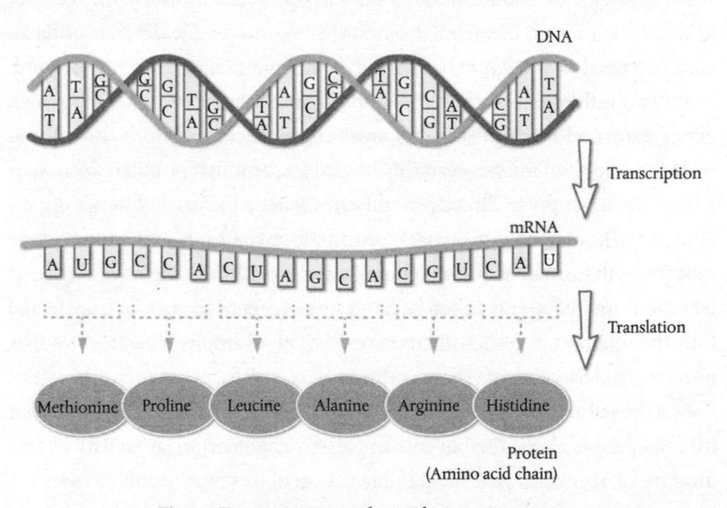

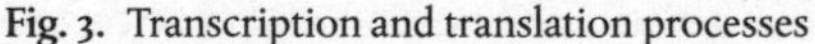
Fig. 3. Transcription and translation processes

Unlike DNA, with its unvarying double helix, each type of protein folds into a unique 3D shape based on its specific sequence of amino acids. It's because of these differences in shape and size that proteins can perform so many different roles in the cell, acting as cellular building blocks, motors, and transporters, as well as carrying out many other functions. The genetic code—the connection between the sequence of letters in the DNA and that of the amino acids in proteins (see Figure 3)—was cracked by the mid-1960s.[107] But knowing how the code worked didn't immediately translate into an ability to manipulate it. That only became possible with the discovery of natural processes in bacteria that provided key tools for genetic engineering.

Engineering Life

The first of these processes allows bacteria to defend themselves from infection. Since we generally think of bacteria being the infective agents, it may seem strange that these microorganisms suffer infection themselves. Yet just as humans can be infected by viruses, so bacteria have their own viruses to deal with—so-called bacteriophages.[108] And just as our own immune system wards off infectious agents, so bacteria have their own miniature form of immunity. This process was discovered by Werner Arber of the University of Geneva, in the 1960s, but Hamilton Smith of Johns Hopkins University in Baltimore, Maryland, worked out its specific details in 1970.[109] He showed that bacteria produce catalytic proteins—enzymes—that recognize a specific DNA sequence in the genome of an invading virus and chop the DNA at that point. The target sequences are small, typically four to six letters, and each bacterial species produces its own set of one or more unique cutting enzymes.

So *Escherichia coli*, more commonly known as *E. coli*—the bacterium that lives inside a human gut but also exists in more dangerous forms—produces an enzyme named EcoRI that cuts within the sequence GAATTC. Since this sequence occurs many times within any typical long stretch of

DNA, this poses the question of why the enzyme doesn't cut the bacterium's own genome into pieces. What prevents this happening is that, just as our immune system can distinguish invading microorganisms from our own cells and tissues, *E. coli* has also evolved a mechanism to protect its genome. This involves the recognition site GAATTC being chemically modified with a methyl ($-CH_3$) group in the bacterial DNA, which prevents the site being cut in the genome. Because of this, the cutting proteins have become known as 'restriction enzymes', since their action is restricted to targeting only foreign DNA.[110]

Many different bacterial species produce their own unique restriction enzyme with a distinct cutting site. So, armed with a variety of such enzymes, it's possible to cut anywhere in a DNA sequence. The first use of such enzymes was demonstrated by Daniel Nathans, also at Johns Hopkins, who employed HindII and HindIII, restriction enzymes Hamilton Smith had purified from the *Haemophilus influenza* bacterium, to cut the SV40 monkey virus into 11 pieces, thereby creating the first 'restriction map'.[111] Long before it became possible to identify a piece of DNA by its sequence, this method made it possible to do so by the precise number and size of fragments it could be cut into, like a kind of molecular fingerprint. For these discoveries, Arber, Smith, and Nathans were awarded a Nobel Prize in 1978.[109]

When told he'd been awarded the greatest prize in science, Hamilton Smith's response was initially one of shock. 'Are you kidding?' he said when a reporter told him about the award. 'I just didn't imagine it would be taken in that light.'[112] This feeling was shared by his family. When Smith's mother heard the news on her car radio, puzzled, she turned to her husband and said, 'I didn't know there was another Hamilton Smith at Hopkins.'[112] The fact was that, before the award, Smith was viewed at Johns Hopkins only as a fairly obscure researcher, known more for his moth-eaten sweaters, shirts worn bare at the elbows, and thick glasses through which he squinted as if he'd just emerged from a cave, than for any sense of impending fame.[112] His studies of restriction enzymes were seen as highly esoteric, if they were recognized at all. However, Smith's discovery was about to make him famous because of its huge practical potential for increasing humanity's capacity to manipulate the genetic code.

The second step on the road to genetic engineering was the discovery, by Martin Gellert at the US National Institutes of Health (NIH) and Bob Lehman at Stanford University, of another type of enzyme called DNA ligase, which can join two DNA fragments together.[113] This enzyme is employed naturally during the process of DNA replication, alongside the DNA polymerase enzyme that actually creates the new DNA strand. By using both restriction enzymes and DNA ligase, it was finally possible to cut and paste DNA sequences in a test tube. At first, it wasn't clear how the new technologies could be used to modify the genes of a living organism, because introduction of restriction enzymes into a cell would cut its genome at multiple points, killing the cell in the process. A chance meeting between two scientists—Stanley Cohen of Stanford University and Herbert Boyer of the University of California in San Francisco—provided the final link in the chain that would make genetic engineering of living organisms a reality.

In 1972 Cohen and Boyer were in Honolulu attending a molecular biology meeting, at which both gave talks. While Boyer spoke about his studies of the precise cutting mechanism of the restriction enzyme EcoRI, Cohen's talk concerned his investigation of a kind of molecular parasite in bacteria—loops of DNA called plasmids that use the host cell's own DNA replication machinery to propagate themselves.[114] This isn't a wholly selfish act though, for plasmids also give something back. Because they contain genes that code for resistance to antibiotics, help the host bacterium to digest unusual substances, and kill other types of bacteria, they provide the bacterium with an extra capacity for survival.[115]

It was when Cohen and Boyer met to discuss their work over a late-night snack at a delicatessen near Waikiki Beach that an idea was born with technological implications that continue to reverberate today.[114] The two scientists realized that plasmids could be used to transport genes of any species into a bacterial cell, where they would not only be replicated alongside the genome of the host cell, but could also be expressed as functional proteins. All that was required was to cut a gene of interest, and also a plasmid, with

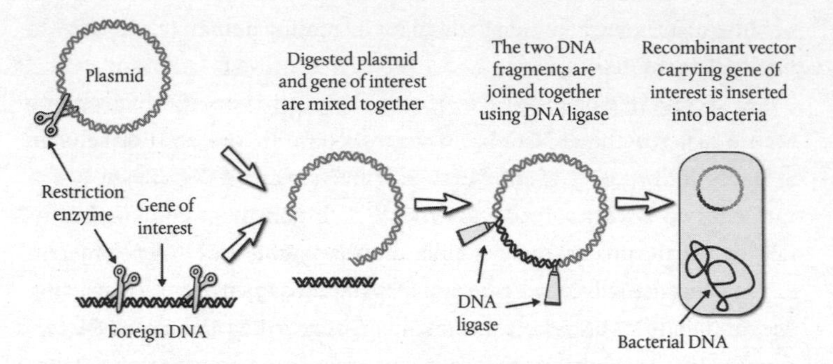

Fig. 4. Procedure for producing recombinant DNA

a restriction enzyme, use DNA ligase to attach the two together, then introduce the resulting gene construct into a bacterium (see Figure 4). Cohen achieved this by using a 'heat shock' to get the bacterium to take up the DNA. Since uptake of the gene construct into the bacterium is highly inefficient, this rare event must be selected for. Such selection makes use of the antibiotic resistance genes that plasmids typically contain. If the experiment is carried out in a solution containing the antibiotic, only those cells containing the plasmid DNA survive.

Birth of Biotechnology

Cohen and Boyer demonstrated their 'recombinant' DNA technology initially by showing that they could join together DNA from two different plasmids and propagate this construct in bacteria. But the real power of the technology was demonstrated by using it to create bacteria that could produce human proteins of clinical relevance. This was achieved first by Genentech, a company set up by Boyer and Robert Swanson in 1976.[114] Swanson was a 28-year-old unemployed banker who recognized that molecular biology had progressed to the point where it could make money—lots of it. So he visited Boyer and proposed a partnership. One of Boyer's colleagues remembered 'standing out in the hallway laughing at this guy in a three-piece suit. We didn't get people like that visiting us.'[116] Yet Swanson's ten-minute pitch was persuasive enough for biologist and banker to go for a beer at a local bar and discuss the proposal further. A deal was struck, but there were still plenty of obstacles ahead; in particular, the fledgling company required start-up funds, as well as a saleable product.

Swanson eventually managed to find investors, but only after a fraught six-month period which he spent drawing unemployment money and living on peanut butter sandwiches. Meanwhile, Boyer had hit upon a product—somatostatin—a hormone that regulates growth and can be used to treat tumours and growth disorders.[114] The biotechnology business was born, and soon other human proteins were being produced, most notably insulin. Previously, people with diabetes had to rely on insulin extracted from the pancreases of pigs slaughtered for their meat. However, slight molecular differences between pig and human insulin meant that this version of the hormone caused adverse immune reactions in some people. But in 1978, Genentech began developing a way to produce human insulin in bacteria in collaboration with the Eli Lilly pharmaceutical company, and in 1982 this became the first biotechnology product to appear on the market.[114]

The growth of biotechnology from university laboratory experiments to billion-dollar industry is one of the success stories of modern science.[117] Yet for a time in the mid-1970s it seemed that biotechnology might never get started at all, because of fears about its safety. And notably, it was scientists themselves, not government officials, political activists, or religious figures, who decided that further development of the technology should cease until the risks had been properly assessed. In particular, Paul Berg, who had been using the technology to study the functional properties of the genes of the simian virus 40 (SV40) virus in a bacterium, became concerned about the health risks if such modified bacteria were to escape into the wild, a reasonable fear given the virus itself had been shown to cause cancer.

Berg and other molecular biologists—notably Sydney Brenner, David Baltimore, Richard Roblin, and Maxine Singer—organized a meeting to discuss such risks and devise ways to reduce them, to be held at the Asilomar Conference Center in Pacific Grove, California, in February 1975.[118] The meeting was to discuss the possibility that, while 'the new technology opened extraordinary avenues for genetics and could ultimately lead to exceptional opportunities in medicine, agriculture and industry... unfettered pursuit of these goals might have unforeseen and damaging consequences for human health and Earth's ecosystems'.[118] In line with such concerns, in the run-up to the meeting a voluntary moratorium was proposed, and despite the commercial potential of the new technology, this was universally observed not only in academia but also in the biotechnology industry. One effect of publicizing the potential risks of recombinant technology was that, in the build-up to the meeting, the media 'had a field day conjuring up fantastical "what if" scenarios'.[118] Some scientists feared this would turn the public against recombinant DNA technology. However, the fact that at the meeting itself there were not only scientists but also lawyers, journalists, and government officials provided an opportunity for members of the public to be informed 'about the deliberations, as well as the bickering, accusations, wavering views and ultimately the consensus' of how to maximize the potential and minimize the risks of biotechnology.[118]

The Asilomar meeting decided that recombinant DNA technology could continue, but only following strict guidelines that regulated the safe disposal of GM bacteria. It also introduced genetic safeguards that limited the bacteria's ability to survive in the wild, should any accidentally escape. Importantly, according to Berg, the meeting introduced the principle that 'the best way to respond to concerns created by emerging knowledge or early-stage technologies is for scientists from publicly funded institutions to find common cause with the wider public about the best way to regulate—as early as possible'. Berg was particularly concerned that 'once scientists from corporations begin to dominate the research enterprise, it will simply be too late'.[118] So, although the biotechnology industry had developed from academic science, already some of the priorities and interests of the two spheres were diverging, in ways that remain relevant today.

A Giant Mouse

So much for introducing a gene into a bacterium. But what about manipulating the genome of a complex, multicellular organism? Such a GM organism—a mouse—was created in 1974 by Rudolf Jaenisch, now at the Whitehead Institute and the Massachusetts Institute of Technology in Boston.[119] At this time he was working as a postdoctoral researcher in Arnold Levine's laboratory at Princeton University. Jaenisch had joined Levine's group because the lab was pursuing an exciting new area of research—the role of certain viruses in causing cancer. Jaenisch's project was to study the mechanism of replication of the SV40 virus mentioned in the section 'Engineering Life'. However, only two months after his arrival, Levine told Jaenisch that 'he was going on sabbatical to Europe and that I should run the lab'.[119] What might have overwhelmed a lesser individual became an important step in Jaenisch's career development, for while continuing his main project into SV40 replication with Levine's graduate students, the absence of his supervisor meant he began exploring areas of research that he might not have otherwise.

In particular, Jaenisch became fascinated by a puzzling fact related to the cancer-causing ability of SV40. When injected into mice, the virus only caused formation of tumours in tissues such as bone, muscle, cartilage, and fat, but not others like the liver. Jaenisch reasoned that this selectivity must arise either because SV40 could not infect liver cells, or because these cells were turning off replication of the virus after infection.[119] To test which scenario was correct, Jaenisch decided to see if he could infect an early mouse embryo with the virus. Since at this stage of life all cells are pluripotent, meaning they can give rise to any cell type, this should allow him to introduce the virus into all tissues of the body. The only problem was the fact that no one had ever tried such an experiment. Undeterred, Jaenisch sought help from Beatrice Mintz of the Fox Chase Cancer Centre in Philadelphia, an expert on isolating and culturing mouse embryos. With her help, Jaenisch injected mouse embryos with SV40, then implanted them into females.

The initial results were disappointing. None of the resulting offspring developed tumours, not even in tissues usually affected by the virus. However, when Jaenisch used a radioactive probe to detect the presence of SV40 genes he found that viral DNA was present within the mouse genome.[119] Here was clear evidence that he had created the first transgenic mouse. But the lack of any tumours suggested that something was suppressing the effects of the viral genes. In fact, such a suppression mechanism makes a lot of biological sense, for without it embryos would be vulnerable to the effects of any virus that happened to infect them. Understanding how such 'epigenetic' effects can alter gene expression is now a central part of Jaenisch's research, and has turned out to be crucial to understanding how embryos develop, and how both the cellular and bodily environment influence which genes are turned on or off.[120]

While Jaenisch had shown that a virus could be used to modify the genome of a mouse, it remained to be seen whether mice could express a gene from another complex species, in the same way that GM bacteria produce human insulin. This feat was first achieved by Richard Palmiter of the University of Washington and Ralph Brinster of the University of Pennsylvania in 1982.[121] They were studying a gene coding for the metallothionein protein. This protein traps metal ions like copper, zinc, and cadmium, and so helps prevent poisoning of the body by such metals. The metallothionein gene is itself turned on by the presence of a metal like cadmium, thereby acting as a sensor (see Figure 5A).

Palmiter and Brinster fused the regulatory region, or promoter, of the metallothionein gene to the rat gene coding for growth hormone, and injected this gene construct into fertilized mouse eggs, which were then implanted in females. Remarkably, if cadmium was included in the diets of the resulting offspring, these animals grew much bigger than normal since the foreign growth hormone gene was now permanently stimulated (see Figure 5B). This showed not only that the rat gene had been inherited by these mice, but that it functioned perfectly too. 'For Brinster: 'The giant mouse experiment...made everybody, including us, stop and say, "This is incredibly powerful"...It's the first time man was able

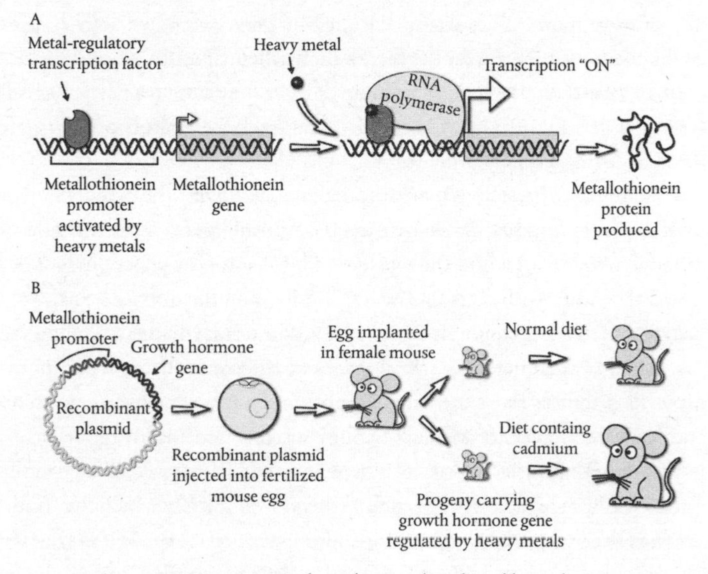

Fig. 5. Transgenic animals with growth induced by cadmium

to experimentally modify the genetic code that will make the next individual.'[122]

Such transgenic mice have been useful to biomedical research in numerous ways. For instance, they have been used to study why genes are turned on in some cell types in the body but off in others. Investigating this issue could help us understand the molecular mechanisms underlying the development of the different bodily tissues, and why sometimes such mechanisms go wrong, leading to a developmental abnormality or cancer. Many different regulatory elements are present in the promoter of any given gene. By fusing each separately to a 'reporter' gene, and then creating transgenic mice expressing such gene constructs, it's possible to uncover the specific contribution of each regulatory element.[123] The first such reporter gene was a bacterial gene called β-galactosidase. The presence of the protein product of this gene can be detected by a chemical reaction that produces a blue colour, thereby labelling the cells in which the gene is normally turned on.

But an even more direct way of visualizing gene promoter activity uses genes coding for fluorescent proteins that, when fused to regulatory elements, signal their presence by the fluorescence they emit in a particular cell type or tissue. By such means scientists can track the expression of a gene product during a bodily process.

This approach has proved very important for studying the progression of disease in mice models. Osteoarthritis is a painful joint disease that affects millions of people around the world.[124] This disorder is generally first detected when painful symptoms occur, but by then the disease has already progressed to a late stage. So there is a lot of interest in understanding the mechanisms underlying the initial stages of osteoarthritis in order to improve its diagnosis and treatment. Recently, researchers at Tufts University and Harvard University Medical Schools used a fluorescent reporter approach to monitor the activity of a gene involved in loss of cartilage in the joint—the key characteristic of osteoarthritis—in mice in which the condition had been triggered by injury. 'The fluorescent probe made it easy to see the activities that lead to cartilage breakdown in the initial and moderate stages of osteoarthritis, which is needed for early detection and adequate monitoring of the disease,' says Shadi Esfahani, one of the researchers.[124] The team believe their approach could be used to study the effectiveness of new osteoarthritis drugs, leading to improved treatments.

Crop Controversy

Transgenic technology has also been used to create GM plants both for research and agriculture. This technology has been employed to produce plants that are resistant to viruses, other infectious agents, and even to insects; to create forms that are more resistant to weedkiller so that the latter can be used more effectively to destroy neighbouring weeds; and to change the appearance, flavour, and nutrient composition of the modified plant.[125] There is now widespread production of such crops, with a recent report claiming that 'around one 10th of the world's planted crops are GM.'[126] There has also been substantial opposition to GM crops, for instance from the Soil Association, an organization committed to 'organic' methods of farming, on three main grounds.[127] First, concerns have been raised about the impact of such crops on the environment, for instance if a herbicide resistance gene were to be transmitted to weeds. Second, some have claimed that GM crops will not help ordinary farmers, but rather allow giant agribusiness companies to further increase their grip on the farming industry. Finally, there are claims that GM crops pose risks to human health.

Assessing the scientific basis of such claims has not been helped by the highly polarized way in which the debate has developed. Science journalist Natasha Gilbert claimed in an article in *Nature* in 2013 that, 'in the pitched debate over genetically modified (GM) foods and crops, it can be hard to see where scientific evidence ends and dogma and speculation begin'.[128] To try to get a clearer picture about the impact of GM crops, Gilbert looked at three key issues: whether the spread of herbicide resistance genes in the USA had helped create superweeds; whether the introduction of GM insect-resistant cotton in India had dramatically increased suicide rates amongst small farmers; and whether transgenes in GM crops imported into Mexico from the USA had contaminated local maize strains.[128]

Gilbert claims that, in only one of these cases, is there an unambiguous answer: namely that the suicide rate amongst Indian farmers does not seem to have been directly affected by the introduction of modified cotton. So a study by Ian Plewis of the University of Manchester in 2012 found that the suicide rate in areas of India that now grow GM cotton did not change significantly following the introduction of such cotton to India in 2002.[129] In contrast, Gilbert did find evidence of an increase in resistant weeds in fields planted with GM herbicide-resistant crops. But this may reflect overreliance on one specific herbicide rather than any actual transfer of transgenes, as the same phenomenon has also been seen in fields planted with standard crops. Finally, Gilbert found evidence both for and against the possibility that transgenes from GM plants can spread to non-modified crops.[128]

One aspect of GM crops that has caused particular controversy is so-called 'terminator' technology. A plan to develop this technology, which creates transgenic plants that produce sterile seeds, thereby forcing farmers to buy new seed for each planting, was announced by the multinational Monsanto in 1998.[130] The proposal generated huge opposition, on the grounds that it would force farmers in the developing world to purchase expensive seed each year from Monsanto, rather than saving new seed to sow the following season as was their normal practice. In fact, many conventional hybrid crops—the result of breeding two different plant varieties[131]—also produce sterile seeds. And, according to Paul Moyes of the European Association for Bioindustries, 'plant breeders and farmers have preferred hybrid seeds for more than 30 years because they were more productive. This means they have to buy their seeds again every year because hybrid seeds can only be used once.'[130] Nevertheless, the fact that terminator technology seemed to represent a deliberate attempt to impose such sterility was seized upon by anti-GM activists. So, for example, Andrew Simms of Christian Aid, the development charity, says, 'Terminator technology was the lynchpin of a strategy to protect corporate royalties in developing countries.'[130] And indeed, opposition was such that, in 1999, Monsanto's chairman Robert Shapiro was forced to promise not to commercialize the technology.

Turning to the third controversial aspect of GM crops, the question of how safe they are for human consumption, there are two potential areas of concern.[132] One is that genetic modification of crops may cause harmful chemical changes in the resulting food product. The second concern is linked to the way GM crops are produced. Because integration of transgenes into the genome is so inefficient, antibiotic resistance is used to select for the rare cases in which integration occurs, a gene that confers such resistance being included in the gene construct. However, this additional gene could, in theory, be transferred to bacteria in soil or the human gut, where it might then create antibiotic resistance in a pathogenic bacterium.

In fact, there is little evidence to indicate toxicity of GM foods, despite widespread media coverage of a study by Árpád Pusztai of the Rowett Research Institute in Aberdeen in 1999.[133] The study appeared to show that rats fed GM potatoes suffered damage to their vital organs and immune systems. But a subsequent review of the study by six toxicologists appointed by the Royal Society, Britain's national science academy, concluded that the study was flawed in many 'aspects of design, execution and analysis', adding that 'where the data seemed to show slight differences between rats fed predominantly on GM and on non-GM potatoes, the differences were uninterpretable because of the technical limitations of the experiment and the incorrect use of statistical tests'.[133]

As for the transmission of antibiotic resistance from a GM crop to a virulent bacterium, this is certainly a concern. But while theoretically possible, it looks likely to be relatively rare. In contrast, a far greater risk in this regard is posed by inappropriate use of antibiotics, partly through overprescription to human patients, but also because of the widespread use of these substances to treat infected livestock in agriculture.[134]

While anti-GM protestors have argued that GM plants are detrimental to human health, recently they themselves have been accused of preventing poor people in the developing world gaining the benefits of one particular GM crop. The plant in question is golden rice, which was engineered to provide vitamin A to counter blindness and other diseases in children in the developing world. 'Vitamin A deficiency is deadly,' says Adrian Dubock, who helped develop the plant. 'It affects children's immune systems and kills around two million every year in developing countries. It is also a major cause of blindness in the third world. Boosting levels of vitamin A in rice provides a simple, straightforward way to put that right.'[135] Yet despite being developed in 1999, cultivation of the plant has been blocked by a vigorous campaign by anti-GM groups like Greenpeace, on the grounds that the crop's introduction in the developing world is part of a general plan to make farmers increasingly dependent on Western industry. Greenpeace also argued that the rice would provide only a tiny amount of the vitamin A required by a person each day, and ensuring that people in the developing world had a normal balanced diet was a much surer way of combating vitamin A deficiency. Yet recently a trial carried out on Chinese children, aged between 6 and 8, showed that a bowl of cooked golden rice could provide 60 per cent of the recommended intake of vitamin A for young people.

Findings like this have led Mark Lynas, an environmental campaigner and one of the founders of the anti-GM crop movement, to recently publicly apologize for opposing the planting of GM crops in Britain. According to him: 'The first generation of GM crops were suspect, I believed then, but the case for continued opposition to new generations—which provide life-saving vitamins for starving people—is no longer justifiable.'[135]

What this discussion of the benefits and risks of GM crops does highlight is that such issues are not purely scientific ones. Instead, they are intimately connected with the development of GM crops in a free market system, with all that implies in terms of the interests of public versus private ownership, and the priorities of profit compared to those of a sustainable agriculture that benefits the maximum number of people.[126] These are issues to which we will return in Chapter 6 when assessing more recent approaches to genetic modification of plants and animals that are already promising to have a major impact on agriculture.

Genes as Therapy

As well as its use in agriculture, standard transgenic technology has been employed in gene therapy in humans. Potential targets for such therapy are conditions caused by absence of a normal gene product. Following a pattern of inheritance established by Mendel in pea plants, these disorders are known as recessive disorders, because a person is only affected when they have two faulty copies of a defective gene.[136] These diseases occur when two unaffected 'carriers', each with a single faulty copy of a gene, have an affected child. Following Mendel's laws, on average this will be the case in a quarter of their children (see Figure 6).

Cystic fibrosis is one such disorder, caused by the absence of a protein, cystic fibrosis transmembrane conductance regulator (CFTR), that forms pores in cell surface membranes and transports chloride ions out. In the lungs, lack of the protein causes a chemical imbalance and subsequent stickiness in the mucus that lines the airways, which causes havoc in the

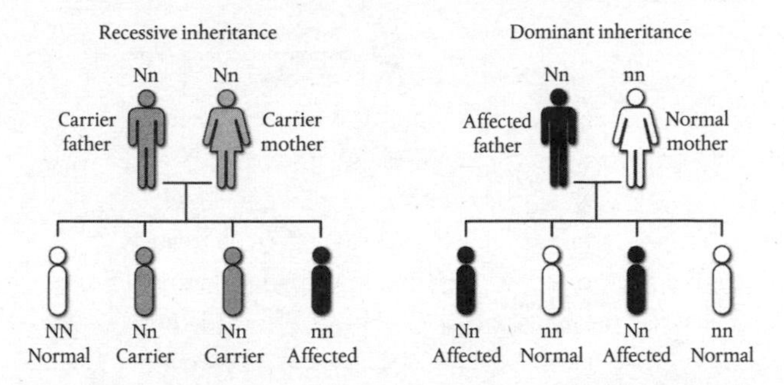

Fig. 6. Inheritance of recessive and dominant characteristics

operation of this organ, leading to lung dysfunction and infection. But this has also made cystic fibrosis a potential target for gene therapy, since lung cells are relatively accessible via the respiratory tract.[137] Diseases affecting the blood cells are also targets for gene therapy, since the stem cells that produce such cells are located in the bone marrow. By extracting a sample of bone marrow, introducing a functional copy of the gene that's defective into the stem cells within, and then replacing the treated bone marrow, it should be possible to treat the disorder.[138]

Unfortunately, gene therapy using standard transgenic technology has been far from a success story. A key challenge is getting a gene construct into an affected tissue. Attempts to do this include wrapping the gene construct in a detergent shell, to help it cross the cell's fatty membrane.[137] However, this is an inefficient process. An alternative strategy uses a virus to carry the gene construct into the cell.[139] This is a potentially attractive route because viruses have evolved to cross cellular boundaries and, in some cases, even integrate their genetic material into the genome of a host cell. This latter property is a particular feature of retroviruses, of which human immunodeficiency virus (HIV) is the most famous member.[140]

A safe, adapted form of HIV was used in a clinical trial carried out in Paris in the late 1990s to treat a disorder called severe combined immunodeficiency (SCID). In this disorder, a gene defect in white blood cells leaves sufferers without a functioning immune system and thereby extremely vulnerable to

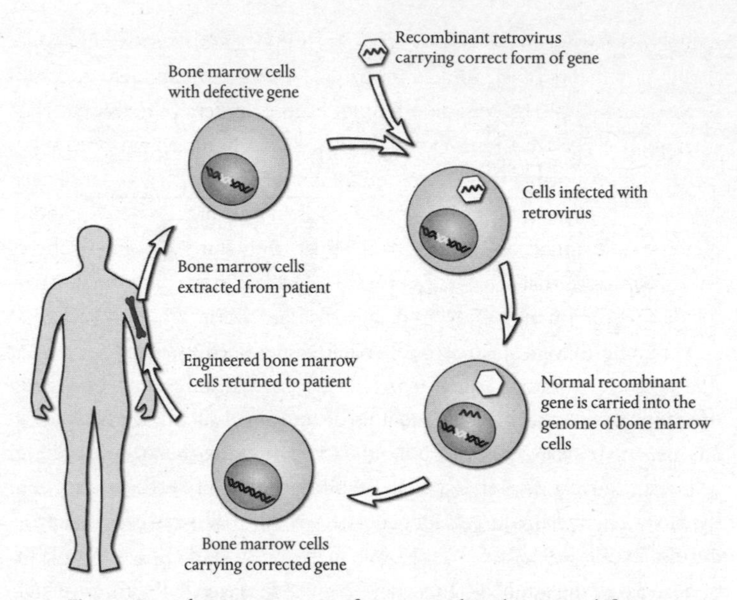

Fig. 7. Gene-therapy treatment of severe combined immunodeficiency

infection. The treatment of bone marrow from such individuals with a retrovirus carrying a normal copy of the defective gene cured the disorder (see Figure 7).[141] However, some treated patients subsequently developed leukaemia. An investigation showed that while the retrovirus had successfully carried the normal gene into the genomes of sufferers' cells, in some cases it also disrupted the action of genes that control cell growth and division, causing cancer. A more recent trial led by Patrick Aubourg at the French National Institute of Health and Medical Research in Paris, has been successful in curing SCID without causing leukaemia in the process. 'The new generation of [viral] vectors is much safer, although the risk is not zero,' says Aubourg.[142]

For a dominant genetic disorder like Huntington's disease, the problem is not the loss of a gene, but rather the fact that the mutant gene product disrupts normal cellular function. In this case, every generation is affected, with the chance of an affected person passing on the condition to their children being one in two (see Figure 6). In Huntington's, what begins as odd, jerky movements of the limbs rapidly progresses into psychosis and full-scale dementia.[143] The condition has not been considered suitable for gene therapy because treatment would involve precisely replacing the mutant gene with the normal version, something not possible with standard transgenic approaches. But at the end of the 1980s, a more precise way to modify genes was identified, and it came from a surprising starting point—cancers called teratomas that can occur in various locations in the body, but are particularly common in tumours of the testicles and ovaries.[144]

Teratomas have a startling property first identified by Leroy Stevens at the Jackson Laboratory, Maine, in 1953. Newly arrived at this mouse breeding institute, Stevens noticed that one mouse breed had a tendency to develop abnormally enlarged testicles. 'We killed it, and looked at the testes, and they had strange things inside,' said Don Varnum, a technician working with Stevens.[145] This was somewhat of an understatement given that the tumour, for this is what it was, comprised a grotesque mishmash of different tissues, including bone, hair, and teeth. It was as if the tumour cells could form any cell type in the body. And the discovery of teratomas in humans showed that this wasn't just a peculiarity of mice. Teratomas were also discovered in ovaries and in other parts of the body, like the brain.[144] Trying to understand the phenomenon would occupy Stevens for the rest of his career. In 1970 he made a major step forward when he discovered that cells from an early mouse embryo, transplanted into the testicles of adult mice, also generated teratomas. Based on this observation, Stevens proposed that teratomas might provide a clue as to how the unspecialized cells of the embryo could develop into all the specialized cell types that make up the body—so-called 'pluripotency'.[145]

Pluripotent Potential

One scientist who became particularly interested in this question was Martin Evans. Working with Gail Martin at University College London in the mid-1970s, Evans began studying teratoma cells in culture and found little difference between them and cells extracted from a normal embryo.[146] Were the tumour cells simply unspecialized cells that only became malignant in the wrong environment? In line with this, while cells isolated from normal embryos turned malignant when injected into an adult mouse, teratoma cells injected into an early mouse embryo became part of the normal tissues of the resulting mouse. This suggested that embryonic cells might be used to generate a whole mouse. And indeed, by taking embryonic cells from one mouse breed and injecting these into the embryo of another, Evans showed that the resulting offspring were 'chimaeras'—products of more than one original embryo and named after those mythical beasts like the Sphinx with the head of one animal and the body of another.[146]

This chimaeric quality was shown graphically by taking embryonic cells from a black mouse and injecting these into the embryo of a white one. This produced mice with patches of black amongst mainly white fur, and genetic analysis showed that other tissues of these animals were a similar patchwork, suggesting the embryonic cells could develop into any cell type. This capacity included an ability to develop into the cells that form the next generation—sperm and eggs—since breeding male and female chimaeras led to some offspring that were totally black, that is, totally derived from the embryonic cells.[146] Reflecting their pluripotent properties, the embryonic cells were named embryonic stem (ES) cells.

The discovery of ES cells immediately suggested a new route for making transgenic mice. Rather than injecting a gene construct into a fertilized egg and hoping it would integrate into the latter's genome, an ES cell could be genetically modified and used to create a transgenic mouse. Nonetheless, the same limitations would apply to such a mouse as with the standard route—unless a way were found to modify a gene in the ES cell more precisely than was possible with standard methods. In the end, such precision was achieved using a process that occurs naturally in cells, called 'homologous recombination'.[147] This occurs when two pieces of DNA that contain sequences which are identical, or very similar, come into contact. Such proximity triggers a cellular mechanism in which the two sequences are exchanged.[148] In fact, this mechanism is central to a process without which you or I would not exist.

This process is sexual reproduction. Although we tend to associate sex with the physical act and the feelings that go with that, in evolutionary terms the primary role of sex is to create new characteristics for natural selection to work on.[149] Novel characteristics also occur in organisms that reproduce asexually, as shown by bacteria that become resistant to antibiotics. Such changes occur through mutation, and although this may be a one in a million event, it only takes one bacterium to develop resistance to an antibiotic for this single organism to pass on this characteristic to its descendants. Mutations are also ultimately the source of variation in complex multicellular organisms, including humans, but these changes are very rare, and occur very slowly in a species like ours that currently produces a new generation every 25 years on average,[150] unlike bacteria which can produce a copy of themselves in less than an hour.[151]

Sexual reproduction, however, can generate novelty in a single generation by mixing and matching the genetic material in the mother's and father's genomes.[152] This is possible because, while cells in the body generally contain two copies of each gene, the sperm and egg only have a single copy. This is a necessary state of affairs given that the union of an egg and a sperm creates a new life, since otherwise the number of copies of each gene would increase with each new generation. However, within a testicle or ovary, sperm or eggs develop from stem cells that contain two copies of each gene—one derived from the original maternal genome, the other from the paternal genome. It is during this process that homologous recombination takes place, swapping similar regions of the maternal and paternal chromosomes (see Figure 8).[152] As a consequence, when eggs or sperm are formed, each has a unique genome. This explains why, although we have many things in common with our siblings, we can also differ greatly in looks and temperament.

A second important role for homologous recombination is in DNA repair. As we saw in Chapter 1, organisms from bacteria to humans have evolved mechanisms for correcting damage to their DNA. Homologous

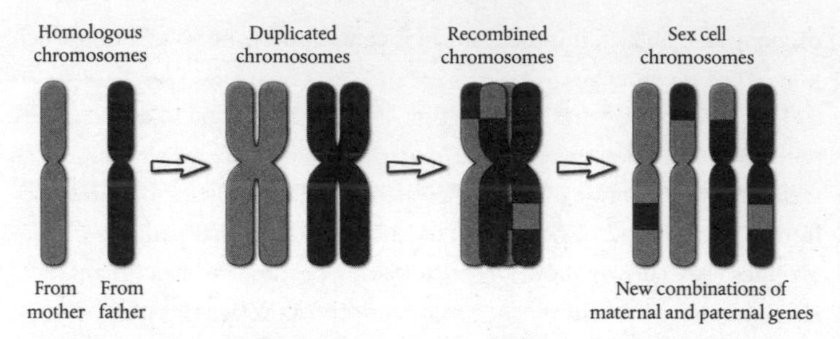

Fig. 8. Homologous recombination during sex cell formation

recombination plays a role here by allowing a sequence in which a double-stranded break has occurred to be repaired using the non-broken copy as a template.[148] The importance of this process is shown by what happens when it's defective. As we saw in Chapter 1, the role of the BRCA genes in cancer susceptibility was highlighted by Angelina Jolie's decision to have a double mastectomy, and subsequent removal of her ovaries, because she inherited a defect in the BRCA1 gene from her mother.[153] In fact, the BRCA genes are central components of the homologous recombination process (see Figure 9).[154]

So why does a gene defect that affects a fundamental process like DNA repair and is present in all cells of the body only cause breast or ovarian cancer? One explanation is that both tissues undergo rapid cell division under the influence of the hormone oestrogen. This leaves them vulnerable to mutations that can occur during the DNA replication that accompanies cell division, creating a greater need for a correctly operating DNA repair mechanism.[155]

The link between homologous recombination and cancer illustrates the importance of this process for normal cellular function. But this capacity of the cell to swap similar stretches of DNA is now a key tool in genetic engineering. The two scientists who recognized that homologous recombination offered a way to precisely modify genes in the ES cell genome were Mario Capecchi of the University of Utah and Oliver Smithies of the University of North Carolina.[156] Although homologous recombination is

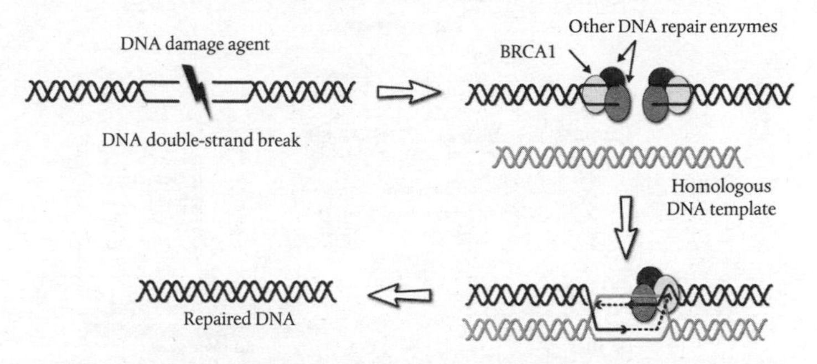

Fig. 9. Involvement of BRCA1 in DNA repair by homologous recombination

very rare in this cell type, in 1989 Capecchi and Smithies independently developed ways to select the cell in which successful gene targeting had taken place by an ingenious drug selection method that selected for the one in a million event in which homologous recombination had taken place, against the far more common random integration of the targeting construct into the genome. In essence, this method selects for a gene that confers resistance to the antibiotic neomycin which is present in the targeting construct, while selecting against the presence of a 'suicide gene' thymidine kinase (TK), that is transferred during random integration of this construct but not during homologous recombination (see Figure 10).

The development of such gene targeting revolutionized biomedicine by providing, for the first time, mice with precisely engineered genomes. Its importance was acknowledged by the award of a Nobel Prize to Capecchi, Smithies, and Martin Evans in 2007.[156] Justifying the award, Göran Hansson of the Nobel Committee, said: 'It is difficult to imagine contemporary medical research without the use of gene-targeted models. The ability to generate predictable designer mutations in mouse genes has led to penetrating new insights into development, immunology, neurobiology, physiology and metabolism.'[157]

Capecchi's part in the discovery was particularly remarkable given that he was lucky to make it to adulthood, never mind scientific success.[158] His grandfather was accidentally gunned down by his own men during World

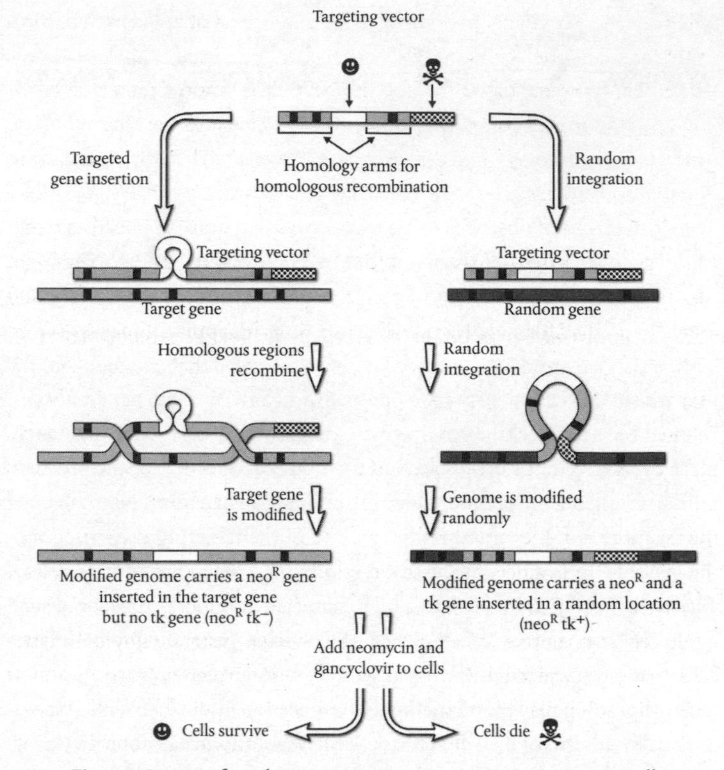

Fig. 10. Strategy for selecting correct gene targeting events in ES cells

War I, while Capecchi's father, a fighter pilot, perished in World War II. His American mother became an anti-fascist activist in Mussolini's Italy during that war, but was caught in 1941 by the Gestapo and sent to Dachau concentration camp. Capecchi was only 4 years old, and his mother's provision for someone to look after her son if she was arrested fell through. So for the next four years Capecchi fended for himself, 'sometimes living in the streets, sometimes joining gangs of other homeless children, sometimes living in orphanages'.[158] Miraculously, his mother survived Dachau, and in 1946, after a year's search, she finally traced her son to a hospital in Reggio Emilia, where he was surviving on a cup of coffee and a crust of bread each day in a ward for abandoned children.[158]

The half-starved Capecchi left for the USA with his mother, and after growing up on a Quaker commune, he eventually made it to the Massachusetts Institute of Technology, and then Harvard University. There he worked with Jim Watson, co-discoverer of the DNA double helix structure, but there were still plenty of obstacles in his way, for his first grant application to develop gene targeting, submitted to the US NIH, was rejected as being 'not worthy of pursuit'.[158] Luckily, Capecchi decided to carry on regardless, and when he applied for an NIH grant a second time, his application was greeted enthusiastically, and with an added note: 'We are glad that you didn't follow our advice.'[158]

Knockouts and Knockins

The first mice produced by gene targeting were referred to as 'knockouts', being engineered to totally lack a particular gene product.[159] Studies of such mice are now routine in biomedical science. For instance, my colleagues and I recently used this approach to show that proteins called two-pore channels (TPCs) play important roles in regulating bodily processes including the formation of new blood vessels, skeletal muscle development, heart contraction, and regulation of glucose levels in the blood.[160] Sometimes knocking out a gene has very clear effects on bodily function. However, in a surprising number of cases, eliminating a gene's function has far less effect than might have been expected.

This lack of effect is thought to be due to the organism compensating for the loss of a gene, by increased expression during embryo development of other genes that can substitute for the missing one.[161] To get around this problem, scientists have devised ingenious ways to delay the elimination of a gene until adulthood. ES cells, for example, can be engineered so that a gene is marked with a molecular tag and thereby becomes disabled when the resulting mouse is exposed to an enzyme called a recombinase that's activated by a chemical fed to the animal.[159] Another complicating factor with knockout mice is that a gene may have effects in multiple cell types and tissues, making it difficult to isolate each effect because of the ways that different parts of the body interact. To address this issue, mice have been engineered so that the recombinase gene is present in the animal but only switched on in specific cell types, making it possible, for instance, to knock out a gene only in the brain and not in other parts of the body, or even only in specific cell types within the brain (see Figure 11).[159]

Although totally knocking out a gene can lead to important insights about its normal function in the body, as well as providing a model for human diseases in which a gene is completely disabled, many genetic diseases are caused by more subtle changes. For example, sickle cell anaemia involves only a single amino acid change in the haemoglobin protein, yet results in a devastating, life-threatening disorder, by affecting the ability of

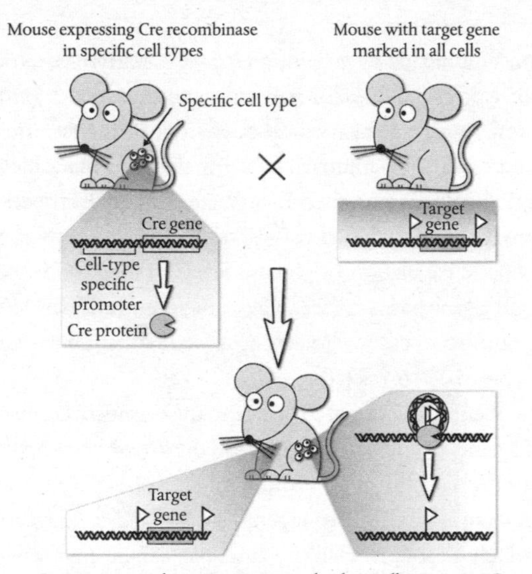

Fig. 11. Cell-type-specific knockout mice

the protein to carry oxygen and giving the red blood cells of sufferers an odd, sickle shape.[162] My colleagues and I recently identified an infertile man whose sperm's inability to activate his partner's eggs to develop into embryos is caused by a single amino acid change in the phospholipase C zeta (PLCζ) protein, which appears to play a central role in the egg activation process.[163]

To address these cases, gene targeting can also be used to create mice with such a subtle change.[159] This 'knockin' approach may also be used to attach a fluorescent tag to a protein to allow its movements to be tracked during particular cellular processes. This can provide clues about a protein's function, and, if a protein only found in a particular cell type is tagged, this also makes it possible to fluorescently 'label' such a cell type and thereby identify it in a living animal so its properties can be studied. This can be particularly important in the brain, where it can be difficult to distinguish different cell types on morphological grounds. In fact, the use of fluorescent tags, and indeed of light itself, in combination with knockout and transgenic technology, has gone way beyond simply identifying the location of a protein in a living cell or of that cell in the body. It's now becoming possible to use light to trigger functional activity in a cell. This is transforming our understanding of how the brain works, but also other organs in the body, in some very exciting ways. So let's now look at what light has to offer as a tool to manipulate life.

3

Light as a Life Tool

Where would life be, without light? The centrality of the Sun's rays to our existence has been recognized since the dawn of humanity, with light featuring strongly in accounts of life's creation in various religious texts. In the Bible, God commands 'Let there be light!', while worship of the Sun was central to religions like those of the ancient Egyptians, the Aztecs, and the Celts.[164] This ancient acknowledgement of light's importance reflects the fact that the world's ecosystem is ultimately powered by solar rays. Plants use photosynthesis to turn the Sun's energy into organic molecules, which animals like ourselves either consume directly by eating plants, or indirectly by feeding on other animals. Solar rays also provide the warmth that allows cellular activities to take place at a reasonable rate, and allow us to see what's around us. So important is light to life that organisms from simple microbes to our own species have evolved cellular and bodily mechanisms to detect the day–night cycle.[165] This body clock tells us when it's time to go to bed, and also regulates our metabolism. It's the reason why we suffer jet lag, and why many shift workers are prone to ill health, with recent studies suggesting this may be due to disturbances in the normal pattern of gene activity.[166]

Organisms have also developed ways to directly sense the Sun's rays, from light-sensitive pores on the surface membranes of single-celled algae, to sensors on the skin of worms that regulate their movement, or the exquisite structure of the human eye that allows us to see with such colour and precision.[167] Some organisms even generate light. Fireflies, with their glowing patterns—signals to attract mates—can make the forests they inhabit look like something from a fairy tale.[168] Other phosphorescent land organisms include some species of fungi, and a tropical snail found in South-East Asia.[169] However, most phosphorescence is found in the sea, being emitted by tiny plankton, but also larger organisms like jellyfish, squid, and various types of fish. This can be used to attract prey, like the glowing filament the anglerfish uses to entice smaller fish into its ferocious jaws, or as a defence mechanism, like the vampire squid, that ejects not ink, but a sticky bioluminescent mucus to startle, confuse, and delay predators.[169]

For years scientists have sought to understand how light is sensed, as well as generated, by organisms. More recently, we've begun to use light as a tool to manipulate and control life processes. Of course, light has been an important tool in biology for centuries. The first light microscopes made it possible to visualize aspects of life invisible to the naked eye. Thus, Robert Hooke, in Oxford, built one of the first microscopes and used it to study the natural world, his findings being published in 1665 in a book entitled *Micrographia*, which became the world's first scientific bestseller.[170] Some people disputed his diagrams, such as one showing the compound eye of an insect, because the miniature world Hooke had uncovered looked too alien! One object Hooke studied was a piece of cork, which he showed to be composed of tiny units; he called these cells, after the spaces that monks, or prisoners, inhabit.[171] We now recognize cells as the basic units of life.

The first person to visualize a living animal cell was Antonie van Leeuwenhoek, who lived and worked in seventeenth-century Delft in the Netherlands, and was a contemporary of the painter Vermeer. An expert lens maker, he produced microscopes whose resolution wasn't bettered until the nineteenth century.[172] In 1677, van Leeuwenhoek used one to study his own semen and observed for the first time a living sperm cell, with its characteristic head and tail, which he described as 'moving like a snake or like an eel swimming in water'.[173] He presented his findings to Lord Somers, President of the British Royal Society, although he was careful to add that 'what I investigate is only what, without sinfully defiling myself, remains as a residue after conjugal coitus. If your Lordship should consider that these observations may disgust or scandalize, regard them as private and to publish or destroy them as your Lordship thinks fit.'[174] In fact, van Leeuwenhoek was rather fond of self-experimentation, once leaving a piece of stocking containing three lice tied to his leg for 25 days to assess their capacity for reproduction.[175] By these means he estimated that two pairs of lice could generate 10,000 young in only eight weeks. Quite what his wife Cornelia thought about the lice and the post-conjugal microscopy experiments, there is no record.

Later development of higher magnification microscopes made it possible to visualize structures inside cells like the nucleus. But there's an inherent limitation to the magnification possible with a light microscope, and that's the wavelength of light itself. To get around this problem, scientists turned to radiation of a smaller wavelength—electrons. With the electron microscope, it became possible for the first time to visualize subcellular structures at a fine level of detail.[176] And recent developments in electron microscopy mean it's now possible to study the structure of important biomolecules like proteins at the atomic level. Such is the importance of this approach that in 2017 a Nobel Prize was awarded to Jacques Dubochet, Joachim Frank and Richard Henderson, for developing this type of electron microscopy.[177] There are disadvantages, though, to this type of microscopy: it needs to be carried out in a vacuum and an electron beam can only pick out structures dense enough to deflect its rays. This means that, using this technique, it's only possible to view cells that are already dead and stained with electron-dense heavy metals.[176]

Cellular processes are controlled by proteins, and one way of understanding a specific protein's function is to study its location inside a cell. For instance, if a protein is located in the nucleus it's likely to be involved in turning genes on or off, or if in the membrane it may regulate entry or exit of substances from the cell, or mediate the interactions between cells. One way of identifying a protein's cellular location uses antibodies—chemically tagged with fluorescent molecules—that recognize specific proteins. The power of this approach was shown by a study my colleagues and I recently carried out at Oxford University. I mentioned in Chapter 2 that one of my research interests is a sperm protein called PLCζ, which we believe is the agent that activates the egg to develop into an embryo at fertilization.[178] One way we confirmed this role was to show, by antibody labelling, that

PLCζ is located in a region of the sperm head that's precisely where an egg activation stimulus should be, being the part that first comes into contact with the egg.[179] What's more, our analysis of sperm from an infertile man in which the PLCζ protein is mutated showed that the mutant protein is incorrectly localized, preventing it from performing its normal role.[180]

While such studies use a light microscope, if an antibody is tagged with a heavy metal like gold it's possible, with an electron microscope, to pinpoint the location of proteins in subcellular structures. Recently, Gregory Frolenkov and colleagues at the University of Kentucky used this approach to uncover the precise interaction between two proteins called protocadherin-related 15 (PCDH15) and cadherin-related 23 (CDH23), in the hair cells of the inner ear, thereby defining their role in the hearing process.[181] According to Frolenkov, the study 'reveals the details of a process that is likely to be vital for the development, maintenance, and restoration of normal hearing'.[182] And since the genes coding for the PCDH15 and CDH23 proteins can be mutated in a type of deafness called Usher syndrome, this information could help devise new treatments for this particular type of hearing disorder.[181]

A Living Palette

Despite these positive features of combining microscopy and antibody labelling, this approach can only be carried out on dead cells in which the cellular structures are immobilized with a chemical fixative like formaldehyde—also used to embalm bodies—and then the cell membrane disrupted with detergent. This treatment is necessary because antibodies cannot cross the cell membrane, but such analysis gives a picture of the cell as a static entity, rather than the dynamic object it really is. What if it were possible to label proteins within a living cell so they could be visualized without killing the cell in the process? This is indeed now possible, and the origins of this new technology are worth relating, for they show that sometimes discoveries with a major impact on biomedicine originate not from thought of practical benefits, but because of sheer curiosity about the natural world.

In this case, it was the curiosity of a Japanese scientist called Osamu Shimomura that eventually led to a new technology. Shimomura was born in Nagasaki, and at the age of 16 he was lucky to survive the atomic bomb that devastated the city in 1945, being only seven and a half miles from the centre of the blast.[183] This demonstration of the lethal application of atomic physics didn't destroy his growing passion for science and, studying chemistry at Nagoya University, Shimomura became fascinated by 'sea fireflies'. These organisms, actually tiny crustaceans that emit a blue light, are particularly abundant in the sea off Kone and Ikuchi islands, near Hiroshima. In 1956, as a graduate student, Shimomura decided to try and isolate luciferin, the luminescent substance in these organisms, despite the fact that US researchers had been trying in vain to do so for over twenty years. For ten months Shimomura also had no success, until one night he 'accidentally' added a strong acid to a sea firefly extract.[184] Next morning he saw that the acid treatment had caused the luciferin to form pure crystals. 'That success offered hope for my future, which had looked grey ever since the end of World War II,' he later recalled. 'I was so excited and happy that I wasn't able to sleep at night.'[185]

Later, having moved to the USA, Shimomura became a researcher at the Woods Hole Marine Biological Laboratory near Cape Cod, where he began studying phosphorescent jellyfish. Shimomura found that their vivid colour is the product of two proteins—aequorin, which generates blue light when it comes into contact with calcium ions,[186] and green fluorescent protein (GFP), which only becomes phosphorescent when in close proximity with light from aequorin. Because aequorin emits light following contact with calcium, scientists recognized that it could be used to detect changes in the concentration of this ion in the cell. Such calcium 'signals' convey information coming from the outside of the cell to effector proteins. These effector proteins carry out important tasks in the body, for instance as regulators of heart contraction, secretion of insulin by the pancreas, and release of neurotransmitters in the brain.[187]

Lionel Jaffe and colleagues, also at Woods Hole, used aequorin to show that calcium signals play a key role during activation of the egg by the sperm during fertilization. By injecting the protein into fish eggs and then adding fish sperm under a microscope, Jaffe's team showed that, as the sperm fuses with the egg, a burst of bioluminescence occurs, beginning at the point of sperm–egg fusion and travelling across the egg like a forest fire.[188] In 2002, my colleagues and I identified the sperm PLCζ protein, and subsequently showed that this triggers the calcium signal in eggs ranging from fish to humans.[189] And Andrew Miller at Hong Kong's University of Science and Technology has used aequorin to study the role of calcium signals during development of the zebrafish (*Danio rerio*).[190] This species is ideal for studying vertebrate development, since its embryos develop outside the mother's body and are transparent, making it possible to perform fluorescent imaging in the living embryo. Miller and his team used this approach to show that calcium signals of many different sizes and shapes regulate key stages of embryo development, from establishment of the main body and tissue layers, to development of specialized tissues or organs like the heart or brain.[190]

Yet despite aequorin's importance for studying calcium signals, it was Shimomura's discovery of GFP that has had the bigger impact on biomedical science.[191] For the isolation of the gene coding for this protein suggested that GFP could be used to make the protein products of other genes 'visible' by fusing the GFP gene sequence to them. In particular, Martin Chalfie of Columbia University and Roger Tsien of the University of California, San Diego, developed GFP in this way. The ability to tag a gene, and therefore its protein product, has revolutionized cell biology, by making it possible to track a protein's movements in a living cell. And Tsien's creation of a range of differently coloured fluorescent proteins, which fluoresce at different wavelengths in the light spectrum, means the localization of two or more proteins in the cell can be studied at the same time by giving them differently coloured tags. For the discovery and development of GFP as a technology, Shimomura, Chalfie, and Tsien received a Nobel Prize in 2008.[191]

Use of fluorescent tags to label proteins is now a standard approach to track the movements of proteins in cells, thereby providing important clues as to their functions. For instance, my colleagues and I have used this approach to study how proteins called two-pore channels, or TPCs for short, defend the body against infection or cancer.[192] Cytotoxic T-cells are white blood cells that recognize infected or cancerous cells and kill them by forming a connection with the target cell, introducing noxious substances that destroy it. By labelling TPCs with a red fluorescent tag and expressing these in human cytotoxic T-cells in culture, we showed that when these cells come into contact with an infected cell, TPCs physically move to the point of contact, where they trigger calcium signals that regulate the killing event.[192] Such information may aid the design of new drugs that enhance this process.

Green Eggs and Sperm

We can learn much from studies of cells in culture, but there are limits to how much they can reflect the complexity of many bodily processes. A particularly powerful use of protein labelling combines this technique with transgenic technology to create a whole animal that expresses a labelled protein in its cells. One study investigated the inheritance of subcellular structures called mitochondria, which produce most of the energy in our bodies. Mitochondria have another distinctive feature. They contain their own DNA genome, distinct from that in the cell nucleus. This reflects the fact that these subcellular structures were originally free-living bacteria that became incorporated into our single-celled ancestors about 1.5 billion years ago, in a relationship that became mutually beneficial: the mitochondria providing energy and the host cell a sheltering environment.[193]

How vital mitochondria are to multicellular life is shown by the effects of cyanide, which blocks energy production by these tiny powerhouses, causing almost instant death.[194] Some people have mutations in their mitochondrial genome, leading to a reduction in the capacity of these subcellular structures to produce energy.[195] This particularly affects processes requiring lots of energy, like vision, muscle contraction, and conduction of impulses in the brain. These gene defects tend to be associated with muscle weakness, neurological problems, and in one type of disorder, blindness that occurs in middle age. The particular symptoms vary depending upon which gene is affected and its specific role in the energy-making process. But what all these disorders have in common is their inheritance through the mother.[195]

This inheritance pattern reflects the fact that a human embryo only inherits mitochondria from the egg, not from the sperm. Why this is so, was for a long time not very clear. For although the egg is substantially bigger than a sperm, and consequently contributes more mitochondria, the sperm also contains these structures; indeed, mitochondria supply the energy that powers the sperm's rapidly beating tail.[196] Moreover, studies of the fertilization process have shown that the whole sperm is engulfed by the egg following the fusion of these two cells.[197] So why isn't any sperm mitochondrial DNA passed on to the next generation?

To find out, Hiromichi Yonekawa and colleagues at Tokyo Metropolitan Institute of Medical Science created male transgenic mice in which a protein only found in mitochondria was labelled with GFP.[198] Since this colours the mitochondrion with fluorescence, it allowed Yonekawa's team to track the movement of the mitochondria in sperm from such mice as they fertilize the egg. Sperm mitochondria are concentrated in a region called the midpiece that lies midway between the sperm head and tail, making it easier to track their movements under a fluorescence microscope. The researchers found that when the sperm unites with an egg, the fluorescent mitochondria are engulfed into the egg with the rest of the sperm body, where they remain visible. But then the fluorescence suddenly disappears.[198] Further investigation revealed that the egg has a mechanism for identifying the male mitochondria and destroying them, although why it does this remains unclear.[199]

Other studies that have used this technology to study the reproductive process may have very important implications for medicine, and indeed society, in the future. For such studies have challenged a long-accepted dogma, namely that women are born with a finite supply of eggs that's depleted throughout life and exhausted at the menopause. Instead, it now seems possible that women may retain a hidden capacity to produce fertile eggs after the normal child-bearing age. This suggestion, first proposed by Jonathan Tilly of Massachusetts General Hospital in Boston in 2004, is based on the idea that stem cells exist in the ovaries of both young and old mammals, with the potential to develop into fertile eggs. In 2009 Ji Wu and colleagues at Jiao Tong University, Shanghai, isolated and cultured such putative stem cells, and infected them with a virus expressing GFP. When Wu's team injected the cells into the ovaries of sterilized female mice, the mice gave birth to green fluorescent offspring, suggesting that the stem cells gave rise to these offspring.[200] Subsequently, in 2012, Tilly and colleagues identified similar stem cells in human ovaries. And after expressing GFP in these cells with a virus, they showed that these stem cells could generate green fluorescent eggs in human ovary tissue transplanted into a mouse (see Figure 12).[201]

Not everyone is convinced of the existence of stem cells that can produce eggs in the ovaries of either mice or humans. Some critics believe the first study's findings resulted from the sterilization procedure not being totally effective, and the virus transferring its GFP fluorescence to normal eggs still remaining in the host ovary.[200] Something similar could account for the apparent generation of green fluorescent eggs from the human ovary stem cells. And at least four other research groups have failed to reproduce Tilly's or Wu's findings. Kui Liu, of the University of Gothenburg in Sweden, has said that 'we immediately repeated the experiment...My lab never got the cells.'[202] Yet a previous high-profile critic of Tilly, Evelyn Telfer, who studies

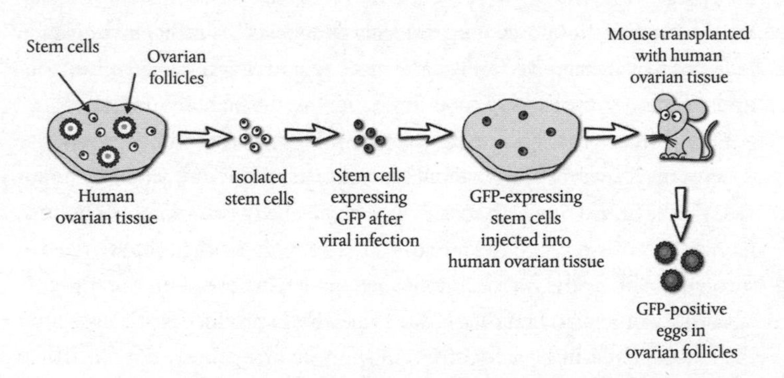

Fig. 12. Ovaries contain stem cells that can produce eggs

egg maturation at Edinburgh University, has become a convert to the idea of the existence of ovarian stem cells.[203] Telfer says she was impressed by data Tilly shared with her, and his 'openness and diligence' when she visited his lab.[202] Both Tilly and Telfer believe the menopause may not be due to a shortage of egg cells as such, but reflects a depletion of the cells within the ovary required to support and nurture eggs.

Might stem cells in the ovaries of post-menopausal women be coaxed to produce fertile eggs? If this were possible, it could help women who undergo the menopause prematurely, and so miss out on having children. More controversially, it could allow women to conceive long after the normal child-bearing age. And since the menopause is associated with health risks such as osteoporosis, heart disease, and cancer, finding ways to artificially sustain the production of eggs and the hormones they produce could be of great general benefit to women's health. 'It's very clear that keeping the ovaries working has tremendous health benefits on the ageing female body,' says Tilly. 'To me, there is a grander golden chalice here which is ageing itself. These cells may provide a way for us to tackle that tremendously important problem.'[203]

Visualizing the Brain

While combining fluorescent tagging and transgenic technology has led to valuable insights into the reproductive process, the most important application of this technology is in studies of the brain. The human brain is the most complex object in the known universe, with its 100 billion nerve cells, or neurons, joined by 100 trillion connections.[204] The mouse brain may contain a thousand times fewer cells, but is still highly complex.[205] There are also many functional similarities between human and mouse brains, with a recent study showing that the two species use their brains in remarkably similar ways while finding their way around a maze.[206] Identifying how the different cell types in a mouse brain are connected together, as well as studying their unique electrical properties—for the brain may be viewed as one huge circuit board—could lead to important insights into human brain function.

If regulatory DNA elements found next to genes only expressed in specific types of neurons are linked to fluorescent 'reporter' genes, these constructs can be used to create transgenic mice that express such tags only in these cell types, giving scientists a vital new resource for brain studies.[207] This technique makes it possible to identify specific cell types by their fluorescence. One approach, called 'brainbow', was developed by Jeff Lichtman and Joshua Sanes at Harvard University. 'In the same way that a television monitor mixes red, green and blue to depict a wide array of colors, the combination of three or more fluorescent proteins in neurons can generate many different hues,' says Lichtman.[208] The brainbow approach, which creates images that wouldn't look out of place in a modern art museum, makes it possible to tag several hundred neurons at once with 90 distinct colours (see Plate 1).

Such studies show that particular types of neurons are located in very specific parts of the brain, but also make it possible to visualize the intricate 3D structure of each neuron. The brain is like a dense forest in which the trees—the neurons—grow near, around, and on top of each other, their branches and roots intertwining.[209] Labelling an individual neuron using fluorescence reveals not just the body of the cell—akin to the trunk of a tree—but also its roots and branches, amongst those of adjacent neurons. Brain researchers have gained insights into the connections, and functional interactions, between nerve cell types by using different coloured fluorescent tags.[207] Use of GFP technology to create a detailed microanatomy of the brain mirrors something of what can be achieved using antibodies that recognize specific neurons by the proteins they produce, except that antibody staining can only be applied to dead cells. Fluorescence tagging allows scientists to label living cells. In the brain, scientists can insert microelectrodes that measure the electrical properties of the glowing neurons—including their roots and branches, which receive signals from other cells and transmit their own responses in return—providing clues about their function.

This approach has been used to investigate the neurons involved in smell and taste. A major insight into the molecular basis of such sensations was made by Richard Axel and Linda Buck at Columbia University in 1991.[210] They were trying to understand how the 5 million or so neurons in the skin of the inner nose convey sensory information to the brain. Each of these neurons possesses hair-like projections that detect the molecules associated with different odours and send a message to the olfactory bulb in the brain. This structure, located at the front of the brain, acts like a clearing house, relaying information about the odours that our nose detects both to the brain's higher cortex, which handles conscious thought, and the limbic system, which deals with emotions.[210]

Until Axel and Buck's discovery, the identity of the 'receptor' proteins on the surface membranes of the neurons in the inner nose had been unclear. To identify these, Buck and Axel decided to look for genes that were only expressed in these neurons, on the basis that some of them would code for the smell receptors. At first this search proved fruitless, because of a problem later acknowledged by Axel: 'there are a large number of odorant receptors, and each was expressed only at a very low level'.[210] But then Buck had the idea that the smell receptors might have similar characteristics to a protein involved in another sensory process. This protein—rhodopsin—is expressed in the rod cells of the eye and allows us to see. And sure enough, a search for genes related to rhodopsin revealed a huge family of different genes, each expressed in a specific neuron in the inner nose. In fact, there are around a thousand different odorant receptor genes, which still begs the question of how this number allows us to detect and remember the 10,000 different odours that a typical person can smell. Buck has likened the process to forming words with different letters. 'Just as you put different combinations of letters of the alphabet together to form words, you put different combinations of receptors together to get different smells,' she says.[211] For the discovery, Axel and Buck were awarded a Nobel Prize in 2004.[210]

Odorant receptors detect smells in the nose, but how is this sensory information transmitted to the brain? It's here that GFP technology has proven particularly useful. Peter Mombaerts, a member of Axel's laboratory who subsequently founded his own group at the Rockefeller University in New York, tagged one of the odorant receptor genes with GFP.[212] He then created

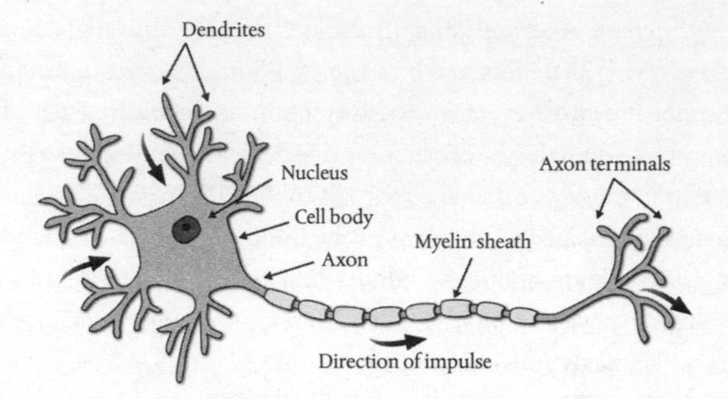

Fig. 13. The nerve cell

a knockin mouse expressing the modified receptor. Studying the pattern of fluorescence in the knockin mouse brain, he made a surprising discovery about the neurons that detect odours. Neurons have multiple inputs—dendrites—and one output—the axon (see Figure 13).

The axon can be shorter than a millimetre or longer than a metre in the case of neurons in the spinal cord.[213] GFP labelling made it possible to track the axon of an odour-sensing neuron as it passed from the inner nose into the brain. But while the labelled cells that expressed a specific odorant receptor were scattered across the surface of the inner nose, in the brain they all converged at a specific spot on the olfactory bulb.[212] This is a remarkable route-finding feat—like a blind person navigating the length of a football field through a crowd of thousands of people by touch alone. Subsequent studies have shown that specific odorant receptors on the developing axon interact with other proteins to guide the axon in the correct direction.[214]

Light-Induced Thoughts

Fluorescently labelling cells in the brain to measure both their electrical properties and anatomical relationships is one way in which light has been harnessed for biomedical research. An even more remarkable technology uses light to activate cells in the living brain. Optogenetics, as this new technology is known, uses genetically encoded switches that allow neurons to be turned on or off with bursts of laser light. To understand optogenetics, we need to step back and consider how the brain and nervous system works. At the most basic level, this system is like a highly complex electrical circuit.[215] Each neuron has various pump and pore proteins in its dendrites that control the ionic composition of the cell. The unstimulated neuron normally has a negative electric charge. But brain chemicals—known as neurotransmitters—modulate the charge of the neuron by their effect on the different pumps and pores in the dendrites. Activating neurotransmitters cause a flow of positive ions into the cell and negative ones out. At a certain point this triggers an explosive change called an action potential, a rapid flow of positively charged sodium ions into the neuron (Figure 14).[215] This causes a chain reaction that propagates itself down the axon to its terminus, where it stimulates the release of further neurotransmitters that cause positive or negative changes in neighbouring neurons. In contrast, inhibitory neurotransmitters make the neuron even more negative than normal, making it harder for an action potential to occur.[215]

Optogenetics uses light to manipulate the positive or negative charge of neurons genetically engineered to respond in this way (see Figure 15). Underpinning this technology was the discovery that certain types of bacteria and algae have protein pores on their surface membranes that allow ions into these microorganisms in response to light.[216] Such pore proteins, known as opsins, are related to rhodopsin which, as we've noted, allows the rods of our own eyes to detect light. In microorganisms, opsins play a variety of different roles. By powering movement, they can orientate an aquatic microbe towards the Sun's light in order to generate energy through photosynthesis, or help it find shade to escape the damaging effects of UV light, in environments ranging from tropical seas to polar regions.[216]

Opsins were mainly viewed as a curiosity of nature, but a few scientists began to recognize that such proteins might make it possible to activate neurons in the brain. In fact, as early as 1979, Francis Crick had speculated that 'this seems rather far-fetched, but it is conceivable that molecular biologists could engineer a particular cell type to be sensitive to light'.[217] Crick

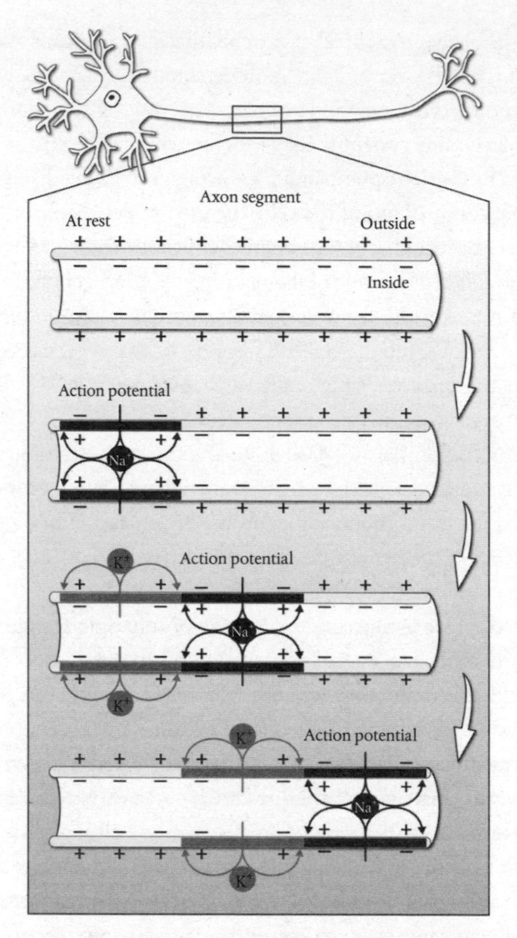

Fig. 14. Propagation of an action potential

offered no specific molecular strategy for such an approach; however, at the turn of the twenty-first century, a few scientists decided to test whether opsins provided a solution. The first was Gero Miesenböck, working at the Memorial Sloan Kettering Cancer Center in New York. In 2002, Miesenböck engineered neurons in the brain of a fruit fly to express the microbial opsins and showed

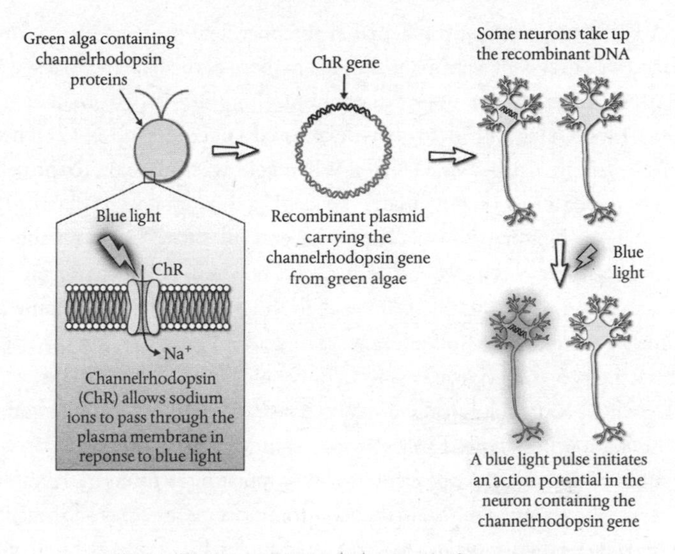
Green alga containing
channelrhodopsin
proteins
ChR gene
Some neurons take up
the recombinant DNA
Blue light
ChR
Na+
Channelrhodopsin
(ChR) allows sodium
ions to pass through the
plasma membrane in
reponse to blue light
Recombinant plasmid
carrying the
channelrhodopsin gene
from green algae
Blue
light
A blue light pulse initiates
an action potential in the
neuron containing the
channelrhodopsin gene

Fig. 15. Optogenetics uses light to stimulate neurons

that they could now respond to light.[218] Few people thought this approach would ever work in mammals, but, undeterred, Karl Deisseroth of Stanford University decided to see if he could develop this technology in rodents.

A practising psychiatrist as well as a neuroscientist, Deisseroth is concerned that psychiatry's ability to treat the most intractable disorders—severe depression, schizophrenia, autism—is limited by a lack of understanding of how the brain works. 'A cardiologist can explain a damaged heart muscle to a patient,' he says. 'With depression, you cannot say what it really is. People can give drugs of different kinds, put electrodes in and stimulate different parts of the brain and see changed behavior—but there is no tissue-level understanding. That problem has framed everything. How do we build the tools to keep the tissue intact but let us see and control what's going on?'[219] Deisseroth decided that using transgenic technology to express opsins in rodents could provide a revolutionary way to explore the functional role of different neurons in the mammalian brain.

A key initial question was whether the microbial proteins would function as well in rodent neurons as they did in those of the fruit fly. To test this, Deisseroth's research team—at that time just himself and two graduate students, Feng Zhang and Ed Boyden—expressed a microbial opsin in rat neurons in culture using a viral carrier. When light was shone on to the cells, this triggered an action potential.[219] So, could a similar effect be obtained in the living rodent brain? In fact it took several years for Deisseroth and his team to achieve this aim because it meant not only using transgenic approaches to modify specific cell types in the rodent brain to become responsive to light, but also finding ways to deliver light deep into the brain. This latter goal was achieved using an ultra-thin fibre-optic wire attached to a laser light source, surgically implanted into the brain. A combination of these approaches allowed Deisseroth's team to make a dramatic demonstration of the power of optogenetics. By stimulating neurons in the motor cortex—the region of the brain that controls movement—they showed that light could be used to make a mouse run in circles, like a remote-controlled toy. 'That's really the moment we knew that it could drive very, very robust behavior,' says Zhang.[219] But what fully convinced the neuroscience community of the power of optogenetics as a tool to study key mechanisms of brain function and dysfunction were a series of studies published in 2009.

First, another of Deisseroth's graduate students, Viviana Gradinaru, published a study with him in *Science* describing the use of optogenetics to define the precise neuron connections in Parkinson's disease.[219] Shortly afterwards, another study involving Zhang and Deisseroth was published in the same journal that examined the cellular basis of pleasure and reward. Such feelings are particularly associated with the neurotransmitter dopamine. By activating dopamine-producing neurons with light, the researchers could drive reinforced behaviour in the absence of any other cue or reward.[219] This study provided important information about the nerve impulses that may underlie addiction or conditions like depression, in which a person is unable to be pleased or excited by events around them. In addition, two other studies published in *Nature* by Deisseroth and his colleagues used optogenetics to identify neurons that regulate brain activities, which were known to be abnormal in schizophrenia and autism. These studies appeared in quick succession, and succeeded in convincing neuroscientists of the revolutionary potential of the technology. 'That was all people needed,' says Deisseroth. 'The world ran with it.'[219]

Since these studies were published, optogenetics has become a vital tool in the arsenal of neuroscientists.[220] Thousands of labs around the world are now using it to better understand the complex wiring of the mammalian brain and the basis of mental disorders including addiction, depression, Parkinson's disease, autism, pain, and stroke. According to Robert Malenka at Stanford University, the technology has 'allowed neuroscientists to manipulate neural activity in a rigorous and sophisticated way and in a manner that was unimaginable 15–20 years ago'.[220] And optogenetics continues to evolve.

One important development is the identification of specific types of opsins with different effects on the neurons in which they are expressed, in response to different wavelengths of light. So the opsin used by Deisseroth and his team in their pioneering studies causes an influx of positively charged ions into the cell in response to blue light, thereby activating them. Another type of opsin allows influx of negatively charged ions in response to yellow light, and this inhibits the neuron's ability to fire.[220] This has made it possible to activate or inhibit neurons purely by shining a different colour light on a specific part of the brain. Meanwhile, a search for other naturally occurring opsins has identified forms that act at different speeds. Scientists expressing these different forms in a mouse brain can now practise exquisite control over the timing and duration of nerve impulses, manipulating brain activity at the same speed as such cells normally communicate with each other.[220]

Yang Dan and colleagues at the University of California, Berkeley, have used optogenetics to investigate the parts of the brain that control sleep.[221] By using light to stimulate neurons in a brain region called the medulla, Dan's team showed that they could induce rapid eye movement (REM) sleep in mice within seconds. This type of sleep is associated with dreaming and involves activation of the brain cortex and total paralysis of the skeletal muscles, presumably so we don't act out the dreams flashing through our mind. 'People used to think that this region of the medulla was only involved in the paralysis of skeletal muscles during REM sleep,' says Dan. 'What we showed is that these neurons triggered all aspects of REM sleep, including muscle paralysis and the typical cortical activation that makes the brain look more awake than in non-REM sleep.'[221] Dan believes that although other regions of the brain have been implicated in the sleep–wake cycle, 'because of the strong induction of REM sleep...this might be a critical node of a relatively small network that makes the decision whether you go into dream sleep or not'.[221] Given that many psychiatric disorders are associated with abnormalities in REM sleep, the researchers hope that such studies might provide insights into the basis of these disorders and possibly future treatments for insomnia.

Making a Memory

Optogenetics has also been used to explore how memories are encoded in the brain. Memory has been a topic of scientific interest ever since Aristotle wrote *On Memory and Reminiscence* in 350 BC.[222] He compared memories to impressions made in a wax tablet, a writing device in use at the time.[223] In the eighteenth century, the English philosopher David Hartley first proposed that memories were encoded in the activity of the brain. It was only in 1904, however, that the German biologist Richard Semon linked memories to changes in specific groups of brain neurons which he called 'engrams'.[223] An important insight into the potential physical basis of memory was made by Tim Bliss and Terje Lømo at the University of Oslo in the late 1960s. They discovered that repeatedly electrically stimulating a neuron in a region of the brain called the hippocampus boosted the cell's ability to talk to a neighbouring neuron.[224] This communication between neurons occurs across tiny gaps called synapses, and Bliss and Lømo realized that such strengthening of synaptic connections—which they called long-term potentiation (LTP)—could be the physical basis of memory. Studies carried out subsequently have shown that synapses are strengthened when rodents run around an enclosure, and that blocking LTP with drugs, or knocking out genes that regulate LTP, can impair memory in mice.[225] Other studies have shown that another process called long-term depression (LTD) has the opposite effect.

Yet despite these pieces of indirect evidence in support of the link between LTP, LTD, and memory, direct evidence for such a link has remained elusive, with Robert Malenka commenting recently that 'proof of causality—that LTP is actually used for the encoding of a memory in a manner that is absolutely required—has been extremely difficult, if not impossible, to generate'.[225] But optogenetics has now supplied what looks like such evidence, in a study led by Roberto Malinow at the University of California, San Diego. He and his colleagues created a virus that expressed microbial opsin and injected this into specific neurons in the brain of a rat. In classical 'conditioning' studies of memory formation, rats can be trained to fear a specific sound by following the sound with an electric shock. After such conditioning the sound alone will make the rats freeze in fear.[225]

By using light to stimulate neurons connecting a brain region involved in processing sound with one that handles fear, and then giving the rats a shock, Malinow's team created the same kind of fearful memory without the rats ever hearing the sound. This demonstrated, according to Malinow, that 'we can make a memory of something that the animal never experienced before'.[225] And an examination of the synapses of the neurons involved showed that the rats had undergone molecular changes that are a hallmark of LTP. What's more, when light was used to induce LTD, the rats no longer cowered after the simulated sound stimulus was triggered in their brains. Subsequent use of light to induce LTP was sufficient to reimplant the fear. 'We were playing with memory like a yo-yo,' says Malinow.[225] The Nobel laureate Eric Kandel, a pioneer of the study of the cellular basis of memory, believes the findings show 'more directly than the indirect evidence that existed before that LTP has a role in memory storage and it can be wiped out by LTD...This is the best evidence so far available, period.'[225]

Susumu Tonegawa of the Massachusetts Institute of Technology has also used optogenetics to explore the cellular mechanisms of memory formation. His studies have shown that memory is stored, not in individual neurons, but in circuits of multiple cells that he calls 'engrams', following the term first proposed by Richard Semon in 1904.[226] Tonegawa's team demonstrated that such memory circuits exist in the hippocampus, and show the synaptic strengthening associated with LTP. Such strengthening was blocked by drugs that inhibit gene expression, indicating that the neurons in these circuits must produce new proteins to consolidate memory. Importantly, Tonegawa's latest studies address an issue that may be central to understanding memory loss in humans, which can occur through concussion, stress, or diseases like Alzheimer's. The issue is whether the amnesia in such conditions is due to inability to store memories, or because access to a memory is blocked, preventing its recall. According to Tonegawa, 'The majority of researchers have favored the storage theory, but...this majority theory is probably wrong.'[226] His conclusion is based on the fact that although inhibiting gene expression in neurons involved in memory made mice forget a learned stimulus, it was possible with optogenetics to reactivate the memory. For Tonegawa, this suggests that 'past memories may not be erased, but could simply be lost and inaccessible for recall'.[226] Clearly, if the brain is severely damaged then the process of memory storage itself may also be defective, but such findings in an animal model offer the exciting prospect that it might be possible to rescue 'lost' memories in human patients suffering from amnesia by targeting the retrieval process.

One of Tonegawa's most remarkable findings is the discovery that optogenetics can activate 'happy' memories to treat a depressive state.[227] This study was stimulated by the discovery of neurons that become active when male mice enjoy a rewarding experience—spending time with a female mouse. The researchers then induced a depressive state by restricting the movements of the male mice, which led to their losing interest in pleasures like a sugary drink. However, light activation of the neurons associated with feel-good activities reversed the depressive state within minutes. While this effect was only short-lived, pulsing the neurons with light twice a day for six consecutive days had an effect that persisted even in the absence of light stimulation. 'We were able to cure the animals' depression,' says Tonegawa.[227] One important finding of the study was that an alternative strategy that involved exposure to females for five days did not cure the depressive state. Amit Etkin, a Stanford neuroscientist, believes 'that's one of the most intriguing aspects of the study. There's something special about the encoding of a positive memory that differs from simply being rewarded.'[227]

But it's not only in the study of brain functions that optogenetics is showing its potential. According to Karl Deisseroth, 'if you had to pick the next logical tissue for work on optogenetics, the heart is a great one'.[228] This is because heart cells are also activated by electrical impulses. And indeed Philipp Sasse and colleagues at the University of Bonn have engineered mouse ES cells to respond to light, and then induced them to turn into heart cells.[228] Shining light on to a patch of cells in a culture dish made the cells beat in unison. In contrast, when the researchers shone light on already beating cells, those cells began beating out of synchrony with the others, in what Sasse calls 'a cardiac arrest in a petri dish'.[228] And by using the modified stem cells to make transgenic mice, Sasse and his team showed that shining light on to the hearts of such mice in different places also made these beat out of step, mimicking the arrhythmias that can trigger fatal heart attacks in humans.[228]

These studies look to be just the start of an expansion in the range of application of optogenetics. There are now moves to stimulate skeletal muscles with light, since they are naturally activated by nervous impulses, with the idea that this might be used to study certain forms of paralysis and ultimately identify ways to overcome them. The technique may be applied to other excitable cells, like those of the immune system and insulin-secreting pancreatic cells, to better understand their properties, as well as conditions such as auto-immune disorders and diabetes.[228]

Optogenetics has also now been extended beyond the initiation of electrical impulses. Some chemicals in the brain do not stimulate pumps or channels but instead activate receptors in the cell surface that regulate important enzymes. By genetically fusing such receptors with opsins, it has been possible to activate such enzymes, and thereby the cellular processes that they regulate.[229] Meanwhile, there have been important new developments in the fibre-optic technologies that deliver light to the deepest reaches of the brain and other parts of the body. Ed Boyden is working on 'multi-wave' arrays that emit light at multiple points, allowing larger areas of the brain to be targeted.[230] What's more, use of different types of opsins that respond to infra-red light, which has a long wavelength and can penetrate deep into living tissue, now make it possible to control brain activity inside a genetically engineered mouse from a device outside its skull, without the need for implanted optic fibres.[231]

Other scientists are working to eliminate the need for external stimulation by light, by creating types of opsins that provide their own illumination. So Jack Tung, Robert Gross, and colleagues at Emory University, Georgia, and Georgia Tech, have taken the enzyme luciferase, which produces a bioluminescent product when exposed to luciferin—the chemical first isolated by Osamu Shimomura in 1956—and fused it to an inhibitory opsin.[232] They then expressed this gene construct in the brain of a rat and showed that when luciferin was injected into the brain, this disabled the brain's ability to respond to an amphetamine drug. The researchers are now using this approach to study ways to halt or prevent seizures—a characteristic feature of epilepsy—in rodents. 'We think that this approach may be particularly useful for modelling treatments for generalized seizures and seizures that involve multiple areas of the brain,' says Tung. 'We're also working on making luminopsins responsive to seizure activity: turning on the light only when it is needed.'[232]

Other research has shown that light itself may eventually be dispensed with as a stimulus to regulate neurons. So Sreekanth Chalasani and colleagues at the Salk Institute in California have pioneered a technique they call 'sonogenetics' that uses ultrasound to control the behaviour of neurons.[233] Admittedly, the demonstration was carried out in nematode worms, not mammals. Chalasani and his colleagues identified a cell surface protein pore called transient receptor potential cation channel C4 (TRPC4), that is naturally sensitive to sound waves. By introducing this protein into worm neurons the researchers could control their behaviour using ultrasound. 'As soon as the ultrasound hits the worm, the neuron turns on,' says Chalasani. 'And when the neuron becomes active, it is telling the rest of the neural circuit, "Hey, I've become active." When that information passes along, the animal then turns, goes back, and goes off in a different direction.'[233] It remains to be seen whether this approach can be applied to mice, but it seems highly likely. And since ultrasound waves can penetrate the skull, it's possible that this approach could be used to study GM rodents in a non-invasive fashion.

Given the powerful ways in which optogenetics has been used in mice and other model animals, would it be possible to use this technology in people? For instance, could a light-emitting device on the skull, use of luminopsins, or even ultrasound, be used to treat a human patient with a brain disorder like epilepsy or Parkinson's disease, or perhaps someone with a psychiatric or mood disorder? Of course, for this to be possible there would need to be some way of precisely genetically modifying the brain cells of such an individual. As we saw in Chapter 2, until now such precision has only been possible in mice, and more recently rats, and then only indirectly by first modifying ES cells, and then using these to create a whole animal knockout or knockin. But all that has begun to change. It's now time to explore in detail the revolution taking place in biology, called genome editing.

4

The Gene Scissors

The words 'revolutionary' and 'breakthrough' can be overused in media reports about new scientific discoveries. Many scientists blame journalists for oversensationalizing research findings, yet while there's undoubtedly some truth in this, it's clearly not the whole story, since a recent report in the *British Medical Journal* found that more than a third of press releases from top British universities contained exaggerated statements.[234] Just as journalists must engage the public's interest in order to sell newspapers—or advertising space on the Internet—so scientists and the institutions they work for are under increasing pressure to demonstrate the practical relevance of their work and may exaggerate its impact. Such hype is damaging not only because it misleads the public about the importance of a specific discovery, but also because it can lead to general cynicism about the value of science to society.

But every once in a while, a scientific discovery is made whose impact on society is likely to be so immense that even an abundance of superlatives may not do it full justice. Genome editing looks set to be such a discovery. This approach has been taken up by laboratories across the world as a research tool, and is drawing billions of dollars of investment from industry because of its potential for medicine and agriculture. Such is the excitement about genome editing that Nobel Prize-winning biologist Craig Mello, of the University of Massachusetts, remarked to me in 2015 that 'there truly is a revolution in genetics going on right now, with gene editing allowing us to modify the genomes of every medically or agriculturally important animal or plant, as well as giving us the potential to genome edit human cells, and potentially, even human embryos'.[235] Only a month after Mello made this statement, a study was published by Junjiu Huang and colleagues at Sun Yat-sen University in Guangzhou, China, reporting the use of this approach to genetically modify human embryos for the first time in history.[236] Subsequently, a team led by Kathy Niakan at the Crick Institute in London also reported genome editing of human embryos.[237] In neither of these cases were the modified embryos implanted into a woman, and Kathy Niakan made a strong case for the importance of such studies in furthering our understanding of the genes involved in human embryo development. However, as mentioned in the Introduction, in December 2018, Jiankui He, a scientist at Southern University of Science and Technology in Shenzhen, China, shocked the world when he revealed that he had edited the genomes of two human embryos that were later born as twin baby girls.[238]

Why genome editing has created such a stir is best understood by comparing it to the genetic engineering approaches considered in Chapter 2. There, we looked at two main approaches to gene modification. One involves the random insertion of a gene construct into the genome of a host cell. The advantage of this approach is that it can be applied to practically any cell. But its usefulness for biomedicine and agriculture has been limited by its inefficiency and also the fact that it involves essentially dumping a piece of DNA randomly into the genome. Also, it only provides the possibility of adding a foreign gene to a genome, not of modifying one already there.

The second approach, using ES cells, has a much greater level of precision. As we saw, it can be used to totally eliminate a gene's action in a mouse or introduce more subtle modifications, like a mutation to model disease or a fluorescent tag to label a gene's protein product. The limitation here lies not in the flexibility of the approach but rather in the gene targeting, which is a complicated procedure that must first be carried out on ES cells. Only subsequently can these be used to create a GM rodent. This use of ES cells is also associated with a more fundamental problem. It has not been possible to isolate, and thereby genetically modify, ES cells from other mammalian species besides mice, and more recently, rats and humans.[239] Cells with many properties similar to mouse ES cells have been isolated from other mammalian species like pigs and sheep.[240] But despite the superficial similarity, for some reason these cells lack the pluripotency required for making GM versions of these other species.[241]

In contrast to such limitations of traditional genetic engineering approaches, the power of genome editing lies in four key features.[242] First, the technique can be applied to practically any cell type from any plant or animal species, ranging from bacteria to humans. Second, it can precisely target any region in a genome, thereby completely knocking out the function of a gene, or subtly modifying it, by introducing a mutation or fluorescent tag. Third, the efficiency of gene targeting is extremely high, so no complicated drug selection to identify a one in a million event is required. Fourth, this type of genetic engineering leaves no trace of foreign DNA in the genome that is being targeted. Finally, the tools for the newest type of genome editing are simple to prepare, being well within the power of any scientist with basic molecular biology skills, reagents, and equipment. This is why laboratories across the world are adopting this technology in their research, whether they're studying bacteria, plants, animals, or human cells in culture. But from the point of view of genetic modification of life, perhaps the most revolutionary aspect of genome editing is the ease with which it can be applied to a fertilized egg, the source of all complex multicellular life.

With genome editing, it is now possible to create a knockout or knockin mouse in several months as opposed to a couple of years for the ES cell approach.[243] And unlike this approach, genome editing can be applied to fertilized eggs of other mammalian species. Over the last few years it has been used to create GM rabbits, goats, pigs, and monkeys. Referring to the technology, the Nobel laureate David Baltimore of Caltech in Pasadena, California, has said: 'These are monumental moments in the history of biomedical research. They don't happen every day.'[244] In plants, genome editing has generated modified versions of species like wheat, rice, potatoes, and tomatoes.[245] 'We are talking about bringing products to market in five to ten years,' says Neal Gutterson, vice president at Pioneer Hi-Bred, part of DuPont's chemicals and biotech seed business. 'That is a pretty damn good time line compared to other technology.'[246] The ease with which it's now possible to create GM species looks set to transform both medicine and agriculture. So what is the scientific basis of genome editing and what makes it so powerful compared to past approaches?

In fact, although the genome-editing revolution is a recent event, much of the science underlying it goes back decades. What sparked the revolution was the recognition that a number of previous discoveries could be brought together to form a new technology that may be thought of as a very precise pair of 'molecular scissors'. Such scissors accurately target a DNA sequence, like the restriction enzymes described in Chapter 2, but unlike the latter, they can be employed in a living cell. Or, as geneticist George Church of Harvard University has put it when describing the new technology: 'You don't even have to take the genome out of the organism...It's like you throw a piston into a car and it finds its way to the right place and swaps out with one of the pistons while the motor's running.'[247] Importantly, these scissors can not only cut a gene in a specific place, but also guide other tools to modify the gene in various ways.

Molecular Scissors

The discovery that a break in the DNA double helix inside a living cell can lead to precise modifications of that part of the genome was first made in the late 1990s by Maria Jasin and colleagues at the Memorial Sloan Kettering Cancer Center, New York. Jasin's main interest was in understanding how such breaks play a role in tumour formation. She was studying the BRCA2 gene, which, as we noted in Chapter 1, plays an important role in DNA repair. Its absence greatly increases the risk of breast and ovarian cancer, since DNA breaks are much less likely to be repaired.[248] An interesting side aspect of this study was the discovery that, in a normal cell, the repair process can proceed in two ways: either by connecting the two broken ends together, but in such a clumsy manner that this adds or deletes DNA sequence; or in a much more accurate fashion by homologous recombination that restores the correct sequence (see Figure 16).[249]

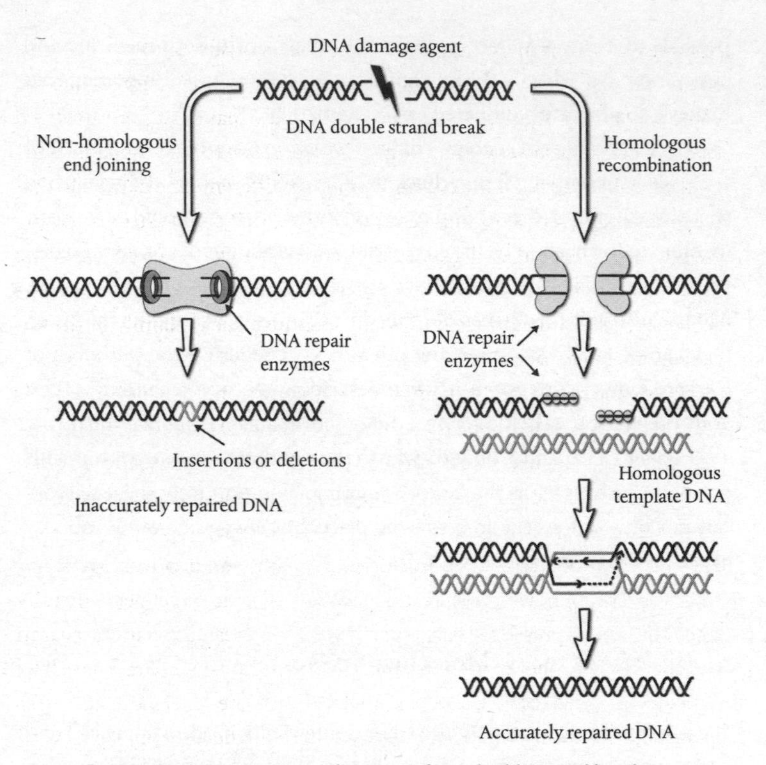

Fig. 16. DNA repair mechanisms after a double strand break

This showed that if a way were found to accurately create a break in the genome at a specific place, the cell's own repair machinery would then create a mismatch join that could result in a knockout of the gene, or if a suitable complementary piece of DNA was available, a substitution that would create a knockin modification. The only problem was that, at this time, no obvious way of creating a sequence-specific DNA break was known.

In fact another decade would pass before such a tool became available. This arose from a discovery made by Srinivasan Chandrasegaran and colleagues at Johns Hopkins University. They were studying a protein called FokI, a restriction enzyme of the sort we looked at in Chapter 2. What became clear from these studies was that FokI is divided into two separate structural regions. One carries out the enzyme's cutting action, while the other recognizes the DNA sequence to be cut. Chandrasegaran realized that, given this clear separation, it might be possible to fuse the part of the FokI gene that coded for the cutting action to a different type of protein that recognized different sites in a genome.[249] If so, this could create a cutting tool that could be targeted against any gene. What was required was some kind of family of proteins that recognized a wide variety of different DNA sequences. And Chandrasegaran finally identified such a family in the so-called 'zinc finger' proteins.

These gene regulatory proteins get their name from the zinc ion at their core and the finger-like appearance of part of their 3D structure.[250] The specific genes they regulate are controlled by steroid hormones. These bodily messengers include the sex hormones testosterone and oestrogen, the pregnancy regulator progesterone, and cortisol, which is released into the body at times of stress. Such hormones work by binding to a particular zinc finger protein, which then attaches to its specific target gene, thereby activating its expression. What particularly interested Chandrasegaran was the large number of these proteins, each specific for a particular DNA sequence. This led to the idea that it might be possible to create a sequence-specific DNA cutting enzyme by fusing different zinc finger proteins with the cutting part of FokI (see Figure 17). And tests with the first zinc finger nuclease, or ZFN as it was named, showed the hybrid protein to indeed have this capacity.

ZFNs made it possible for the first time to precisely genetically modify the genomes of cells from various different species.[251] So Dana Carroll of the University of Utah first used the technology to modify a gene in fruit flies. 'It showed for the first time that we could hit a real gene at its normal genomic site in a real organism,' says Carroll.[252] One species in which genetic engineering was greatly enhanced by ZFNs was the zebrafish. As we saw in Chapter 3, this species has been used to investigate the molecular changes during embryogenesis; it is also proving to be important for studying processes in the adult vertebrate. What's been lacking, however, is a way

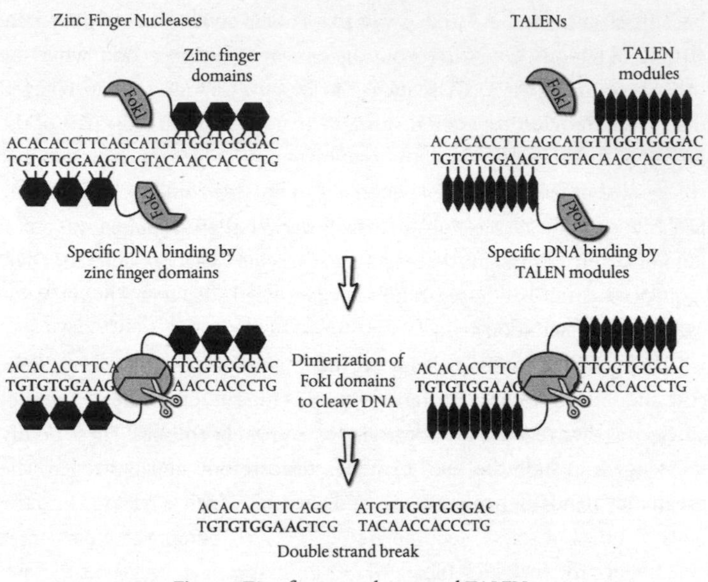

Fig. 17. Zinc finger nucleases and TALENs

to make knockout or knockin versions of this species. Use of ZFNs made it possible for the first time to create GM zebrafish to study the role of specific genes during development and those that regulate important bodily processes in the adult.[253] One study used this technology to knock out a gene, antithrombin 3 (AT3), involved in regulating blood clotting. This resulted in fish with a strongly increased tendency towards thrombosis—the disorder in which excessive blood clotting occurs—as the fish grew older.[254] Since thrombosis can lead to heart attacks and stroke in people, this zebrafish mutant could help scientists to understand the molecular basis of thrombosis. It may also be used to test drugs that can inhibit excessive clotting.

More recently, the cutting part of FokI has been fused to the DNA recognition part of a different protein family, called transcription activator-like effectors, or TALEs. These proteins are secreted by bacteria that infect plants; they activate expression of particular genes in the plant cell and thereby make it easier for the pathogen to establish itself in its host. Once fused to FokI, these are referred to as TALE nucleases, or TALENs (see Figure 17).[255] Importantly, there are far more TALEs than zinc finger proteins, making it possible to target a much larger variety of regions in the genome with TALENs than with ZFNs. And as such, these new tools have been used to successfully modify genomes from species as diverse as yeast, fruit flies, zebrafish, pigs, and cress, as well as human cells in culture.[255]

Despite the power of ZFNs and TALENs, they remain a relatively cumbersome tool. The fact that these proteins must each be created anew to generate a novel cutting specificity has limited the ease with which they can be produced and employed in biomedicine. Because of this, after the development of these cutting tools, the search continued for better ones. In the end, a major breakthrough on this front came with the discovery that a sequence-specific cutting enzyme existed in bacteria that was different from ZFNs and TALENs in one key respect: the recognition device that guides this cutting enzyme to a specific sequence in the DNA genome is the chemical cousin of DNA, namely RNA.

A CRISPR Cut

The cutting system that proved so revolutionary is known as CRISPR/CAS9. The first part of the name comes from the acronym 'clustered regularly interspaced short palindromic repeats', CAS9 being the enzyme that cuts the DNA. In fact, the CRISPR repeats have been known about since 1987, when they were first noticed in the genome of the bacterium *E. coli* by Yoshizumi Ishino and colleagues at Osaka University. The researchers were studying a gene called inhibitor of apoptosis (IAP) and, as well as sequencing the coding region of this gene, Ishino's team also sequenced the adjacent part of the genome to identify regulatory elements that switched the gene on or off. However, instead the researchers found something odd. Near the IAP gene lay five identical segments of DNA, each separated by 'spacer' regions; unlike the repeats, each of these had a unique sequence. While describing these 'unusual' DNA structures, Ishino and his colleagues concluded, almost with a shrug, that the 'biological significance of these sequences is not known'.[256]

So things remained until 2002, when Ruud Jansen and colleagues at Utrecht University in the Netherlands carried out a computer-based 'bioinformatic' survey of different bacterial genomes and made a surprising finding. The researchers noticed that, far from being unique to *E. coli*, the odd genetic sandwiches were present in a staggering number of bacterial species.[257] It was Jansen and his colleagues who christened the sequences CRISPR. They also found that these sequences were generally found next to CRISPR associated (CAS) genes that coded for proteins with a similar structure to enzymes which interacted with DNA. The researchers concluded that this association suggested a 'functional relationship' existed between the CAS genes and CRISPR sequences; however, the role of this relationship remained unclear.[257]

Then, three years later, in 2005, several research groups independently made an intriguing discovery. They noticed that the CRISPR spacers looked a lot like the DNA of bacteriophages—the viruses that infect bacteria which we came across in Chapter 2. Crucially, when Eugene Koonin, an evolutionary biologist at the National Center for Biotechnology Information in Bethesda, Maryland, heard about this discovery, 'the whole thing clicked'.[258] Koonin suggested that bacteria use CAS enzymes to grab fragments of DNA from viruses. They then insert the viral DNA fragments between their own CRISPR sequences. Later, if a virus of the same type comes along, the bacteria use CRISPR to recognize the invader, creating what science journalist Carl Zimmer has called a 'molecular most-wanted gallery'.[258] In other words, just as our own immune system remembers past infections and responds rapidly if they reappear, which is why we're generally only infected once by viruses like chicken pox, so bacteria possess an analogous system. Far from being useless pieces of junk DNA, it was now looking as if CRISPR sequences played a key role in bacterial immunity. And this discovery was to have an immediate practical application in an unexpected area.

In particular, Rodolphe Barrangou, a microbiologist working for Danisco, recognized that the CRISPR sequences might provide an important defence for the bacteria that converted milk into yogurt at his company, for sometimes entire cultures would be lost to outbreaks of bacteriophage.[258] To test Koonin's hypothesis, Barrangou and his colleagues infected the milk-fermenting bacterium *Streptococcus thermophilus* with bacteriophage. The viruses killed most of the bacteria, but some survived. Analysis of the resistant bacteria confirmed that they had inserted DNA fragments from the bacteriophage into their spacers. When the researchers deleted the new spacers, the bacteria lost their resistance. This discovery led many manufacturers to select for customized CRISPR sequences in their cultures so the bacteria could withstand virus outbreaks. 'If you've eaten yogurt or cheese, chances are you've eaten CRISPR-ized cells,' says Barrangou.[258]

So much for the functional role of the CRISPR sequences, but what remained unclear was the underlying mechanism of this defence system. In the end, the big breakthrough on this front was made by Jennifer Doudna, of the University of California, Berkeley, and Emmanuelle Charpentier, a French scientist working at Umeå University, Sweden. Doudna was brought up in Hilo, a region of Hawaii with dramatic waterfalls, fertile rainforests, and blooming tropical gardens.[259] It sounds like an ideal place in which to grow up, but with her blonde hair and blue eyes Doudna felt out of place amongst the other children, who were mainly of Polynesian and Asian descent. 'I think to them I looked like a freak,' she recalls. 'And I felt like a freak.'[259] This feeling of isolation contributed to a kind of bookishness that led to an interest in science from a young age. Doudna found her calling at school after hearing a female scientist discuss her cancer research studies. 'I was just dumbstruck,' she says. 'I wanted to be her.'[259]

Spells as a graduate student working for Jack Szostak at Harvard University, and then postdoctoral research with Thomas Cech at the University of Colorado, provided Doudna with impeccable mentorship, for both men went on to receive Nobel Prizes for their work. Doudna was already recognized as an expert in the structures of different RNAs—increasingly seen as a key player in the cell alongside DNA—when she was asked by the Berkeley environmental researcher Jillian Banfield for help with sequencing the genome of a bacterium isolated from an abandoned

Californian mine. Doudna recalls, 'I remember thinking this is probably the most obscure thing I ever worked on.'[260] The analysis revealed CRISPR sequences, and, while analysing them, Doudna became fascinated by this bacterial defence system and determined to understand its molecular basis. In particular, given her interest in RNA, Doudna was intrigued to discover that this molecule appeared to play an important role as an intermediary in the CRISPR system, although how it did so remained unclear.

Such was the situation when Doudna met Emmanuelle Charpentier, at a conference. Charpentier, a microbiologist, was also studying CRISPR, but in the context of its role in *Streptococcus* bacteria, a group that includes bacterial species that can cause sore throats as well as those that have alarming—and potentially fatal—'flesh-eating' properties.[261] The particular type of *Streptococcus* bacteria that Charpentier was studying produced a CAS protein called CAS9. As the two scientists talked about the different approaches they were using to study CRISPR, it became clear that combining their complementary skills would be far more powerful than working alone. And indeed, in less than a year, the two made a major discovery. 'We found that CAS9 has the ability to make a double-stranded break in DNA at sites that are programmed by a small RNA molecule,' says Doudna. 'What was so important was that we could really show how the CAS9 protein worked.'[262]

Doudna and Charpentier had discovered that CRISPR works by producing an RNA copy of a virus DNA and this 'guide RNA' then directs the CAS enzyme to a specific site in the viral genome, which is cleaved (see Figure 18). In a way, it's a bit like a word search function combined with a cut-and-paste facility, with the guide RNA carrying out the search and the CAS9 enzyme doing the cutting. However, the truly revolutionary aspect of this discovery was not only that it provided the mechanism of a key process in microbes, but that it also suggested the system could be adapted as a form of genetic engineering.

'One day... we realized, gosh, this could be a very powerful technology,' says Doudna.[263] She and Charpentier recognized that it might be possible to reprogramme the system to recognize new DNA sequences: 'If it could be made to work in eukaryote systems—plants and animals—then you'd have

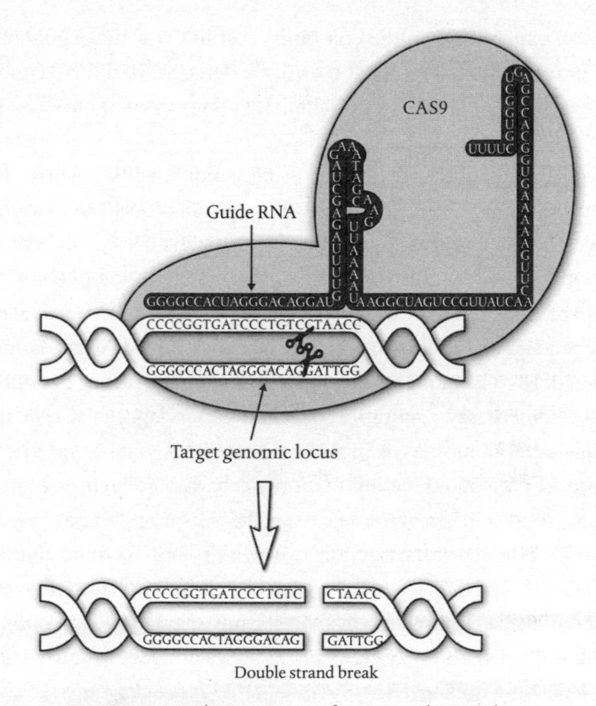

Fig. 18. CRISPR/CAS9 system for targeted DNA cleavage

a system where you could effectively decide where to produce a double-stranded break in that cell's genome.'[262] Of course, this is also possible with ZFNs and TALENs. But as Charpentier pointed out: 'All the other tools, each time you want to target DNA at a specific site you have to engineer a new protein... With CAS9, anyone can use the tool. It's cheap, it's fast, it's efficient, and it works in any size organism.'[264] Unlike ZFNs and TALENs, CRISPR always uses the same enzyme—CAS9. Instead, it's the guide RNA that varies. Since the guide RNA can be generated in a few days for a small sum, CRISPR can be performed in a fraction of the time and cost of the other technologies. Indeed, James Haber of Brandeis University recently estimated that ZFNs, which typically cost US$5,000 or more to order, are 150 times more expensive than the combined cost of CAS9 and a guide RNA, which comes to about $30. 'That effectively democratized the technology so that everyone is using it,' says Haber. 'It's a huge revolution.'[265] So what form is this revolution taking?

The most straightforward way of using genome editing is to create a knockout cell or organism. When a cutting enzyme cuts DNA at a specific sequence, the cell responds by seeking to repair the break. But when this happens the repair is often a botched affair, as we noted in the section 'Molecular Scissors'. If this occurs in a gene coding for a protein, it can disrupt the genetic code and ensure that the protein is not produced. However, a segment of DNA identical to the one surrounding the cutting site but with a subtle difference, say a mutation or a fluorescent tag, can also be introduced into a cell alongside CAS9 and the guide RNA. In this case, the cell's own natural homologous recombination mechanisms can create a knockin change. As we saw in Chapter 2, such specific gene targeting has been possible in mice for some years, but only through the indirect route of modifying ES cells first, and then using these to create a whole modified mouse. But genome editing can be applied to any cell from any species, and unlike the ES cell strategy, it's extremely efficient, and does not need drug selection to identify a one-in-a-million event. Because of this, research groups around the world are applying this approach to their cell type and species of interest.

Life in a Dish

One key system used in biomedical research is human cells grown in culture. Such cells may be obtained from biopsies or organs donated to research by people when they die. However, such 'primary' cells have an inbuilt limited lifespan, so that, if allowed to divide in a culture dish, after 40 to 60 cell divisions (the figure varies depending on cell type), the cells eventually stop dividing. This is called the 'Hayflick limit' after Leonard Hayflick, the man who discovered it in 1962, and it is thought to be connected to the natural ageing process.[266] Cancer cells have no such limit to their capacity to divide, having gained immortality through mutations that overcome the natural barriers to unlimited division that normal cells possess. The first and most famous of such 'immortalized cell lines' is the HeLa cell.[267]

This was isolated from the highly malignant ovarian tumour of Henrietta Lacks, a poor black woman, in 1941, without her knowledge or consent. Lacks soon succumbed to the cancer, but HeLa cells continue to multiply in the incubators of laboratories across the world, having been used to develop the polio vaccine, aided research into cancer and AIDS, and used to assess the effect of toxins and radiation on human cells.[267] Subsequently, many other immortalized human cell lines were isolated from tumours, while infection of normal human cells with tumorigenic viruses can immortalize such cells.[268] Some immortalized cell lines retain characteristics of the cell types from which they originated in the body. This means that they can be used to study the specific properties of such cell types in a culture dish. For instance, some immortalized cell lines isolated from the pancreas will secrete insulin in response to a glucose stimulus, just like normal pancreatic beta cells.[269]

As we saw in Chapter 2, the discovery of ways to precisely modify the genomes of mouse ES cells, and then the use of these to create a GM mouse, revolutionized biomedicine by making it possible to dissect gene function in a living mammal. Ironically, humans are the only other species besides mice and rats in which it has been possible to isolate ES cells.[270] Clearly though, there are ethical reasons why it would be impossible to use such cells to create, first, living human chimaeras and then breed these to produce GM humans. So the question was whether genome editing could be applied to immortalized human cell lines, or indeed primary cells generated from biopsies, in order to study the roles of specific genes in cellular processes.

In fact, even prior to the development of genome-editing technology, there has been a way to modify the expression of genes in human cells using an approach called 'RNA interference'.[271] This is a natural process found in cells of species ranging from petunias to people, whereby certain types of RNA act to suppress gene expression. As we saw in Chapter 2, the DNA sequence of letters in a gene acts as a linear code that's translated into another linear code, the unique string of amino acids that makes up each protein.

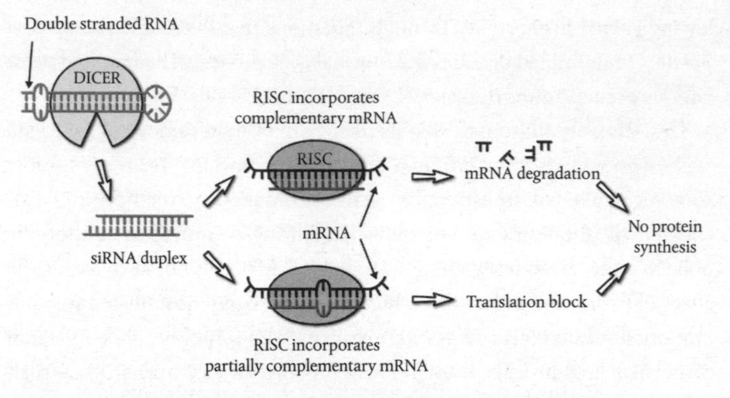

Fig. 19. RNA interference inhibits gene expression

However, this isn't a direct translation; rather it requires an intermediary—messenger RNA—essentially a copy of the protein-coding sequence of each gene. In RNA interference, small regulatory RNAs either trigger destruction of the messenger RNA or block its translation into protein. The process is regulated by proteins that exist naturally in cells—one called DICER generates the small regulatory RNAs, also known as siRNAs; while a protein complex named RNA interference silencing complex, or RISC, mediates the RNA destruction or the block to translation (see Figure 19). The discovery of this process by Craig Mello and Andrew Fire of the University of Massachusetts led to their receiving a Nobel Prize in 2006 and was important for two reasons. First, it revealed a far more central role for RNAs in the control of gene expression than had been suspected. Second, it provided a tool to inhibit the expression of specific genes in various cell types from different species; in particular the approach could be used to 'knock down' gene expression in human cells.[271]

This made it possible for the first time to explore the function of specific genes in human cells. So, for instance, in 2009 my colleagues and I used this approach in a human cell line to show that a protein named TPC2, which forms a pore on the surface of a cellular structure called the lysosome, plays a key role in generating calcium signals.[272] Formerly, the lysosome was thought merely to be the 'dustbin' of the cell, gobbling up its discarded waste, but studies like ours show that it also has another important role to play in regulating a variety of cellular processes via such calcium signals.[273]

Despite these important features of RNA interference, the approach does have some limitations. One is that the inhibition of protein production is often incomplete, in contrast to knockout at the DNA level, which completely blocks expression of a gene. Second, the approach can only be used to inhibit gene expression, not introduce subtle alterations like the knockin approach discussed. Because of this, when CRISPR genome editing was first developed, an obvious question was whether it could be used to make knockouts or knockins of genes in cultured human cells. And the person who showed that this was indeed the case was a scientist we have already come across—Feng Zhang.

As we saw in Chapter 3, as a PhD student Zhang worked with Karl Deisseroth, who would later say of Zhang that his 'skills were absolutely essential to the creation of optogenetics'.[274] However, making his mark in one hot new area of biotechnology was clearly not enough for Zhang, who has been called the 'Midas of Methods' because of his pioneering role in different areas of biotechnology.[274] As a postdoctoral researcher working with Paola Arlotta at Harvard University, he devised a way to use TALEs not to cut, but to activate genes artificially. Arlotta has subsequently praised Zhang's capacity for highly inventive problem solving. 'He has an ability to see the simplicity of things,' she says. 'That's a gift that not everybody has.'[274] Subsequently, having set up his own group at the Broad Institute, itself part of the Massachusetts Institute of Technology, Zhang heard about CRISPR from a speaker at a scientific advisory board meeting. 'I was bored,' he says, 'so as the researcher spoke, I just Googled it.'[275] Then he went to Miami for a conference, but while there he spent his time reading papers on CRISPR and filling his notebook with ideas about how to use it to modify the human genome. 'That was an extremely exciting weekend,' he says.[275] Back in Boston, Zhang quickly set about seeing whether the technology could be adapted to edit human cells in culture. In early 2013 he published his findings, showing he could knock out not just one gene, but several simultaneously, in such cells. In fact, he was not the only scientist to pursue this goal; in the same issue of *Science* in which Zhang's paper was published, George Church reported that he had also edited human cells with CRISPR/CAS9.[275]

The ability to knock out genes in a human cell line is proving important in uncovering the functions of those genes. The Human Genome Project showed that there are just over 22,000 genes in our genome, but our understanding of their functions is still very incomplete. And the molecular basis of many important cellular processes is far from clear. So there is much work still to be done to connect particular genes with specific processes. One way to do this is to knock out genes individually and assess the effect on a particular process. But it could take many years to screen a whole genome by this route. A more powerful approach is to carry out a so-called 'genome-wide screen'.[276] In such a screen, cells are cultured in thousands of individual 'wells' arrayed in a grid. Into each well a CRISPR/CAS9 construct is introduced that inhibits the expression of a specific gene. Given there are over 22,000 genes in the human genome, an equivalent number of wells need to be screened. Then all these wells are subjected to a test of the cellular process of interest.

Such a screen, by a group led by Feng Zhang, identified genes involved in the developing resistance of human melanoma cells, growing in a culture, to the drug vemurafenib, also known as Zelboraf®, which is used to treat skin cancer.[277] Since such drug resistance is a major reason why some cancers start regrowing following initial successful treatment, this could provide important information that allows clinicians to devise ways to combat such resistance. Another study led by Zhang and Philip Sharp, also at the Massachusetts Institute of Technology, used genome editing to screen for genes involved in tumour formation in a mouse model. According to Sharp, it's important to study cancer in a living animal because 'tumour evolution is an extremely complex set of processes, or hallmarks, controlled by networks of genes'.[278]

The study's aim was to identify genes involved in metastasis, the process in which malignant cells escape their tissue of origin and travel around the body in the blood, spreading cancer where they go. First, genome editing was used to knock out genes in cultured mouse lung cells across the whole genome (see Figure 20). These cells were then injected into live mice, and in

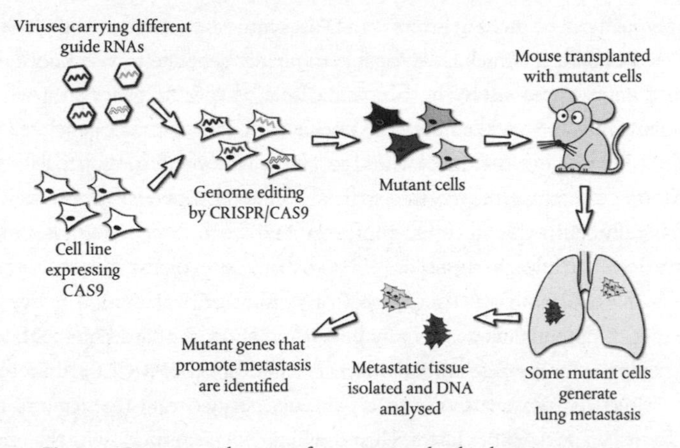

Fig. 20. Genome-wide screen for genes involved in lung metastasis

some cases this created metastatic tumours. By isolating these tumours and sequencing their genomes, it was possible to identify the genes knocked out and establish a role for these genes in metastasis. Zhang believes the study 'represents a first step toward using CAS9 to identify important genes in cancer and other complex diseases'.[278]

Remote-Controlled Genes

In these studies, genome editing was used to knock out genes. However, the same approach can also be used to introduce more subtle knockin changes, such as single amino acid modifications like those found in human diseases, or the addition of a fluorescent tag. We saw in Chapter 2 how such subtle genetic modifications can be introduced into the genome of a mouse using ES cells as an intermediary. However, with genome editing, it's possible to create a knockin mouse in a single step simply by including a segment of DNA similar to the region of the gene to be modified.[279] Compared to the complicated construct required for the ES cell method, such a DNA fragment can be made in hours on a DNA synthesizer. As such, the CRISPR/CAS9 approach is making it feasible to rapidly generate mouse knockouts and knockins to study the functional role of specific genes in a whole animal model. So while standard knockout and knockin mice generated via ES cells typically take 18 months and cost up to $20,000, according to Douglas Mortlock, a transgenic mouse expert at Vanderbilt University, 'now we can basically squirt this stuff into mouse embryos and three weeks later mice are born that have the mutation … at a cost of $3,000 or less. It's stunning.'[280]

A particularly interesting aspect of genome editing though is that its range of potential uses goes way beyond creating standard knockout, and knockin, cells or whole animals. So, especially for CRISPR/CAS9, the ability to direct the enzyme CAS9 to a particular sequence in the genome has opened up possibilities beyond just disabling a gene or modifying the properties of its protein product. Instead, the capacity to position CAS9 in this sequence-specific manner can also be used to switch genes on or off in a controlled fashion. In this case, instead of acting like a pair of scissors, CRISPR/CAS9 works more like a dimmer switch. To understand how, it's worth looking in more detail at how genes are expressed.

Genes in organisms as diverse as bacteria and humans are regulated by gene regulatory proteins known as transcription factors.[281] These proteins bind to regulatory DNA sequences adjacent to the gene (the so-called gene promoter) and thereby affect the activity of the RNA polymerase that produces the messenger RNA that acts as the intermediary between a gene and its protein product (see Figure 21A). Given that each gene is controlled by a large number of regulatory sequences, which can activate but also inhibit the polymerase, this allows an exquisite level of control depending on which transcription factors happen to be in any particular cell type.

The gene constructs used to create a standard transgenic mouse generally include a regulatory DNA sequence adjacent to the gene that binds a powerful activating transcription factor, thereby ensuring the gene is always turned on. Alternatively, a transgenic mouse may be engineered so that the regulatory region binds the transcription factor only when the mouse is injected with a specific chemical, allowing reversible control of gene expression in the

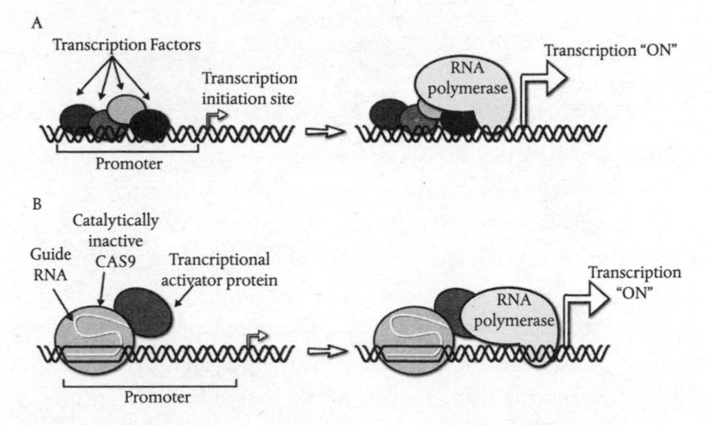

Fig. 21. Regulation of gene expression by transcription factors and CAS9

animal.[282] A limitation of this approach is that the transgene is a foreign gene inserted randomly in the genome. But genome editing makes it possible to reversibly control the expression of a cell's own genes. A guide RNA directs a version of CAS9 engineered not to cut DNA, but instead to attract a specific transcription factor to a site in the genome (see Figure 21B). While one version of this approach uses a transcription factor that can be induced to bind by a chemical injected into the mouse, there may be other potential ways to turn gene expression on or off. One would be a cross between the optogenetic approach we looked at in Chapter 3 and CRISPR/CAS9.

Moritoshi Sato's team at the University of Tokyo recently demonstrated how light can be used to regulate CRISPR/CAS9 genome editing. For Sato, one limitation of conventional CRISPR/CAS9 technology was that 'the existing CAS9 does not allow us to modify the genome of a small subset of cells in tissue, such as neurons in the brain . . . we have been interested in the development of a powerful tool that enables spatial and temporal control of genome editing'.[283] To achieve this, Sato and his colleagues split CAS9 into two inactive halves, each tagged with light-sensitive tags. When the two halves are expressed in a cell with a guide RNA they are unable to edit their target gene; however, when blue light is shone upon the cell, the two halves unite to form

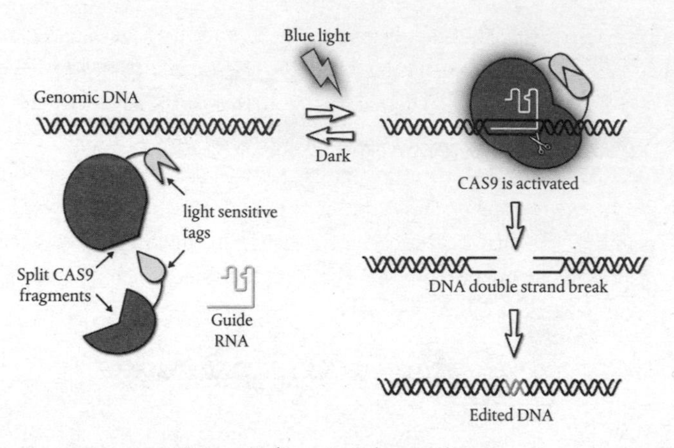

Fig. 22. Light-activated gene editing

an active enzyme, which cuts the DNA (see Figure 22). The study impressed Paul Knoepfler, a biologist at the University of California, Davis, who said that 'this is an effective new system for extremely precise control of gene editing via light'.[283] And although it has only been tested in cultured cells, given the speed of development of genome editing it's surely only a matter of time before such an approach is developed in a mouse model. In such a model it could, for instance, be used to make precise knockouts of specific genes in specific cells in the brain. An adapted version of such an approach might involve a light-activated CAS9 that could regulate gene expression. And the effect could be reversible, for as soon as the light was switched off the two CAS9 halves would fall apart and the target gene would stop being expressed.

In fact, this is only one way in which gene expression might soon be controlled at a distance. A study led by Jeffrey Friedman of the Rockefeller University and Jonathan Dordick of Rensselaer Polytechnic Institute, Troy, New York, used a magnetic field to activate expression of a gene in a living mouse.[284] The researchers engineered mice to express a version of the insulin gene that's activated in the presence of calcium signals. The mice were also engineered to express a pore protein, transient receptor potential cation channel V1 (TRPV1), on their surface membranes, which allows entry of calcium into their cells but only in response to a magnetic device. When the mice were placed in the magnetic field, the presence of insulin was detected in the blood. Dordick believes that this approach is a 'big advance in remote gene expression because it is non-invasive and easily adaptable. You don't have to insert anything—no wires, no light systems—the genes are introduced through gene therapy.'[284] While in this case an indirect approach was used, it seems possible that we'll soon see a situation in which a gene engineered to respond to a regulatory protein is directly activated by a magnetic field.

Problems with Patents

Something the pioneers of CRISPR/CAS9 agree upon is the importance of basic scientific research. So Jennifer Doudna has stressed that often scientists 'don't set out with a practical goal in mind beyond the goal of understanding how things work…the idea of working on a bacterial immune system and studying how bacteria deal with viral infection was a fun, fascinating project that I certainly never anticipated would lead to something like this'.[285] And while acknowledging the importance of 'some kind of big hypothesis or direction or interest' in scientific research, Emmanuelle Charpentier believes the discovery shows that 'sometimes you just hit something, certain components that you want to put together in a way that maybe does not make sense but you just want to do it. Scientists need to have the support to be able to do some crazy experiments to see where they go.'[286]

Yet despite its origins in basic research, the great practical potential of genome editing for both medicine and agriculture has led to tensions between the different researchers involved in the CRISPR/CAS9 discovery about who played the biggest role and therefore who has the biggest claim to patenting the technology. The first signs of such tensions came in April 2014 with the news that Doudna and Charpentier had split into rival teams. So Charpentier established a consortium called CRISPR Therapeutics that included Craig Mello, co-discoverer of the RNA interference technique mentioned in the section 'Life in a Dish', which was promised $25 million by venture capitalists to commercialize the invention for medical use. Meanwhile, Doudna joined Feng Zhang and George Church in a rival consortium named Editas Medicine with $43 million in venture capital promised to develop the technology for the clinic.[287]

Subsequently Doudna broke off relations with Editas, and focused her efforts on another company, named Caribou Biosciences.[288] The reason for this second parting of the ways was a clash about patents, with Feng Zhang filing an application that claimed he had been first to develop the genome-editing technology.[289] Yet doesn't the timescale of discovery we outlined in the section 'A CRISPR Cut' show it was Doudna and Charpentier who not only first discovered the basic mechanism of CRISPR/CAS9 but also recognized its potential as a genome-editing tool? Certainly that was the opinion of the committee who, in November 2014, awarded the two scientists each a Breakthrough Prize. These prizes, worth $3 million, twice the value of a Nobel Prize, were set up by Yuri Milner, a Russian entrepreneur who left a PhD in physics to make $1 billion through investments in internet companies, Mark Zuckerberg of Facebook fame, Sergey Brin, who co-founded Google, and Jack Ma, the Chinese internet magnate.[290]

However, Zhang seems intent on challenging the idea that his contribution to the genome-editing discovery was a secondary one.[291] Backing this position, Robert Desimone, who heads the McGovern Institute for Brain Research, where Feng has an appointment, has challenged *The Economist*'s account of how CRISPR/CAS9 was invented. In a letter to the magazine he wrote that Doudna and Charpentier's paper studied 'a purified protein in a test tube: it involved no cells, no genomes and no editing. Rather, the paper simply highlighted the potential that genome editing might be possible.'[292] And the Broad Institute, part of the Massachusetts Institute of Technology, where Zhang also has an appointment, released a statement saying the institute 'was not the first to file a patent request related to CRISPR. However, [it] was the first to file a patent that described an actual invention—experimental data regarding a successful method for mammalian genome editing.'[292] Meanwhile, in January 2016, Eric Lander, the head of the Broad Institute, came under fire for a review article he wrote for *Cell*, entitled 'The

Heroes of CRISPR'. Although the article was supposed to be an objective history of genome editing, some critics argued that it skewed the history of CRISPR to favour Feng Zhang's contribution, with Berkeley professor Michael Eisen calling it 'science propaganda at its most repellent' and 'a deliberate effort to undermine Doudna and Charpentier's patent claims.[293] Yet despite all the harsh words and legal wrangling, the lengthy, and highly costly, nature of this type of dispute was shown by the fact that in 2020, five years after the dispute began, it was still far from clear which side, if any, had won any ascendency.[294]

The controversy surrounding the dispute may be one factor behind the failure so far to award a Nobel Prize for the CRISPR/CAS9 discovery. In October 2015, Doudna and Charpentier were tipped to win the prize for chemistry by Thomson Reuters but instead it went to Tomas Lindahl, Paul Modrich,and Aziz Sancar, for their studies of DNA repair.[295] Subsequently, other discoveries have been awarded the prize. In fact, it would be very unusual for a Nobel Prize to be awarded for a discovery as recent as the CRISPR/CAS9 one, but it's possible that the Nobel Committee would also prefer to see a clearer resolution of each individual's role in the discovery.[292] And, demonstrating that some senior geneticists have concerns about this patent battle, the late Nobel laureate John Sulston, who played a key role in the Human Genome Project, drew attention in 2015 to the dangers of patenting as fundamental a technology as genome editing.[296] 'This is not just a philosophical point of view,' he said. 'It's actually the case that monopolistic control of this kind would be bad for science, bad for consumers and bad for business, because it removes the element of competition.'[296]

In fact, the very speed with which CRISPR/CAS9 genome-editing technology is evolving may itself be a factor that could scupper attempts to monopolize ownership of the technology. Thus, a recent report concluded that, 'given the pace of innovation in gene editing, today's legal fights could end up serving little purpose. Improved versions of CRISPR/CAS9 have already been invented, and entirely new methods are likely.'[297] And, indeed, recent studies of natural CRISPR systems are revealing a vast menagerie of different types. Tapping that diversity could lead to more effective genome editing, or open the way to applications no one has thought of yet. 'You can imagine that many labs—including our own—are busily looking at other variants and how they work,' says Doudna. 'So stay tuned.'[298]

A Step Too Far?

While arguments about who should reap the commercial benefits of CRISPR/CAS9 have been one source of controversy, a far more potent issue in the public eye has been the news that scientists have used CRISPR/CAS9 to modify the genome of a human embryo for the first time in history (see Plate 2). The news first appeared in the form of a rumour. On 12 March 2015, a comment piece appeared in *Nature* by Edward Lanphier, president of Sangamo BioSciences and chairman of the Alliance for Regenerative Medicine, and four co-authors, all experts in genome editing.[299] This called on scientists not to use CRISPR/CAS9 to modify human embryos, even for research. It transpired that Lanphier and his co-authors had heard that unspecified scientists had already used this approach to create GM human embryos and were seeking to publish their findings. Normally, when a paper is submitted to an academic journal, it is sent to scientific experts who send back comments and a recommendation about whether to publish the study. In this 'peer review' process, the reviewers remain anonymous to the researchers submitting the paper, while they themselves should not reveal to colleagues or friends that they are reviewing a particular study.[300] In this case though, the reviewers clearly felt it was in the public interest to raise the alarm.

The call for a 'voluntary moratorium' on attempts to use genome editing to modify human embryos had clear similarities with the situation in the run-up to the Asilomar conference in 1975. At that time, as we saw in Chapter 2, fears about the safety of the new recombinant DNA technology led to a halt of any further development of the technology until potential risks had been discussed and safety guidelines agreed. Yet while some scientists agreed with Lanphier's call for a moratorium, others were less convinced. Lanphier himself said: 'We are humans, not transgenic rats. We believe there is a fundamental ethical issue in crossing the boundary to modifying the human germ line.'[301] However, George Church was more qualified in his thoughts about the new development, saying that there should be a moratorium on embryo editing, but only 'until safety issues are cleared up and there is general consensus that it is OK'.[301] In April 2015, the study that had been the main object of the rumours was published in a relatively obscure journal called *Protein & Cell*—having reportedly been rejected by reviewers for both *Nature* and *Science* on 'ethical grounds'—and it was finally possible to examine what had been achieved indetail.[302] The study was led by Junjiu Huang, of Sun Yat-Sen University in China, and the report revealed that Huang and his team had used CRISPR/CAS9 to correct a gene defect that underlies the disease β-thalassaemia, a potentially fatal blood disorder. The defect is in the β-globin gene, which codes for one of the components of the haemoglobin protein that carries oxygen around the blood.

In an attempt to counter potential ethical objections, Huang and his team used embryos created by the accidental union of two sperm with a single egg; such embryos were obtained from a local *in vitro* fertilization (IVF) clinic, which would normally have discarded them. These embryos can undergo the first stages of development but will never result in a live birth. The study showed that CRISPR/CAS9 could correct the gene defect, but with low efficiency and accuracy. So only a fraction of the treated embryos were successfully modified, and there were also a number of 'off-target' effects upon other genes in the genome. 'If you want to do it in normal embryos, you need to be close to 100%,' said Huang. 'That's why we stopped. We still think it's too immature.'[303] However, George Church was less convinced, pointing out that the Chinese researchers did not use the most up-to-date CRISPR/CAS9 methods and that many of the problems of efficiency and accuracy could have been avoided if they had.[303]

Following the publication of the paper, scientists remained deeply divided about the wisdom of seeking to apply genome editing to a human embryo. For Edward Lanphier, the result 'underlines what we said before: we need to pause this research and make sure we have a broad based discussion about which direction we're going here'.[302] However, George Daley, a stem-cell biologist at Harvard University, supported editing of human embryos for research purposes, noting that employing genome editing to modify the genomes of human embryos grown in culture could be used to study the role of certain genes in early development. 'Some questions about early human development can only be addressed by studying human embryos,' he said.[303]

Meanwhile, philosophers of ethics were equally divided on this issue. John Harris, a bioethicist at Manchester University, said that although the poor performance of the technique in Huang's study 'should be a stern warning to any practitioner who thinks the technology is ready for testing to eradicate disease genes,' he believed Huang's use of non-viable embryos was 'no worse than what happens in *in vitro* fertilization all the time'.[304] However, Tetsuya Ishii, a professor of bioethics at Hokkaido University, Japan, feared that if genome editing of human embryos became permissible as a form of preventative medicine, it could be the start of a 'slippery slope to designer babies in countries with lax regulations'.[305]

In fact, the different points of view in this debate were soon rapidly tested by subsequent events. First, a scientist called Kathy Niakan, of the Crick Institute in London, obtained approval from the UK Human Fertilisation and Embryology Authority (HFEA), to carry out her own use of CRISPR/Cas9 to genetically alter human embryos. In 2017, Niakan published her first exploration in this area in *Nature*; the study looked at the effects of altering the expression of a gene called OCT4, on human embryo development. Importantly, the study showed that OCT4 plays a subtly different role in human embryos than it does in those of mice.[306] For Dieter Egli of Columbia University in New York, the findings of Niakan and her colleagues underscored the limitations of relying on animal models. For as he put it, 'If we are to truly understand human embryonic development and improve human health, we need to work directly on human embryos. We cannot rely only on inference from model organisms.'[306] And in general, perhaps because Niakan's studies had carefully sought ethical approval, were carried out with a rigour lacking in those of Huang and his team, and were clearly advancing scientific knowledge, they drew far less criticism than had the latter.

The situation was soon to blow up once again, however, and this was because of the shock revelation in December 2018 that a scientist had decided to ignore ethical considerations and even the laws of his own country, and use CRISPR/Cas9 to edit a gene in human embryos that were then implanted into their mother, resulting in the first genome-edited human babies in history. For me, this news came just a day before I was due to give a talk about genome editing at a conference in Rotterdam, and it meant that all my carefully prepared discussion about the pros and cons of editing human embryos for research purposes, now had to be altered to acknowledge the new development, and its implications for medicine, and indeed the likely future consequences of our newfound ability to edit genomes.

The scientist who decided to make this leap into the unknown was Jiankui He of Southern University of Science and Technology in Shenzhen, China, who announced that had used CRISPR/Cas9 to edit a gene called CCR5 in human embryos to make them resistant to HIV; after implantation of two of the embryos into their mother, they were subsequently born as twin baby girls.[307] Later in this book, we will examine how loss of CCR5 expression can protect against HIV. The rationale for creating the genome-edited babies in this way was that their father had HIV, and He claimed that this was the only way he could have children. In fact, it is possible to prepare sperm so that they are not infected with HIV, so this was a piece of false information. But this was far from the only problem with what He had done; he had also flouted ethical guidelines and broken laws of his country, including those that prevented 'germline' genome editing. In fact, He has now not only been sacked from his university job, but he has also been jailed for three years.[308]

In fact, these issues are far from the only controversial aspects of genome editing. Another issue likely to generate future debate is whether this approach can be employed safely and effectively as a form of gene therapy to treat disease in adult humans. But there's also the fact that genome editing is likely to have a major impact on agriculture, as it can be used not just to create new types of GM crops, but GM animals too. And ultimately the question arises whether, in the future, genome editing may be a legitimate method for 'enhancing' the human species. For instance, John Harris believes that 'if we could become more resistant to disease, more resilient to injury, improving our cognitive powers, or increasing our life expectancy, I don't see why we would not do that'.[304]

Clearly, these are questions that need to be examined in detail, and in the following chapters that's exactly what I'll be doing. But while, ultimately, medicine is supposed to be about treatments and cures, we can only hope to develop such therapies on the basis of a proper understanding of the human body, both in terms of normal function but also when it's in a pathological state. And in the pursuit of such understanding, the availability of other species that can act as 'model organisms' in which to study human health and disease has been a vital aspect of the biomedical sciences. So, as the initial step in this discussion about the applications of genome editing, we should turn to look at the varied ways that this technology is beginning to transform the study of such model organisms, large and small.

5

Next Year's Models

The use of other species to model human health and disease has been central to biomedical science since its inception. As early as the seventeenth century William Harvey was using dogs to demonstrate the circulation of the blood. Nowadays, the details of such vivisection seem barbaric, with the animals being cut open while they writhed in agony on the operating table, with no anaesthetics or painkillers to alleviate their distress.[309] And currently in Britain, all experiments on live animals, whether involving surgery or a GM mouse that will spend its whole life being monitored in a cage and may never undergo any invasive procedures, must be carried out under the strict regulations of a Home Office licence.[310] Anaesthetics or painkillers must also be used wherever possible to alleviate distress in the animal, although there will be experiments in which the whole point of the exercise is to study the pain response. Similar regulations apply to research on animals in the USA, Japan, Australia, China, and other countries.[311] Despite such guidelines, it's not surprising that, since most people are sensitive to the suffering of other creatures, animal experimentation remains controversial.

Yet if we really want to understand how the human body works, and what happens when things go wrong in disease, then animal experimentation will continue to be central to biomedical research.[312] Opponents of such experimentation often point to alternatives like biochemical analysis carried out in a test tube, study of cells in culture, or modelling of bodily processes on a computer. In fact, such approaches form a normal part of biomedical research; so, for instance, my colleagues and I carried out more than half of our recent studies into the molecular basis of calcium signals on cells in culture.[313] But when researching the biology of the heart, the liver, or the brain, it's impossible to get a picture of the true complexity of these organs from such studies. That's because the structure of organs, and the way different cell types interact within that structure, have so far proved impossible to mimic accurately in culture.[312] This is particularly the case with the brain, with its billions of cells, themselves existing as hundreds of different cell types, connected by trillions of nerve connections; although, as we'll explore in Chapter 8, ways in which it may be studied could be starting to change with new advances in human stem-cell research. Another reason why cultured cells cannot mimic the complexity of many processes occurring in a living animal is that organs communicate with each other and the rest of the body via hormones, growth factors, and other chemicals.

Of course, animal experiments do not necessarily have to involve mammals. So much of our knowledge of the mechanisms by which genes are turned on or off originally came from studies of bacteria,[314] while the genes regulating the cell cycle—the process by which cells reproduce and divide—were originally identified in yeast.[315] We saw in Chapter 1 how studies of fruit flies laid the foundation for modern genetics, and how, more recently, this organism has furthered our understanding of embryo development, particularly that of the nervous system and brain. The lowly nematode worm has also been of great importance in this respect. Studies of the nematode identified the phenomenon of controlled cell death, the process whereby cells commit suicide, which is important in precise modelling of the embryo during development and the prevention of cancer during adulthood.[316] And, as we saw in Chapter 3, the zebrafish has also been an important organism for studying embryogenesis.[317]

The reason we can gain valuable insights from these different organisms is that all life evolved from the same original genetic stock, and also because evolution has an innate conservatism. It tends to adapt what is already there, rather than totally fashioning things anew, with the consequence that there is a remarkable degree of similarity between many processes occurring in such organisms and those in the human body.[318]

The reasons for studying such non-mammalian lifeforms are both ethical and scientific. Ethically, it's seen as more acceptable to carry out experiments on 'lower' organisms because, having a less developed nervous system, they are judged less likely to feel pain or distress. To some extent this is a subjective judgement. In Britain it was only in the mid-1980s that it became necessary to have a licence to study zebrafish or frogs, perhaps because studies on warm-blooded furry animals like mice excite more interest than ones on cold-blooded, slimy ones. Yet frogs and fish are still complex organisms, and there's increasing recognition that they also need to be treated in ways that minimize any pain or distress they may suffer. Studies of invertebrate species, apart from the octopus, still require no licence.[319]

Invertebrate species also have features that make them valuable for research purposes.[320] With their short lifespans and large numbers of offspring, fruit flies and worms have been particularly suitable for genetics, because it's possible to dose them with radiation or mutagenic chemicals, and then screen resulting offspring for mutant forms.[321] For studies of embryogenesis, the fact that fly and worm embryos develop outside the mother makes them much more amenable for study than the mammal in its mother's womb. The zebrafish embryo also develops outside its mother's body, yet being a vertebrate it shares many specific features of development with mammals. That zebrafish embryos are transparent is important because it allows sophisticated imaging approaches to be used to study what's going on at the molecular level inside the cells of a living embryo.[322]

Genome editing looks set to have a big impact on the study of these non-mammalian organisms because of the sheer speed at which genes can be targeted using this technology. For instance, Shawn Burgess of the US NIH has recently shown that CRISPR/CAS9 can be used to knock out genes in the zebrafish on a mass scale. 'What we have done is to establish an entire pipeline for knocking out many genes and testing their function quickly in a vertebrate model,' he says.[323] Burgess's team successfully mutated 82 genes, about 50 of which are similar to human genes linked to deafness. Such mutants can now be screened to assess exactly how such genes are involved in hearing. But Burgess has even bigger ambitions. 'We've shown that with relatively moderate resources, you can analyse hundreds of genes,' he says. 'On the scale of big science, you could target every gene in the genome with what would be a relatively modest scientific investment.'[324]

Mice as Models

Undoubtedly though, a principal way that genome editing will affect biomedical research is through its capacity to precisely modify the genomes of different mammals. Despite the importance of research in worms, flies, and fish, major differences exist between mammals and other multicellular organisms. Such differences include not just the development of the mammalian embryo inside the womb, but also the particular challenges posed by maintaining a constant body temperature in a warm-blooded animal. In addition, in mammals there has been a trend in some species towards larger brains and a greater role for learning compared to instinct, with this feature being particularly pronounced in the primate group of which our own human species is a member.[325]

We saw in Chapter 2 how the discovery of ways to create knockout and knockin mice using ES cells was a key development in biomedical science. Yet the mouse is far from being the best mammalian model for many aspects of human health and disease.[326] This is partly due to its much smaller size compared to people, and also because of the shortness of the mouse lifespan. In my own laboratory, CRISPR/Cas9 allowed us to rapidly create mice with a defect in the PLCζ gene. As I mentioned in Chapter 2, PLCζ is a sperm protein that we believe plays an important role in the activation of the egg to develop into an embryo. Creating PLCζ knockout male mice allowed us to definitively prove this was the case as sperm from such mice lacked the ability to activate the egg in the normal fashion.[327] But mice also differ from humans in the biology of some cell types and tissues that limit the usefulness of this species as a model. Take, for instance, the heart. Understanding the molecular and cellular basis of cardiac function, and why hearts fail, is a pressing matter given the increasing numbers of people succumbing to cardiovascular disease.

A recent study by Emanuele Di Angelantonio and colleagues at Cambridge University found that a combination of heart disease and diabetes can shorten a person's life dramatically. According to Di Angelantonio, 'an individual in their 60s who has both conditions has an average reduction in life expectancy of about 15 years.'[328] This is particularly a problem given the recent dramatic global rise in diabetes, itself a product of increasing obesity. Clearly, one important way of stemming the rise in cardiovascular disease is preventative. As Jeremy Pearson, a medical director at the British Heart Foundation, which part-funded this particular study, said: 'the results of this large study emphasise the importance of preventing diabetes, heart attacks and strokes in the first place, through encouraging patients to live a healthier lifestyle'.[328]

Unfortunately, this message is not getting through; instead the obesity 'epidemic' in the developed, and increasingly the developing, world is getting worse, not better. Some recent reports highlight the scale of the problem. The National Obesity Forum found that a previous estimate that half the British population would be obese by 2050 actually underestimates the scale of the obesity crisis.[329] Another study, funded by Cancer Research UK (CR UK), revealed that more than a third of obese or overweight teenagers think their size is perfectly normal.[330] Finally, a study by researchers at King's College London, that screened the electronic health records of 279,000 people in Britain, concluded that existing weight loss programmes are 'not working for the vast majority of obese patients'.[331] Similar problems exist in many other developed countries, with more than two-thirds of US adults now being classified as overweight or obese, compared to less than half in the 1970s.[332]

Despite the clear link between heart disease and environmental factors such as junk food, lack of exercise, plus increasing stress within society, there is still much we can learn from animal models about this disorder. And valuable insights into the biology of the heart have come from studies of knockout and knockin mice. By interfering with the expression of specific genes and then assessing their effects, such studies allow scientists to build up a picture of the molecular mechanisms that underlie heart function. This is important because although the environment has an important impact on a person's susceptibility to disorders of the heart and circulation, so apparently do genetic factors.[333] And GM mice have made it possible to explore the way genetic differences identified in people affect heart function at the level of basic biology.[334] At the same time, such mice can be placed on different diets or exercise regimes to explore the way that genetics and environment can combine to cause a particular problem within the heart or circulation.

Another important use for GM mice is to develop and test new therapies. For while basic scientific research aims to uncover how molecules, cells, tissues, and organs combine to make a living, functioning human being, the ultimate goal of biomedical research is to devise new ways to treat disease. A central aspect of modern medicine is its arsenal of drugs—generally small molecular weight chemicals that can either be ingested or injected and which have a beneficial effect upon the body. Because drugs may be ineffective, or have nasty, even fatal, side effects, testing new drugs on animals prior to their use in human clinical trials is a vital part of drug development.[335] Such testing allows the clinical usefulness of a drug to be assessed, but also its potential for adverse side effects. Testing can be carried out on an unmodified animal if the aim is purely to assess the potential toxic effects of a drug. But given that the point of drugs is to treat disease, drug testing in animals should also test the capacity of a drug to relieve the symptoms, or even better remove the underlying cause, of an illness. In this respect too, knockout and knockin mice models of disease have been important.[336]

But there are also some limitations to the usefulness of the mouse as a model for understanding human heart function and disease.[337] One major difference is the speed at which the hearts of the two species beat. A typical mouse pulse is 600 beats a minute, whereas our own is around a tenth of that speed. This reflects significant molecular differences between the mouse and human heart. The protein responsible for the heart's contractile force is different, as are the cellular pumps and channels that produce the chemical signals that regulate heart contraction.[337] And because of these differences, trying to model human heart disease in a mouse by knocking out or subtly altering specific genes may result in misleading information about the importance of such genetic factors for humans.

Matters of the Heart

Because of such concerns, scientists are exploring the possibility of modelling heart function and disease in mammalian species more similar to ourselves. The pig, in particular, has much to offer in this respect. As we saw in Chapter 1, pigs have a long and intimate connection with our species through our cultivation of this animal for meat. And the fact that humans and pigs share many common features—such as our size and tendency to enjoy a good feed—has attracted the attention of writers and dramatists throughout the ages, most famously George Orwell in his political parable *Animal Farm*, in which the originally revolutionary pigs become so similar to their former human masters that it is 'impossible to say which is which'.[338]

In fact, the similarities between pigs and humans are more than superficial, and the potential of the pig for modelling human health and disease is increasingly recognized. Bhanu Telugu of the University of Maryland believes that 'from a biomedical standpoint, the pig is really one of the most important animals'.[339] He points out that other large animals like cows or sheep don't have digestive systems, diets, or physiology similar enough to ours to provide insight into human diseases. In size, structure, contractile proteins, and the electrical and chemical signals that regulate it, the pig heart is much more similar to a human heart than that of a mouse.[337]

The importance of the pig for modelling human heart function goes beyond the similarities of the organ in these two species. When it comes to testing drugs designed to treat heart disorders, the pig has other similarities to us that make it a more suitable model than the mouse.[340] An important aspect of how drugs work is not only the effect they have on their target in the body, but also how they get into the bloodstream, and from there to the target cell or tissue. How a drug is processed and excreted from the body by the liver and kidney are also important factors in designing a drug regime that will successfully treat a disorder while resulting in minimal side effects. But here factors like an organism's size, diet, and metabolism all play an important role, and have meant that some findings about a drug's usefulness and side effects gleaned from mouse studies can be misleading.

In contrast, the similarities in size and also diet between ourselves and pigs makes the pig a better model in this respect.[340] That pigs live much longer than mice is important for modelling heart disease, since heart disease is a condition that generally affects older people.[337] The similarity in size means that pigs are important models for surgery, since similar procedures can be used as in human surgery.[340] It also means the various devices developed for sophisticated imaging of the human body are directly applicable to the pig, which can help in the comparison of pathologies in the two species.

Pigs may also become important as a source of heart transplants. This might seem odd, given that another human heart would seem to be the most obvious choice for such a transplant. Here though, we face a number of problems. The first is that human donors are in short supply because of the lack of people volunteering to donate an organ after their death, the difficulty of getting consent from grieving relatives, and a reduction in most countries of the rate of fatal road accidents (the most reliable source of healthy organs).[341] 'It's a cruel situation currently, that someone who needs a heart transplant has to pin their chance for a healthy life on the untimely death of another person,' says David Dunn, an expert on transplantation at the State University of New York in Oswego.[342] Even if an organ is available, it's highly likely to be rejected by the recipient's immune system. Each of us has proteins called major histocompatibility complexes (MHCs) on the surface of the cells that make up our major organs. These tell our immune systems that these organs are part of us, and subtle differences in MHCs in a transplanted heart will cause the immune system to generally attack it as a foreign body.[343] This can lead to severe blood clots and failure of the new organ.

With thousands of different MHC variants, the chance of finding a donor with the same profile as the person requiring a transplant is slim indeed. Because of this, and the shortage of donors, there is much interest in transplanting pig hearts into people whose own heart has failed, so-called 'xenotransplantation'.[343] Transplanting a normal pig heart would also lead to rejection because of the mismatch between the MHC proteins. But, unlike human donor hearts, there is a potential solution: to genetically modify the pigs so that the MHC proteins on their hearts are no longer recognized as foreign by the human immune system. And the same would apply to other organs too, so such pigs could be the source of a new liver, kidney, pancreas, or lungs.[343]

Despite the potential usefulness of the pig for biomedicine, whether to model human health and disease or as a source of organs, the main obstacle until recently has been the lack of an effective technology for producing knockout and knockin pigs. Now, genome editing offers such a technology. Recently, Bruce Whitelaw and colleagues at the Roslin Institute near Edinburgh—the birthplace of Dolly the cloned sheep—successfully used ZFNs and TALENs to create GM pigs.[344] And in February 2015, Hong Wei and his team at the Third Military Medical University in China used CRISPR/CAS9 to create pigs with a knocked out version of a gene named Niemann-Pick disease C1-like (NPC1L1).[345] Previous studies showed that this gene plays a key role in regulating the uptake of cholesterol by the intestine and liver. Ezetimibe, also known as Ezetrol®, a drug that targets the expressed protein, lowers blood cholesterol in humans.[346] Inhibiting the expression of this gene in pigs will provide an important model for studying the beneficial effects of low cholesterol levels on the heart and circulation. My colleagues and I also recently used CRISPR/Cas9 to create pigs with a defect in TPC2, a gene that I mentioned in Chapter 4, which codes for a protein that helps to transmit chemical signals inside cells.[347] Since we have already identified important roles for this protein in heart function and metabolism in our mice studies, we are excited to extend these studies using the pig as a model. Such studies could help to identify new drugs for treating heart disease and diabetes.

As for creating GM pigs as a source of transplantable organs, Craig Venter, who led one of the teams that first sequenced the human genome, has announced that his company, Synthetic Genomics, is developing such a strategy.[348] In collaboration with another company, United Therapeutics, the plan is to use genome editing to create pigs with 'humanized' lungs. If successful, this could help the 400,000 people in the USA alone who die each year from various forms of lung disease. 'We're going to start with generating a brand new super-accurate sequence of the pig genome, and then go through in detail and compare it to the human genome,' says Venter.

'The goal is to go in and edit... the pig genes that seem to be associated with immune responses. We want to get it so there is no acute or chronic rejection.'[348] In fact, Venter is far from alone in such a goal, since, in October 2015, George Church reported that his team had modified more than 20 genes in pig embryos, including ones that code for MHCs. 'This is something I've been wanting to do for almost a decade,' he said.[349] And a biotech company named eGenesis that he has co-founded in Boston is now trying to engineer pigs that can be used for organ transplantation as cheaply as possible.

Although Venter and Church both believe they'll soon be able to deliver modified pigs whose organs will not be rejected by a human recipient, a big challenge will be confirming that such organs are safe to transplant. It will not only be necessary to show that the pig organ won't be rejected by the recipient's body, but also that the organ is compatible with the rest of the human body in a more general sense. Although a pig organ may function perfectly well as an isolated entity in a human being, recent studies indicate that interactions between organs also play a key role in bodily function. Such studies show that the heart has other functions besides its vital role in pumping the blood around the body; for instance, it also produces hormones that send signals to other organs.[350] It will be important, therefore, to confirm that a transplanted pig heart can also fulfil these other functions.

Another concern is the possibility of pathogens like bacteria or viruses being transferred from the donor heart to the recipient human. Careful breeding in sterile conditions can eliminate most problematic microorganisms from the bodies of donor pigs. However, there is one class of viruses that poses a problem in this respect—retroviruses. As we saw in Chapter 2, these viruses, of which HIV is the most famous member, can integrate their genetic material into the genomes of the cells they infect. This is one reason why a person infected with HIV can go for years without showing any symptoms. And pigs, through exposure to such viruses in the past, also contain retroviral DNA in their genomes, called porcine endogenous retroviruses, or PERVs, to use their rather unfortunate acronym.

One concern is that such retroviruses might become activated in a human host's body and cause a serious disease in that individual.[351] An even more serious concern is that such a transplanted virus could jump to other human individuals and cause an epidemic. Since HIV seems to have first arisen in humans by jumping species from chimpanzees,[352] the potential consequences of the spread of a virus from pigs to humans needs to be taken very seriously. One particular concern is that viruses that jump species can have far more serious effects in their new host than in the previous one. The precursor of HIV has few adverse effects in chimpanzees. Jonathan Allan, a virologist at the Southwestern Foundation for Biomedical Research in San Antonio, USA, remarked on this: 'African primates all carry their own little viruses. In some species, the viruses have been there for thousands of years. And the natural host never gets sick.'[353] However, in a species that has not evolved to tolerate a particular virus the consequences can be devastating, as seen with HIV in humans.

Until recently, there seemed little that could be done about the presence of PERVs in potential donor pigs. 'They were part of these animals' genomes,' says Jay Fishman, associate director of the transplant centre at Massachusetts General Hospital.[342] Instead, those in favour of transplanting pig organs into humans pointed to studies like one in which pig hearts transplanted into baboons failed to show any activation of pig retroviruses once inside the baboon's body.[354] None the less, the potential risks of allowing such viruses into a human body have always been one of the objections levelled against xenotransplantation. So, according to George Church, billions of dollars were invested in xenotransplantation research in the mid- to late 1990s, but funding fizzled out because of the inability to find a way to remove the viral sequences.[355] Yet in October 2015, Church and his team used CRISPR/CAS9 to delete all the PERVs in the genomes of pig kidney cells in culture.[356] 'Fast-forward 15 years later, we got rid of them in 14 days with CRISPR and a lot less money,' says Church.[355]

Church's team first analysed the genomic DNA of the pig cells. This revealed 62 PERVs located at different sites in the genome; however, these were practically identical in sequence, reflecting their origin from a single ancestor that invaded the pig genome millions of years ago. The researchers then used a form of CRISPR/CAS9 that didn't just cut the viral DNA but deleted it from the genome. Amazingly, in some cells this treatment eliminated all 62 copies of the virus from the pig genome. And those edited cells showed up to a thousandfold reduction in their ability to infect human kidney cells with PERVs in a culture dish.[356] While this initial phase of the work only removed PERVs from pig cells in culture, in 2017 Church and his colleagues reported the birth of 37 healthy piglets whose genomes had been edited to be free of PERVs.[357] Commenting on these studies, and the plans to alter pig MHCs, David Dunn believes that 'this work brings us closer to a realization of a limitless supply of safe, dependable pig organs for transplant.'[342]

The Complex Brain

The potential usefulness of pigs for modelling human health and disease, and as a source of replacement organs, is due to the great similarities in basic physiology between the two species. But one human organ that's peculiar in both size and complexity compared to other mammalian species is the brain. Human beings are unique in possessing a self-conscious awareness, a language capacity that allows us to communicate with other humans as well as think, talk, and write conceptually, and an ability to develop and use new tools and technologies in each new generation.[358] Such characteristics must ultimately be based on the distinctive biology of the human brain, although obviously the shape of the mouth, throat, and hand are also vital in allowing us to articulate thoughts vocally and use tools.[359] This poses the question of how to model the human brain, both in terms of its normal function but also degenerative diseases like Parkinson's and Alzheimer's and personality disorders like schizophrenia, depression, and bipolar disorder.

The problem that such disorders pose for society is shown by recent figures from the US National Alliance on Mental Illness, which indicate that about one in four adults in the USA experiences a diagnosable mental disorder in a given year.[360] Figures from the UK Mental Health Foundation reveal a similar situation in Britain.[361] Of course, one might wonder what type of 'illness' affects such a large proportion of the population, and some experts criticize what they see as the increasing overmedicalization of individuals whose actions lie within the spectrum of normal human behaviour. Peter Kinderman, head of Liverpool University's Institute of Psychology, notes that: 'Many people who are shy, bereaved, eccentric, or have unconventional romantic lives will suddenly find themselves labeled as mentally ill. It's not humane, it's not scientific, and it won't help decide what help a person needs.'[362] This is particularly a concern given the increasing diagnoses of children as suffering from conditions with names like 'disruptive mood dysregulation disorder'.[363] Some critics argue that this is not only inappropriate labelling since such children may simply be bored or naturally boisterous, but also dangerous, since such conditions are increasingly treated with drugs whose long-term effects remain far from clear.

Despite such concerns, it would be a mistake to underestimate the misery caused by conditions like schizophrenia, major depression, or bipolar disorder—which affect one in 17, or about 13.6 million of the US population.[360] A similar proportion of people are affected by these disorders in Britain.[364] An insight into the devastating impact of such disorders, and the difficulties in successfully treating them, was recently provided by Robert Desimone, who, as we saw in Chapter 4, heads the McGovern Institute for Brain Research, part of the Massachusetts Institute of Technology.[365] As a first-year psychology student, aged 18, Desimone spent a month at a state mental hospital, living alongside and observing patients with schizophrenia and bipolar disorder. He befriended a schizophrenic patient in her 30s, formerly a promising student but now trapped in a world of delusion and failing to respond to any of the available drug treatments. 'What was so frustrating was that here was an intelligent person who I could have imagined as a fellow student in college, and yet here she was stuck in this mental hospital because she had these delusional ideas,' recalls Desimone. 'That really opened my eyes to how terrible some of the serious mental illnesses were, and how frustrating it was for the physicians trying to treat them.'[365]

This incident happened over 40 years ago. Yet the same frustrations persist, due to the continuing lack of effective, precise ways to diagnose and treat serious mental disorders. This was particularly brought home to Desimone in 2000 when his own 16-year-old son was diagnosed with bipolar disorder, and he had to face the stark reality that the treatments available were essentially unchanged from those used in the psychiatric ward where he had worked more than three decades earlier. Desimone believes the reason is that neuroscientists still don't really understand what's happening in the brains of those with mental disorders. 'I felt that we really needed to go back to the lab and concentrate more on fundamental knowledge of brain mechanisms and genetics,' he says.[365] And a chance to pursue such a goal arose when he was asked to direct the McGovern Institute in 2004.

Under Desimone, the institute has recruited top researchers working in the most cutting-edge aspects of neuroscience. Feng Zhang, who, as we saw in Chapters 3 and 4, helped pioneer both optogenetics and CRISPR/CAS9 genome editing, joined the McGovern Institute in 2011. Zhang is also critical of the current lack of precision that underlies our understanding of the biology of mental disorders and the drugs used to treat them. 'Traditionally when we think about developing drugs to treat brain diseases, it's all about this hypothesis that there's some kind of chemical imbalance,' he says. 'All the cells in the brain live in this milieu of chemicals, and if there's an imbalance in the composition of the chemicals, then the brain has problems. But that's a very gross and inaccurate way of thinking about how the brain functions.'[365] Instead Zhang believes the focus should be on understanding the 'abnormal signalling between different cells in specific neural circuits… probably underlying many of the neurological or psychiatric diseases that we know today'[365]—features highlighted by the optogenetic studies being carried out by himself and others.

Optogenetics is only one of the technologies being developed at the McGovern Institute. Another approach is the use of genome editing to assess the functional relevance of genomic regions shown to be associated with mental disorders in humans. On the completion of the Human Genome Project in 2003, some influential figures predicted that we would soon identify the genes associated with such disorders. Consequently, Daniel Koshland, editor of *Science*, declared that the basis for 'illnesses such as manic depression, Alzheimer's, schizophrenia and heart disease' would all be unravelled, with new drug treatments for these conditions sure to follow.[366] Unfortunately, the reality has proved rather more complex.

The main strategy in trying to identify genes associated with human disorders has been so-called 'genome-wide association studies' (GWAS).[367] These studies typically analyse the genomes of a large number of individuals with a disorder and compare them with people lacking the disorder. And studies of schizophrenic individuals have indeed identified links with a number of genomic regions. But far from identifying a few clear links, these studies have found associations with over a hundred different regions.[368] And each of these regions of the genome individually appear to have only a minor impact. What's more, the same complexity is emerging from genetic studies of other mental disorders such as bipolar disorder, depression, and autism.[369] Currently there is a debate about whether this indicates that genetic predisposition to such disorders requires changes in a large number of different genomic regions, or whether rare differences have a much more significant impact, but only in a few individuals.[370] Either way, this still leaves the question of how to assess the functional importance of these genetic differences.

Up till now, the main way to do this has employed ES cells to create knockout or knockin mice with defects in genes associated with human mental disorders. This makes it possible to explore the underlying biology of such disorders, as well as test drugs that may be used to treat them. Here genome editing can play a vital role by dramatically reducing the time and cost required to produce such mice. Guoping Feng, a scientist at the McGovern Institute, is pioneering this 'high-throughput' approach. 'We have models of obsessive-compulsive disorder and autism,' he says. 'By studying these mice we want to learn what's wrong with their brains.'[371] In one such model, mice show obsessive self-grooming, and Feng has shown that this behaviour ceases when the missing gene is reintroduced, even in adulthood. 'The brain is amazingly plastic,' says Feng. 'At least in a mouse, we have shown that the damage can often be repaired.'[371] So far he has only created mice with differences in single genes, but human mental disorders appear to involve many contributing genetic differences. And here the capacity of CRISPR/CAS9 for editing multiple genes simultaneously will be very important.

Such a focus on mice does, however, raise the question of whether we can gain a meaningful idea of the role of specific genes or neurons in human mental disorder purely from studying rodent models. One concern is that many potential drugs for brain disorders have been tested successfully in mice, only to prove ineffective in subsequent human trials. Thus, according to Robert Desimone, 'a lot of the treatments that are tried out in mice seem very promising, and then they go into clinical trials and don't go anywhere. You hear the expression all the time that this is a great time to have Alzheimer's disease if you're a mouse. You could say the same thing about autism or any number of disorders.'[365]

Modify My Monkey

Because of such limitations there is increasing interest in using GM primates to complement rodent studies. For while human brains are particularly large and complex compared to the size of our bodies, this is only the culmination of a trend in primates as a whole.[372] Primates therefore offer an excellent model for studying the human brain and its various functions. Of particular interest are brain regions like the prefrontal cortex, critical for many higher human brain functions but not well developed in mice. In contrast, the monkey prefrontal cortex is much closer in size and structure to our own (see Figure 23).[372] Another aspect of primates that makes them potentially better models for studying human brain disorders is the fact that we and other primates interact with the world around us in similar ways. So while rodents mainly rely on smell to negotiate their way around, primates, including humans, are far more dependent on visual cues. This makes it easier to devise tests to assess primate behaviour that have relevance to those used on humans.[373]

Despite this, most neuroscience studies have traditionally used mice and rats, not only because such rodents are much cheaper to breed and

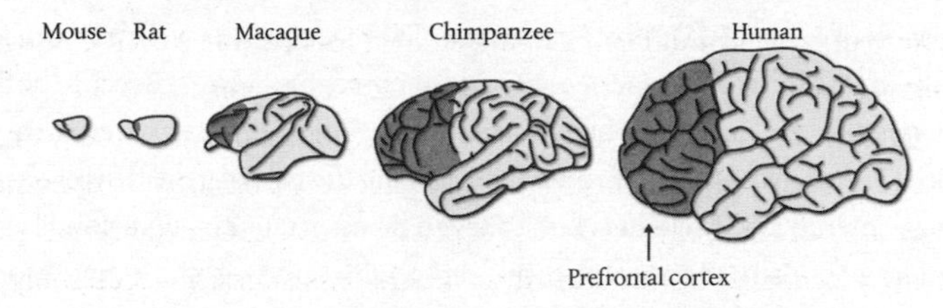

Fig. 23. Mammalian brains and prefrontal cortex

maintain, but also because of ethical concerns about work involving primates.[374] And of course, a major attraction of mice, and more recently rats, has been that these were the only mammalian species from which scientists could create knockout and knockin versions. But now that genome editing makes it possible to create primates that are precisely genetically modified, there is likely to be increased interest in using such primates to study the brain, notwithstanding the potential controversy this may generate in terms of ethical issues.

That genome editing can be employed in primates was first demonstrated in January 2014 by Weizhi Ji and colleagues at the Yunnan Key Laboratory of Primate Biomedical Research in China. They used CRISPR/CAS9 to knock out the genes peroxisome proliferator-activated receptor gamma (PPARγ) and recombination activating gene 1 (RAG1) in cynomolgus macaque monkeys.[375] Ji and his team targeted the genes in fertilized macaque eggs, then implanted these into surrogate mothers. Two resulting offspring, named Ningning and Mingming, had knockouts in the genes. PPARγ regulates metabolism and RAG1 is involved in immunity, and the researchers are now investigating what effect loss of these genes has upon these bodily functions. Ji's team subsequently targeted dystrophin—the gene defective in people with Duchenne muscular dystrophy (DMD)—in rhesus macaque monkeys, and found that this caused severe muscle degeneration similar to that observed in humans with this condition.[376] The macaque may therefore be a useful model for studying such diseases that affect humans.

But the main interest in GM primates is for studying brain disorders. With this in mind, scientists at the McGovern Institute are now in the process of using CRISPR/CAS9 to create GM macaques, as well as marmosets—a smaller primate species with a faster breeding cycle.[371] Because both species are very social with highly structured communication patterns, they should offer a valuable new route for assessing the role of genes involved in social interaction. The aim is first to try to breed primates with a condition similar to autism, and then to move on to schizophrenia and other disorders. Such primate models will be important for understanding not only the basic biology of mental disorders, but also for testing new drugs. Robert Desimone hopes that 'the primate models will give us better testing and treatment platforms'.[365]

Such is the potential of gene editing in primates, but the application of the technology in this manner is creating controversy. Troy Seidle, of Humane Society International, believes there should be an outright ban on the genetic manipulation of primates: 'You can't genetically manipulate a highly sentient non-human primate without compromising its welfare, perhaps significantly. GM primates will be just as intelligent, just as sensitive to physical and psychological suffering as their non-GM counterparts, and our moral responsibility toward them is no less.'[377] Such opposition may have an impact in countries that would otherwise be well placed to develop the use of GM primates in brain research. In the USA, commercial airlines have already ceased all primate shipments by air within the country, making it difficult for researchers to transport animals.[378] Many airlines in Europe have taken similar steps, although Air France still provides a service. And a recent report showed that a compromise European Union initiative, which was supposed to balance the welfare of primates while allowing some research on this animal group, is already in danger of unravelling because of political lobbying by animal rights activists.[378]

Consequently, future development and study of GM primates may well shift to countries where there are fewer ethical concerns about such research. As we mentioned earlier in this section, the first successful use of genome editing to create such modified primates was in China. And China not only has more relaxed attitudes to primate research but is also investing significant funds in this area. The dynamics of the situation are illustrated by the changing fortunes of the Yunnan Key Laboratory of Primate Biomedical Research in Kunming, the institute in which Weizhi Ji and his colleagues announced the birth of the first genome-edited primates. Ji recalls that when he first began his career at the Kunming Institute of Zoology in 1982, 'we did not have enough funding for research. We just did very simple work, such as studying how to improve primate nutrition.'[379]

The subsequent expansion of the Chinese economy has led to a growth in the country's scientific ambitions, and major funds are being invested in transforming the primate research institute at Kunming. It now boasts 75 covered homes for primates, housing more than 4,000 animals, which spend their time swinging on hanging ladders and scampering up and down wire mesh walls, while 60 trained animal keepers look after them full time.[379] The institute also has extensive facilities for creating genome-editing constructs, microinjection systems for injecting these into fertilized monkey eggs, incubators for culturing the resulting embryos, and facilities for implanting these into surrogate mothers. So, despite the ambitions of well-established US centres of science like the McGovern Institute to develop GM primates for studies of brain function, big future breakthroughs in this area may well be at Chinese institutes like the one at Kunming. Indeed, Guoping Feng, of the McGovern Institute in Botson, whom I mentioned in the section 'The Complex Brain', now travels to China several times a year to carry out research on genome-edited monkeys that he cannot do in the USA. Research currently taking in China using such monkeys, includes studies of genes linked to severe autism and Parkinson's disease.[380]

The Language Gene

The use of GM primates to study mental disorders is likely to further our understanding of such disorders but also generate controversy. However, an even more controversial question is whether primates might ever be used to investigate the biological basis of human uniqueness. To reiterate, humans are unique among species in our capacity for self-conscious awareness and our ability to change the world around us with tools. And ultimately this must reflect differences in the human genome compared to other species. The question then is whether genetic modification and the study of primates might be one route to gaining key insights into the biological basis of human uniqueness, and to what extent it would ever be deemed ethically acceptable to pursue such research.

For instance, could GM primates be used to explore the biological basis of human language? Until recently, scientists had no idea about the genes that underpin the unique capacity of our species for communicating through a system of abstract symbols—words. But in 2001, a team led by Anthony Monaco and Simon Fisher at Oxford University studied a British family with several members who had trouble with pronunciation, putting words together, and understanding speech, and showed that these individuals had a defect in a gene called forkhead box protein 2 (FOXP2). For Fisher, now at the Max Planck Institute for Psycholinguistics at Nijmegen in the Netherlands, the discovery 'opened a molecular window on the neural basis of speech and language'.[381] Yet although FOXP2 was initially hailed as a 'language gene', the question of its precise role in human language has been complicated by the subsequent discovery that the gene is present in many other animal species.[381]

Instead, attention has focused on the fact that the human FOXP2 protein differs from the chimpanzee version in only two amino acids, while the mouse version has a further amino acid difference, suggesting that such differences could be one of the reasons for our species' capacity for language compared to other mammalian species, including our ape cousins.[381] Recent studies suggest that one of FOXP2's functions is to regulate the number of nerve connections—synapses—in the brain, a role that it fulfils partly through its regulation of another gene, named sushi repeat containing protein, X-linked 2 (SRPX2), which is involved in synapse formation.[381]

Yet exactly how all this might relate to a role in language capacity in humans remains unclear. To explore this question further, in 2009 Fisher and his colleagues created a knockin mouse in which the human FOXP2 gene was substituted for the mouse version.[382] Initial studies of such mice showed that they produce more frequent and complex alarm calls, while a more recent study demonstrated that they are better at responding to cues that require repetitive learning while learning to negotiate a maze.[383] One interpretation of such findings is that they confirm that the specific FOXP2 gene difference found in humans is central to the complexity of human language, as well as our species' unique ability to learn such complexity as children. But some critics are sceptical about the conclusions that can be drawn from studies carried out on mice, which have considerable differences from us in brain size and structure, as well as in the way in which they interact with the environment.[383] Faraneh Vargha-Khadem, a neuroscientist at University College London, for example, has pointed out that mice depend on visual cues to work out what to do, whereas human infants respond to audio cues. 'If you really want to deal with the right [brain] circuit,' she says, 'you have to work with the right stimuli.'[383]

It would be interesting to assess the effect of introducing a 'humanized' version of FOXP2 into a primate, as well as investigate the effect of altering the activity of other genes, such as SRPX2, in a primate model. And indeed Bing Su, of the Kunming Institute of Zoology in China, has said he plans to use genome editing to make such a change to FOXP2 in macaque monkeys; he has stated that 'I don't think the monkey will all of a sudden start speaking, but will have some behavioural change.'[384] Moreover, showing the pace in this area, Su has already introduced human versions of a gene called SRGAP2, thought to endow the human brain with processing power by allowing the growth of connections between neurons, and of MCPH1, a gene related to brain size, into macaques.[384] As a human baby's brain develops after birth, MCPHI is expressed in abundance, but much less so in non-human primates, and it has been suggested to play a role in the human brain's specific development. A recent report from Su and his team provided evidence that the brains of monkeys with the human version of MCPH1, took longer to develop, and the animals performed better in tests of short-term memory as well as reaction time compared to normal monkeys.

Su claims that this is the 'first attempt to understand the evolution of human cognition using a transgenic monkey model.'[385] However, the study has been criticized on ethical grounds. Jacqueline Glover, a University of Colorado bioethicist, believes the study is likely to raise concerns about 'Planet of the Apes' type scenarios 'in the popular imagination' and has added that 'to humanise [the monkeys] is to cause harm. Where would they live and what would they do? Do not create a being that can't have a meaningful life in any context.'[385] But Larry Baum, a researcher at Hong Kong University's Centre for Genomic Sciences, has downplayed sci-fi comparisons, stating that 'the genome of rhesus monkeys differs from ours by a few percent. That's millions of individual DNA bases differing between humans and monkeys. This study changed a few of those in just one of about 20,000 genes. You can decide for yourself whether there is anything to worry about.'[385]

It is not surprising that such studies are controversial, or that the use of genome editing to generate new animal models or modify human cells for research, tends to dominate media coverage of this technology. But it is important to note that this is far from the only type of application for this technology that is likely to transform our lives in the future. In particular agriculture and farming are becoming a hot new area for the application of genome editing. It's time to enter the molecular farm.

6

The Molecular Farm

What could be more natural than the British countryside? Whether your fancy is strolling through valleys filled with daffodils in spring, taking a dip in burbling rivers in high summer, or hiking across hillsides dotted with sheep in autumn, it's not hard to see why Britain's green and pleasant land attracts so many visitors throughout the seasons. Yet for all its idyllic qualities, one thing the British countryside cannot claim to be is natural. Not only was the landscape itself produced by human activity—and this includes moors and heaths as well as the patchwork fields, for both are products of the deforestation that began as early as 6,500 years ago[386]—but the domesticated plant and animal species that populate England's rural regions are themselves products of thousands of years of selective breeding, as we saw in Chapter 1. It was in Britain that the industrialization of agriculture was first pioneered. With the establishment of the country as one of the first capitalist nations following the English Civil War, and subsequent upheavals of the seventeenth and eighteenth centuries, the way was open for a revolution in farming.[387] Through enclosure of the common land that poor people had farmed for centuries, albeit under a feudal system that claimed much of their output, the new masters of the British countryside gained access both to land to develop as they wished and an army of landless labourers to assist them in this goal. Yet equally important for the agricultural revolution that continued in the 18th and 19th centuries was its harnessing of technology, via mechanization of crop production and livestock handling, and the development of transport networks—canal, road, and rail—to take its products to the growing cities of the nation.

In the twentieth century, the most advanced methods of industrial farming were pioneered in the USA. Before the Second World War, most US farmers cultivated a variety of crops, along with livestock. As agriculture became more industrialized, farmers abandoned diverse farming systems in favour of highly specialized operations that separated crops from animals.[388] Today, US crop production is characterized by fields planted with a single crop over a given season, typically over a very large area. The meat supply chain, meanwhile, is separated into many specialized industries: breeding animals, growing feed crops, fattening the animals, slaughtering them, and processing their meat. And as food production and processing has become more specialized, this has allowed the mechanization of routine tasks, such as sowing seeds and harvesting. Meanwhile, agriculture has become more dependent on resources manufactured off the farm, such as agricultural chemicals and fossil fuels.[388]

Today, in an increasing number of countries, vegetables and fruits are grown under cover in greenhouses with sophisticated lighting and irrigation systems, while animals are reared intensively, with warehouses containing rows and rows of battery hens crammed into tiny cages or milking cows that never see the light of day. Such intensive farming uses large amounts of chemicals, whether fertilizers to boost plant growth and herbicides to prevent the growth of weeds, or antibiotics to keep control of the infections that can spread rapidly amongst animals in close proximity. It's thanks to the adoption of such intensive farming practices that China now produces a third of the world's meat, and why the meat consumption of its vast population jumped from 4 kg to 61 kg per person between 1961 and 2010.[389]

Apart from concerns about animal welfare, modern farming practices also create problems. One is the pollution of the environment by herbicide and fertilizer run-off. To complicate matters, legislation about genome-edited crops varies in different parts of the world.[390] So in 2018 the US Department of Agriculture announced that it was not planning to regulate gene-edited crops as it does transgenic crops created by inserting foreign gene constructs. According to a statement issued by the Department, this was because it sees genome-edited crops as indistinguishable from those developed through traditional breeding methods.[390] However, the European Union at this time confirmed that genome-edited crops would not be exempt from its very strict anti-GM legislation.[390] Another is the infection rife amongst intensively farmed animals, posing a health risk because of contaminated meat, while the continuous use of antibiotics to control such infection is one factor behind the development of antibiotic-resistant bacteria, with potentially disastrous future consequences for human health. Recently, China's Ministry of Environmental Protection warned that 'Chinese policymakers see the US intensive pork production model as the solution to China's food safety problems. Yet it is precisely this system of factory farming that has led to drastic environmental, public health and animal welfare problems in the US.'[389] Another aspect of modern farming is that, increasingly, giant multinational companies control food production, from the development of new varieties of plant seed, to their planting and harvesting in the ground, to the processing of plants and animals into food.

Today, living in Oxford, England, I take it for granted that I can go to my local supermarket and buy beef from Bolivia, spinach from Spain, or tiger prawns from Thailand. So varied are the nations that supply such products, it's doubtful whether many people even register the point of origin of most of what they buy. In some developed countries there is an increasing reaction against the industrialization and globalization of agriculture; so many British restaurants and 'gastro-pubs' now draw attention to the local, seasonal, and 'organic' nature of their ingredients.[391] However, since such foodstuffs are more expensive, this increasingly means two tiers of customers, not just in terms of eating out but also at supermarkets—with more 'ethical' and expensive food distinguished by labels like 'Fair Trade' or 'Finest'.

Another trend is the increasing numbers of supermarket aisles devoted to processed food. Recent years have seen a dramatic rise in food prices, which in Britain rose 32 per cent in the period 2007–12 according to official UK figures. In the USA, a 2012 report showed that food inflation was the highest recorded in 36 years.[392] Such increases have meant that poorer people have focused their stretched budgets on cheaper processed products at the expense of fresh fish, meat, and fruit.[393] And the availability of cheap 'junk' food is a key factor in the obesity 'epidemic' now affecting Britain, the USA, and other Western countries, itself leading to a dramatic rise in diabetes.[394] Meanwhile, the adoption of diets high in meat and processed dairy products has led to an emerging epidemic of diabetes in China, with 50 per cent of the population showing a prevalence of pre-diabetes and 11 per cent already diabetic (up from 1 per cent in 1980).[389]

Feeding Humanity

While it's important to recognize the problems associated with the industrialization and globalization of agriculture, it would be equally mistaken to underestimate its success in feeding the world's growing population—albeit with huge continuing inequalities in food distribution. But can these methods continue to feed the planet in the future? The ability of food production to keep pace with a world population that rose from 3 to 7 billion between 1960 and 2011 was primarily due to the remarkable increase in crop yields with only a slight expansion in land cultivated that occurred in this period, thanks to the 'Green Revolution'.[395] Spearheaded by Iowa-born plant geneticist Norman Borlaug, the Green Revolution introduced more productive varieties of wheat, corn, and rice into many parts of the world, combined with greater use of fertilizer and irrigation. The resulting increase in crop yields fed people directly, as well as the livestock that provide meat. But the success story of the Green Revolution shouldn't blind us to the problems that still exist, with an estimated 795 million people—one in nine people on the planet—lacking sufficient food to lead a healthy active life.[396] And many problems lie ahead. Since at least the turn of the millennium, increases in wheat, rice, and other cereal crop yields have begun showing signs of slowing.[397, 398] This isn't only worrying because such a rise is necessary to keep pace with a growing world population, predicted to reach at least 9 billion by 2050, but also because recent evidence suggests that the fall in yield is linked to global warming.

A study by David Lobell, an environmental scientist at Stanford University, and Wolfram Schlenker, an economist at Columbia University, found that, from 1980 to 2008, climate change depressed yields of wheat and corn; yields still rose during that time but overall production was 2 to 3 per cent less than it would have been in the absence of global warming.[397] This finding suggests that climate change, widely scientifically accepted as being the result of human activity, has not only already adversely affected food production, but that this negative impact will intensify as the planet heats up.[398] A particularly worrying aspect of the findings is that agriculture hasn't adapted to the increased frequency of hot days. 'What surprised me most and should inform us going forward,' says Schlenker, 'is that there has been tremendous progress in agricultural breeding ... but if you look at sensitivity to extreme heat, it seems to be just as bad as it was in the 1950s. We need to have crops that are better at dealing with hot climates.'[397]

One important question is what role genome editing might play in such a situation. In particular, is this technology likely to have a bigger impact than previous transgenic approaches? The latter were originally promoted by companies like Monsanto, Bayer, and DuPont as part of a new green revolution that would greatly increase food production. Yet, as we saw in Chapter 2, the application of transgenic technology to agriculture has been far from the success story predicted. Not that we should underestimate the impact of standard GM crops. Recent surveys estimate that more than 170 million hectares of such crops are grown worldwide, representing one twelfth of global arable land.[397] In the USA, most corn, soybeans, and cotton are now engineered to repel insect pests or withstand herbicides used to treat weeds. And in India, insect-resistant cotton represents 96 per cent of the cotton grown in the country. Yet it's far from clear whether GM crops have resulted in either increased food production worldwide or lower prices for consumers. Little transgenic corn and soybeans is used directly for human consumption; instead much is used for animal feed and biofuels. And such is the commercial success of these GM crops that US farmers are now substituting them for a major human food source—wheat—with adverse effects upon ordinary consumers. So 56 million acres of wheat were planted in 2012, down from 62 million in 2000. And as supply has fallen, the price of a bushel of wheat has risen, from $2.50 in 2000 to almost $8 in 2012.[397]

To some extent, the commercial application of transgenic technology to only a few species with a limited range of new characteristics, such as insect resistance and herbicide tolerance, is a reflection of the continuing controversial status of GM crops. Public opposition and regulatory requirements have made transgenic plants expensive to develop. That's why nearly all GM plants so far are lucrative, big-acreage crops like soybeans, corn, and cotton, and why only giant conglomerates have been willing to invest in such crops, and then only for a limited range of products, particularly those not destined for direct human consumption. Moreover, the continuing political controversy over GM crops means that, for some parts of the world, development has been at a virtual standstill. So in Europe, since the first GM crop was produced in 1994—a tomato whose ripening was delayed, giving it a longer shelf life—the European Union has granted just two licences to cultivate GM crops: one for maize plants engineered to resist corn borers and another for a starchy potato used to make paper.[399]

The limitations of previous GM crops are far from just political though. As we saw in Chapter 2, one problem with standard transgenic technology is that it involves randomly inserting a foreign gene into the genome of the host cell. Not only does this mean the technology is incapable of making the sort of subtle alterations such as those in a knockout or knockin mouse, but it also means the inserted gene may disrupt some other gene in the host genome, leading to unwanted, and possibly adverse, consequences. At the same time, this is a very inefficient process, so usually an antibiotic resistance gene is included in the transgene construct in order that plant cells that have taken up this construct can be selected.[400] And, as we also saw in Chapter 2, this has led to fears about such resistance being transferred to harmful bacteria.

Subtlety and Speed

In contrast, genome editing of plant cells, as with other types of cells, can be used both to delete specific genes or introduce more subtle changes, like the substitution of one type of amino acid for another, similar to what happens in a disease like sickle cell anaemia in humans.[401] A sign of how important genome editing is seen to be by agribusiness was shown by the announcement in 2015 that DuPont had signed an agreement with Jennifer Doudna's company Caribou Biosciences to develop CRISPR/CAS9 as a key new technology for crop production.[402] Indeed, DuPont says it's already using this technology to make drought-resistant corn and wheat that is modified so it breeds like a hybrid, rather than through self-pollination, as it typically does.[402] Hybrid plants are vigorous, and yields can be raised by 10 or 15 per cent. In theory, unlike previous transgenic approaches, genome editing should leave no mark on the host cell genome apart from the precise engineered change.

In practice, initial studies have introduced CRISPR guide RNAs and CAS9 protein not directly, but in the form of DNA constructs that express the guide RNA and CAS9 enzyme within the recipient plant cell. To introduce these DNA constructs into the cell, the approach used is that formerly employed to create standard GM crops, namely that the gene construct is carried into the cell by the bacterial pest *Agrobacterium tumefaciens*.[403] As a result, *Agrobacterium* DNA can end up in the plant's genome. Even if the pest is not used, introduction of CRISPR/CAS9 DNA constructs into a plant cell using other methods means that the construct DNA may be incorporated into the plant's genome. However, Jin-Soo Kim and colleagues at Seoul National University have assembled the CAS9 enzyme together with a guide RNA, and used solvents to get the resulting complex into the plant. Recently, Kim and his team reported that their technique can knock out selected genes in tobacco plants, rice, and lettuce. 'In terms of science, our approach is just another improvement in the field of genome editing,' says Kim. 'However, in terms of regulations and public acceptance, our method could be path-breaking.'[403] Since solvents are used to introduce the construct into plant cells, there's no need for selection for antibiotic resistance. And, given that genome editing can be applied to practically any plant species, this technology has far greater potential for agriculture than previous approaches.

One way that genome editing of plants has already demonstrated its potential is in combating disease. Potato blight is a fungus notorious for its role in the Great Famine that devastated Ireland in the late 1840s.[404] Eyewitness accounts from the period tell of infected plants emitting a horrible stench as they blackened and withered in front of the disbelieving eyes of Irish peasants. Although initially a natural disaster, it was the laissez-faire ideology of the British state at the time that greatly exacerbated the human tragedy developing on England's doorstep.[404] On the one hand, this meant that the government refused to provide the necessary massive food aid because of fears that English landowners and private businesses would be harmed by resulting food price fluctuations. On the other, it led to large quantities of native-grown wheat, barley, and oats being shipped from ports such as Limerick and Waterford to England throughout the Famine years, even though the local Irish were dying of starvation. It was policies such as these that meant that, by the end of the Famine, at least a million people had died—nearly one eighth of the population—and a further 2 million were forced to emigrate. Within five years of the Famine, the Irish population had been reduced by a quarter.[405]

Today, blight is still a major problem for potato farmers. Although fungicides are now available to treat the infection, considerable amounts are required to keep the fungus at bay. So in Britain alone farmers spend £60 million on fungicides each year to combat the pest, and, in a bad year, losses and control measures combined can account for half the total cost of growing potatoes. Because of blight, potatoes are one of the crops most treated with pesticides. In northern Europe, farmers typically spray a potato crop 10 to 15 times a year. But recently, Jonathan Jones and colleagues at the John Innes Centre in Norwich used genome editing to create a GM potato resistant to the fungus.[406] To do this, the researchers introduced a gene difference found in a blight-resistant wild potato from South America into the popular Désirée variety. 'Breeding from wild relatives is laborious and slow and by the time a gene is successfully introduced into a cultivated variety, the late blight pathogen may already have evolved the ability to overcome it,' says Jones. 'With new insights into both the pathogen and its potato host, we can use GM technology to tip the evolutionary balance in favour of potatoes and against late blight.'[406]

In fact, there are already moves to use genome editing to introduce disease resistance into a variety of crop plants. Consequently, Jin-Soo Kim is planning to use this form of CRISPR/CAS9 to make disease-resistant bananas. Currently, the most popular variety of this plant, the Cavendish, is threatened by a devastating soil fungus and may soon be totally wiped out. According to plant pathologist George Mahuku of the International Institute of Tropical Agriculture, this prospect poses 'a serious threat to livelihoods and food security...In Africa, bananas are critical for food security and income generation for more than 100 million people.'[407] Kim wants to edit the genome of this variety to knock out the receptor that the fungus uses to invade the cells of the banana plant. 'We will save the banana so that our children and grandchildren can still enjoy the fruit,' he says.[403]

A team led by Caixia Gao at the Institute of Genetics and Developmental Biology in Beijing recently produced wheat resistant to a fungus called powdery mildew, a major pest of the world's top food source.[408] The researchers deleted genes in the wheat genome coding for proteins that block defences against the mildew. This proved a particular challenge since the wheat genome contains three copies of most of its genes. Yet using both TALENs and CRISPR/CAS9, Gao reported that 'we now caught all three copies, and only by knocking out all three copies can we get this [mildew]-resistant phenotype'.[408] Meanwhile, Daniel Voytas and colleagues at the University of Minnesota in Minneapolis recently used CRISPR/CAS9 to target the geminivirus—a common crop pest that infects plant species ranging from beans to beetroot.[409] This is important as plant viruses cause enormous losses in worldwide crop production each year. What's more, global warming is aggravating viral infection of crops by promoting viral growth and the migration of virus-transmitting insects.[410]

The use of genome editing to introduce characteristics from a wild species into a domestic variety is likely to have much wider applications than introducing disease resistance. 'Wild plants tend to manage much better under harsh conditions than their cultivated relatives,' notes Michael Palmgren, a plant biologist at the University of Copenhagen. 'Many important properties of wild plants were unintentionally lost during thousands of years of breeding.'[411] In fact, the reintroduction of wild characteristics is a practice that has been used by farmers for many years. Traditionally, such 'reverse breeding' is carried out by crossing crops with wild versions of the plant that have the desired characteristics. But the resulting hybrid may also end up with other qualities that breeders had intentionally removed. 'Wild plants are seldom tasty, nutritious, and easy to harvest,' says Palmgren.[411] The process of perfecting the hybrid plant is time-consuming and difficult to control. With genome editing, there's no need to go through extensive crossing to induce the characteristic, which makes the whole process a lot faster.

Palmgren has even suggested that 'all crops would benefit from rewilding'.[411] This would not only protect them against pests and disease, but also allow them to draw nutrients from the soil more efficiently. However, not everyone agrees with this conclusion. 'I find their whole premise to be rather flawed,' says Clay Sneller, a plant breeder and geneticist at Ohio State University in Columbus. 'They appear to think that during breeding we have accumulated negative mutations, and if we got rid of those mutations then the crop would be better. They reviewed no evidence that this occurs on a wide scale in genes that truly matter.'[411] Such concerns show that there will be many divergences of opinion about the best way to employ genome editing in agriculture. However, the precision of the new technology means that different strategies can now be put to a practical test.

Improving a plant's resistance to disease or its capacity for growth is only one way in which genome editing is likely to have an impact on crops. The other important impact will be on the food produced from such crops. For instance, recently a potato plant was engineered so that it doesn't accumulate sweet sugars at typical cold storage temperatures.[412] This modification will let it last longer and, when fried, it won't produce as much acrylamide, a suspected carcinogen that can accumulate in some fried food. The modified potato was created by Daniel Voytas in collaboration with a biotechnology company called Cellectis Plant Sciences. The genome editing disabled a gene that turns sucrose into glucose and fructose and took only about a year to achieve. 'If you did it via breeding it would take five to 10 years,' says Voytas.[412] Luc Mathis, CEO of Cellectis, claims that developing the potato cost a tenth of that needed to create a standard transgenic plant and bring it to market.

A particularly beneficial development for human health could be the use of genome editing to create crops that don't trigger dangerous food allergies. Allergy to peanuts is particularly common, affecting more than 5 million people in the USA and Europe, including 2 million children and adolescents. Indeed, my daughter suffers from such an allergy. And the prevalence of this allergy is rising: a recent study of US children found that peanut allergies have risen threefold, from 0.4 per cent in 1997 to 1.4 per cent in 2010.[413] The reason for the increase remains unclear, with explanations ranging from rising consumption of processed foods to the idea that increased standards of hygiene have led to abnormal immune responses because people are no longer exposed to as many germs in childhood. Whatever the reason, the anaphylactic shock that can result from such an allergy can prove fatal. Acute attacks can be treated with adrenaline, but there are cases of children dying even after multiple injections.[413] Because of this, research is underway in engineering peanuts so they don't trigger an allergic response. Analysis of peanut proteins has identified seven that can cause an allergy.[414] And an important question is whether genome editing can now be used to delete or modify the genes coding for these proteins in the peanut genome.

The economy and speed of genome editing compared to traditional transgenic approaches means that small companies, not just giant corporations, can now develop GM plants. But the ease of modification also means that we may start to see changes to crops that some people might consider frivolous. For example, in 2012 a company called Okanagan Specialty Fruits used RNA interference technology to engineer an apple that doesn't turn brown when sliced or bruised.[415] Neal Carter, the founder and president of the company, which is based in British Columbia, Canada, believes the non-browning apples could increase the consumption of fresh apples, which has fallen from about 20 pounds a year for each person in the late 1980s to 16 pounds today, according to the US Department of Agriculture. A whole apple is 'for many people too big a commitment,' says Carter. 'If you had a bowl of apples at a meeting, people wouldn't take an apple out of the bowl. But if you had a plate of apple slices, everyone would take a slice.'[415] However, showing the continuing political sensitivity of GM crops, many apple industry representatives are wary of the product. 'We don't think it's in the best interest of the apple industry of the United States to have that product in the marketplace at this time,' says Christian Schlect, president of the Northwest Horticultural Council that represents the industry around Washington State, and produces about 60 per cent of the apples in the USA.[415] While many industry representatives don't believe genome editing is dangerous, they think that it could undermine the fruit's image as a healthy and 'natural' food that 'keeps the doctor away', and is as American as, well, apple pie.

The fact is that GM technology applied to crops for human consumption remains a sensitive topic. And ironically, the very subtlety of genome editing that distinguishes it from traditional transgenic approaches is itself controversial. It could mean that food companies will argue that some genome-edited plants no longer need to be classified as GM. Take the modifications that introduce resistance to fungus infection into a commercial potato or banana plant. Because such modifications are based on mutations found in a wild plant, does such a change need to be classified as GM? Not according to Chidananda Nagamangala Kanchiswamy of the Istituto Agrario San Michele all'Adige in Trentino, Italy, who believes that 'the simple avoidance of introducing foreign genes makes genetically edited crops more "natural" than transgenic crops obtained by inserting foreign genes'.[416]

Similarly, Neal Gutterson, vice president at DuPont Pioneer, believes that genome-edited plants 'are basically comparable to what you get from conventional breeding. We certainly hope that the regulatory agencies recognize that and treat the products accordingly.'[417] A recent survey of rice, wheat, barley, fruit, and vegetable crops found that most plants created by genome editing may be outside the scope of current GM organism regulations, including those of the influential US Federal Food and Drug Administration (FDA). Tetsuya Ishii, of Hokkaido University, who carried out the survey, believes that since 'genome editing technology is advancing rapidly; therefore it is timely to review the regulatory system for plant breeding... Moreover, we need to clarify the differences between older genetic engineering techniques and modern genome editing, and shed light on various issues towards social acceptance of genome edited crops.'[418] Certainly, this could be a key demand of GM opponents, on the grounds that genome edited plants may have unintended adverse effects on the environment and human health. To complicate matters, legislation about genome-edited crops varies in different parts of the world.[418] So in 2018 the US Department of Agriculture announced that it was not planning to regulate gene-edited crops as it does transgenic crops created by inserting foreign gene constructs. According to a statement issued by the Department, this was because it sees genome-edited crops as indistinguishable from those developed through traditional breeding methods.[418] However, the European Union at this time confirmed that genome-edited crops would not be exempt from its very strict anti-GM legislation.[418]

Surviving the Extremes

Will GM crops ever be seen as widely socially acceptable in countries across the world, whatever the subtleties of the engineering used to create them? One factor that could be important here is if such crops are seen as making a major contribution to feeding the planet, rather than creating novelties like non-browning apples or simply increasing the profits of giant companies. How humanity responds to the threat posed by global warming is likely to be key. Recent evidence shows the seriousness of the problem, with a joint report in January 2020 by the US organisations NASA and the National Oceanic and Atmospheric Administration (NOAA), noting that the previous decade had been the hottest on record.[419] Commenting on the report, Noah Diffenbaugh, an earth science professor at Stanford University, said that 'we now have clear evidence that people and ecosystems are being impacted across the world, from the equator to the poles, from both in the ocean and on land, from the coastal areas to the high elevations.'[419] Global warming is helping to fuel wildfires from Australia to California, melt permafrost from Alaska to Siberia and fuel more intense storms and floods. It is also altering marine ecosystems from Canada to South America to the African coast, threatening wildlife and the livelihoods of those who depend on the sea.[419] Even Pope Francis has weighed into the debate with his statement that 'leaving an inhabitable planet to future generations is, first and foremost, up to us. The issue is one which dramatically affects us, for it has to do with the ultimate meaning of our earthly sojourn.'[420] Even if the rise in emissions were halted tomorrow, all the evidence indicates that the effects of global warming caused by current carbon dioxide levels will affect us for many years to come. Yet the political will to radically address the root of the problem, namely our continued reliance on fossil fuels and the carbon dioxide this releases into the air, seems to be lacking, despite a succession of meetings of world leaders that have discussed and debated the issue over recent years.[421]

Although the main future trend will be rising global surface temperatures, a recent study by Daniel Horton and colleagues at Stanford University found 'statistically significant' evidence that shows global warming is also causing 'weather whiplash—wild swings from one extreme to another'.[422] This explains not only the recent record high temperatures in places like California but also the equally severe cold conditions that have affected many parts of the USA in recent winters, due to the partial breakdown of the polar vortex that typically confines cold weather to the Arctic region, allowing cold-air breakouts to the south. While such temperature swings can cause chaos to cities and their transport networks, undoubtedly the most serious impact will be on the plant and animal species that provide food for the Earth's 7 billion humans. As we saw in the section 'Feeding Humanity', global warming is already having a worrying effect on the yields of staple crops like wheat and corn. The central highlands of Mexico, for example, experienced their driest and wettest years on record back to back in 2011 and 2012, according to Matthew Reynolds, a physiologist at the International Maize and Wheat Improvement Center in El Batán. Such variation is 'worrisome and very bad for agriculture,' he says. 'If you have a relatively stable climate, you can breed crops with genetic characteristics that follow a certain profile of temperatures and rainfall. As soon as you get into a state of flux, it's much more difficult to know what traits to target.[397]

It's here that genome editing may aid the development of crops able to withstand the extremes of temperature, drought, wind, and snow predicted to be coming our way.[423] To deal with such extreme changes, scientists may need to do more than tinker with existing crop plants; a major transformation of plants' basic biology may be required. Daniel Voytas, working with researchers at the Rice Research Institute in Los Baños, in the Philippines, aims to rewrite the physiology of rice, by focusing on the process of photosynthesis.[397] This is the cellular mechanism, common to all plants, that traps energy from the Sun and uses it to convert carbon dioxide and water into glucose, as a preliminary to producing the other complex molecules of life. Photosynthesis can occur both as the C3 form, found in rice and wheat and other grain-producing plants, or the more complex C4 form, used by plants like corn and sugarcane.[424] The two types are distinguished by the initial molecule formed, with either a 3-carbon or 4-carbon sugar being produced. An important difference between the two processes is that C4 photosynthesis is far more efficient at high temperatures and dryness, so if a way were found to create C4 versions of rice and wheat, their yields could be increased in regions becoming hotter and drier due to climate change.[417]

Other important work along these lines is being pioneered by Eduardo Blumwald, of the University of California, Davis. By introducing genetic differences that provide tolerance to heat, drought, and high soil salinity into rice and other plants, he hopes to create crops that can survive at periods in their growth cycle when they are most prone to stress. 'There's no cure for drought. If there's no water, the plant dies. I'm not a magician,' he says. 'We just want to delay the stress response as long as possible in order to maintain yields until the water comes.'[397] Whether such radical changes to the world's main staple crops are both technically possible and will prove politically acceptable remains to be seen, but at least genome editing now makes such engineering a tangible reality, not just a pipe dream.

Enviropigs and Frankenfish

If genome editing has the potential to vastly expand the possible range of GM plants that can be grown, the technology may have an even bigger impact on farm animals, which, as we saw in Chapter 2, have been little affected by standard transgenic techniques. Political opposition has also played a role here, similar to the situation with GM crops. GM pigs created by standard genetic engineering approaches in the mid-1990s at first looked like they might have a chance of commercial success, only to meet opposition because of their artificial origins. The pigs were produced by Cecil Forsberg and his team at the University of Guelph, Canada, using standard transgenic methods to engineer them to express an enzyme, phytase, normally generated by the gut bacteria of cows and other ruminant animals.[425] Phytase breaks down phosphorus-containing phytate in plants, but normally pigs do not have such bacteria and require a phosphate supplement in their feed. The gene construct used to create the GM pigs contained the gene coding for phytase linked to a tissue-specific mouse gene promoter, which meant the bacterial enzyme was only produced and secreted by the pig salivary glands.

Forsberg named the GM pig variety he had created the 'enviropig', because it produced manure with lower levels of phosphorus, notorious for leaching into groundwater beneath pig farms and fuelling algal blooms in local streams and lakes.[425] He also claimed that the pigs would be more economical, since they would not require the extra cost of adding mineral phosphorus or commercially produced phytase to their feed to ensure that they obtained the nutritional phosphorus they require. But attempts to obtain approval to develop the pigs commercially foundered after opposition from anti-GM activists. They argued that, because of their low-phosphorus manure, if the enviropigs were approved for human consumption, 'agriculture [would have] an excuse to put them in even more concentrated facilities', as Alison Van Eenennaam of the University of California, Davis, who studies the commercial application of GM technologies, put it.[425] Such opposition meant that, in 2011, the Ontario Pork Producers Marketing Board, funder of the project, withdrew its support. Unable to find another industry backer, the researchers decided they had no alternative but to terminate the project. 'These pigs were healthy pigs that did perform as they were designed to perform. They just didn't meet the social requirement,' said Forsberg.[425]

Another transgenic animal project that produced a potentially important commercial product, but then stalled subsequently, is a super-sized salmon. This fish, named the AquAdvantage salmon by its creators Garth Fletcher and his colleagues at Memorial University of Newfoundland, was produced in 1989. They introduced a gene construct combining a growth hormone gene from Chinook salmon and a highly active regulatory DNA element from the eel-like ocean pout into fertilized Atlantic salmon eggs.[425] The resulting GM salmon grow to market size in half the time required for a conventional farmed variety, while consuming 25 per cent less food overall. In order for the fish to be approved for sale, the researchers had to demonstrate that the salmon's meat had the same composition as a normal fish. And to prevent escape and interbreeding with wild fish, the GM salmon are all sterile females and their watery enclosures at Prince Edward Island have physical barriers between them and the sea. In recognition of these safeguards, a 2010 environmental impact assessment found that breeding the AquAdvantage salmon shouldn't harm wild fish. In 2015, after years of delays, the FDA finally approved the salmon for sale.[426] Yet, due to a vigorous campaign of opposition, 65 supermarkets, along with seven seafood companies and restaurants, have already signed a pledge not to carry it. 'We don't want Prince Edward Island to be known around the world as the home of the Frankenfish,' says Leo Broderick, a retired schoolteacher on the island who travelled to an FDA meeting in Maryland to campaign against the salmon's approval.[427]

As we have seen, a major difference between genome editing and previous transgenic approaches is the capacity of genome editing to introduce precise changes to one or more genes in a way that doesn't cause other changes to the genome. It is also cheap, highly efficient, and can be applied to practically any animal species. So could these features allow the new technology to have a major impact on the development of livestock for the industry in a way that previous approaches have not?

Hornless Cows and Beefy Bulls

One way that genome editing is being applied to livestock can be illustrated by a recent study in which scientists used this technology to produce Brazilian Nelore cattle with increased muscle mass. This study, led by Scott Fahrenkrug of the University of Minnesota, in collaboration with the Roslin Institute in Edinburgh and Texas A&M University, introduced a mutation found in Belgian Blue cattle—hulking animals that provide unusually large amounts of prized, lean cuts of beef—into rangy Nelore cattle which are heat-tolerant.[428] The mutation, which inhibits production of a muscle-supressing protein, myostatin, increases muscle mass. The introduction of this mutation into the Nelore cattle has created animals that provide prime meat cuts but can be kept in hot countries like Brazil, unlike Belgian Blues. And such is the interest from the livestock industry that British-based company Genus, the world's largest breeder of pigs and cattle, has funded some of the study's research. Jonathan Lightner is the head of research and development (R&D) at Genus. 'We haven't realized the opportunity for genetic engineering in animals to any degree,' he says. 'But these new approaches that let us move traits around could be transformational.'[428]

Another one of Fahrenkrug's projects that Genus is backing is the introduction of a natural mutation found in Angus beef cattle that means they never develop horns, into a breed with a capacity to produce lots of milk, such as black-and-white Holsteins (see Figure 24). Fahrenkrug began this project after seeing a video showing a young Holstein heifer moaning and bucking as a farmhand burned off its horn buds with a hot iron.[428] This operation is routinely carried out because of the dangers posed to farmhands handling the cows. Introducing the mutation from Angus cattle into the Holstein breed would render such an operation unnecessary. Douglas Keeth, an investor in the project, says his great-grandmother was gored to death by a dairy cow. 'When I was a young man working on a farm, we'd dehorn cattle with mechanical means. You do a hundred steers and, well, it's a bloody mess,' he says. 'You wouldn't want to show that on TV.'[428] And Lightner believes that, because of its potential for eliminating such animal

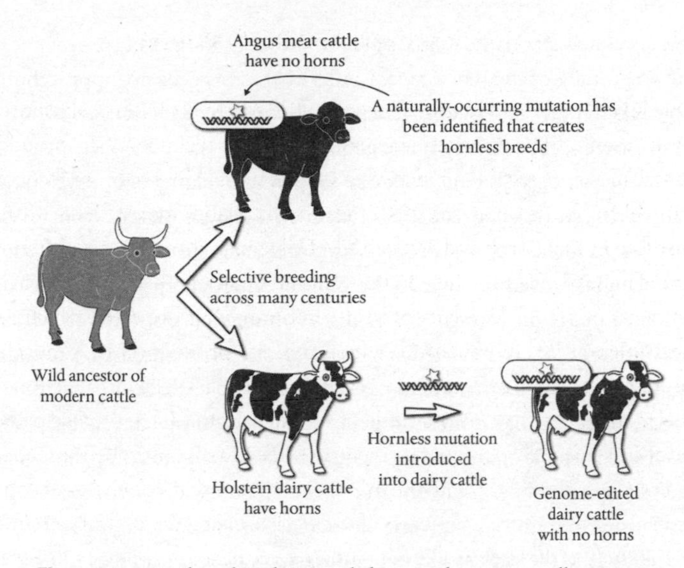

Fig. 24. Creating hornless dairy cattle by introducing naturally occurring mutations

suffering, the project could be viewed much more positively by the public than previous GM experiments. 'There may be an opportunity for a different public acceptance dialogue and different regulations,' he says. 'This isn't a glowing fish. It's a cow that doesn't have to have its horns cut off.'[428]

As for dairy farmers, they are interested, but also cautious about the potential of genome editing. Tom Lawlor, head of R&D for the Holstein Association USA, thinks that the technology 'is very cool'.[428] But he also believes many milk producers are afraid of genetic engineering. 'The technology definitely looks promising and seems to work, but we would enter into it slowly as opposed to rapidly for fear the consumer would get the wrong idea,' he says. 'We get scared to death, because our product is milk, and it's wholesome.'[428] Lawlor also points to other scientific initiatives that provide an alternative route to a hornless dairy cow, such as the 1000 Bull Genomes Project. This has decoded the genomes of 234 dairy bulls, including Swiss Fleckviehs, Holsteins, and Jerseys, and means that breeders can now accurately size up an animal's genetic profile at birth. As a consequence, a few naturally occurring hornless bulls of these breeds are approaching top-ranked status, and such genetic selection may be seen as a less controversial strategy than direct genetic engineering.[428]

Another potentially important use for genome editing is in developing animals that are resistant to disease. One such disease is African swine fever, caused by a highly contagious pig virus, and characterized by high fever, loss of appetite, haemorrhages in the skin and internal organs, and death in two to ten days on average.[429] The disease remained restricted to Africa until 1957, when it was reported in Lisbon. Subsequently, the disease became established in the Iberian Peninsula, and sporadic outbreaks occurred in France, Belgium, and other European countries during the 1980s. Both Spain and Portugal eradicated the disease by the mid-1990s only through a mass slaughter policy. But from 2012 onwards, outbreaks were reported in East European countries including Lithuania, Ukraine, Poland, and Latvia,[429] and since 2018 the disease has had a devastating impact on pigs in China, Vietnam, and other countries in the Far East.[430] In contrast to the severe reactions of domestic pigs to the virus, wild warthogs are far less affected. This is because of a difference in a gene called nuclear factor kappa B3 (NFKB3) that is involved in immune responses. This gene is less active in warthogs compared to domestic pigs, and, ironically, this turns out to be beneficial following infection by African swine fever. So, according to Simon Lillico of the Roslin Institute, while in domestic pigs, 'the immune system grossly overreacts to something that itself isn't that harmful', in contrast 'warthogs still get infected, they just don't drop dead'.[431]

Such is the natural situation; however, recently Lillico's colleague, Bruce Whitelaw, is trying to use CRISPR/CAS9 to create domestic pigs with the genetic difference found in warthogs. Whitelaw believes the welfare aspect of creating disease-resistant animals may prove more acceptable to the public than creating bigger, meatier farm animals. 'We're not trying to make huge pigs, we're trying to make healthier ones,' he says. 'I'd be staggered if anyone says "No, I don't want my animal to be healthier."'[431] Whitelaw claims that farmers have welcomed the focus on disease resistance, because, unlike size and fertility, resilience to disease is almost impossible to breed in with traditional methods. At a recent talk he gave, Whitelaw said the first question from a Lithuanian farmer was 'When can I get these animals?'[431] And if the genome editing is a success, the next step is to apply to the FDA for commercial approval. 'We need these animals to deliver something that could be a product,' says Whitelaw. 'If these pigs show resilience, we will go to the regulators. The limitations are no longer technical, they're legal.'[431]

One final potential application of genome editing is worth mentioning because it shows that this technology might be used to improve livestock farming in more indirect ways. Robert Speight, from the Queensland University of Technology in Brisbane, has been using CRISPR/CAS9 to 'supercharge' the yeast commonly used to supplement livestock feed.[432] 'It's the same yeast that would be added normally, but what we're trying to do is up regulate one of its own enzymes which can also serve to aid digestion,' he says. 'Really, what we want to see is as much of the energy and nutrients in the feed going into the animal to produce meat protein.'[432] There is a particular current relevance to Speight's research, as Queensland farmers are increasingly turning to supplementary feeds to keep their remaining stock alive due to the impact of a severe drought in this region in recent years.[433]

The Pressure to Patent

The examples mentioned in this chapter could be viewed as benign, or even beneficial to the welfare of the animal, and this may help make genome editing of farm animals acceptable to the public and policymakers. But other approaches being planned may be viewed as more problematic. For instance, Pablo Ross at the University of California in Davis wants to use genome editing to design cattle that produce only male offspring. 'Males grow faster than females, and in beef production they are more desirable,' he says.[434] And a gene modification that Scott Fahrenkrug is planning to introduce into beef cattle would prevent them ever reaching sexual maturity.[428] This would make it quicker to fatten animals for slaughter, but could be viewed as a disturbing development that would increasingly turn cattle into meat-producing machines without a thought for their welfare. It could also be viewed negatively by farmers as a way for big companies to gain even more control over the farming process since such sexually immature animals could only be obtained from the company and not used to produce more offspring in the normal fashion. And indeed, the patent filed to cover this application states that this approach will allow genome-editing companies to keep selling animals without the risk of 'uncontrolled breeding of the animals by the buyers'.[428]

In fact, the issue of ownership is likely to become significant if genome editing does become a regular aspect of livestock production. So the filing of patents to cover genome-editing projects has alarmed some farmers already concerned about the way such patents have been used to restrict access to crop seeds. 'They could take semen from my bull, gene-edit it, patent it, and the farmer will get totally screwed,' says Roy MacGregor, who breeds hornless cattle in Peterborough, Ontario. 'They should not be allowed to.'[428] On the other hand, pioneers of genome editing in livestock like Scott Fahrenkrug point to the need to get a return on projects in which they have invested huge amounts of time and money. No doubt this is a debate that will only intensify as the technology becomes established in the industry.

Another likely point of future controversy is whether this technology will only be used to carry out what Fahrenkrug calls 'molecular breeding'—that is, the introduction of natural mutations from one breed into another of the same species, or whether more radical changes are in the pipeline.[428] Recently, a team led by Jin-Soo Kim of Seoul National University in Korea and Xi-Jun Yin of Yanbian University in China created pigs with double the muscle mass of normal animals.[435] In fact, the deletion of the myostatin protein is precisely that found in cattle breeds like the Belgian Blues. But this mutation isn't normally found in pigs. Kim and Yin believe their genetically jacked-up pigs could be of commercial interest, particularly given the increasing consumption of pork in China. Preliminary investigations show that the pigs have many of the beefy cattle's benefits—such as leaner meat and a higher yield of meat per animal. However, the abnormal size of the piglets resulted in birthing defects.[435] What's more, only 13 of the 32 GM pigs survived to 8 months old, with just two remaining alive and only one being healthy, suggesting that there are other adverse effects of knocking out the myostatin gene in this species. The researchers believe that if they sell sperm from this pig for insemination of normal female sows, they can produce offspring with only half the extra muscle mass, and therefore none of the health problems. Whether this will satisfy ethical, health, and safety concerns in China remains to be seen, but the project shows that even if a mutation is found in the wild in one species, its introduction into a different type of animal needs to be considered carefully with regard to potential adverse effects in the host species.

Despite the potential political obstacles that may lie ahead for the practical application of genome editing in farming, Scott Fahrenkrug is sure that the technology represents the future of agriculture. As he puts it, 'People will say to me, "You realize this changes everything, don't you?" Because it does. The genome is information. And this is information technology. We have gone from being able to read the genome to being able to write it.'[428] And it's because of such unprecedented power to manipulate life that the impact of genome editing looks set to extend into every corner of society in the future. Already, its effects are being felt in medicine.

7

New Gene Therapy

In the developed world it's easy to take for granted the wonders of modern medicine. Vaccines and antibiotics, anaesthetics and painkillers, laser and keyhole surgery, drugs to treat diabetes or heart disease, whole body imaging devices—the list could go on. In fact, the situation is very different in far too many parts of the world, with a staggering third of the Earth's population still lacking access to even the most basic health provision.[436] For several billion people, this lack of health care is compounded by inadequate food and access to clean water and sanitation facilities. Little wonder that malaria, tuberculosis, cholera—and even simple malnutrition—remain major causes of death in much of the developing world. The additional terrible burden of HIV is heavily concentrated in Sub-Saharan Africa and is now the world's biggest infectious killer.[437] This is the reason why, even as life expectancy rises globally, it remains stubbornly low in the developing world. So while US average life expectancy is 79, in Zambia it's 55.[438]

Even in the developed world, gross inequalities in health still exist. Half a century after Martin Luther King Jr said that 'of all forms of inequality, injustice in health care is the most shocking and inhumane',[439] millions in the USA lack access to proper health care because they can't afford medical insurance. And ethnic minorities are particularly vulnerable, with a World Health Organization (WHO) report showing that infants born to African-American women are 1.5 to 3 times more likely to die than other infants in the USA.[440] Even in Britain, with its publicly funded National Health Service (NHS), there is still a substantial link between income and health. A recent study by the UK Institute of Health Equity found that the average gap in life expectancy across Britain between the best- and worst-off is seven years, but in London this difference rises to 17, and in Glasgow there is a 28-year difference in life expectancy between rich and poor.[441]

Such statistics show that the amazing technological advances in medicine over the past century mean little without an equal focus on the social inequalities that can cause ill health. Nevertheless, it would be equally wrong to belittle the importance of developing new medical technologies. For all the progress, there is still much that remains backward about modern medicine. So, despite important advances in our understanding of the molecular basis of cancer, the main treatment for this disease remains crude surgery, or chemical or radiation therapies that have serious, sometimes life-threatening effects in normal cells and tissues.[442] As for degenerative disorders of the brain like Alzheimer's and other forms of dementia, which affect one in six people over 80, despite recent advances in our understanding of such disorders, they remain essentially untreatable.[443]

And, as we saw in Chapter 3, although there are now a large variety of drugs for treating personality disorders like schizophrenia, bipolar disorder, or depression, these remain blunt tools that, as many psychiatrists acknowledge, treat some of the symptoms but not the underlying causes of these disorders. Finally, there are concerns about how long we can rely on one of the most powerful tools of modern medicine—antibiotics—with Sally Davies, Britain's chief medical officer, recently warning that antibiotic-resistant bacteria represent a 'ticking time bomb' that threatens to take us back to nineteenth-century standards of health care.[444]

Single Gene Disorders

What is the likely impact of our new-found skills in genome editing on this situation? This technology has led to great excitement because of its potential to advance biomedical research, as we saw in Chapter 5. So genome editing not only makes it possible to study gene function in human cells in culture but has greatly accelerated the development of knockout and knockin mice, and also made it possible to develop models of human health and disease in other mammals, ranging from pigs to primates.[445] In addition, new approaches to studying the brain, like optogenetics, are revolutionizing our understanding of how the brain functions and what happens when it goes wrong in such animal models.[446]

Such new developments increase our understanding of the human body and its various ailments by modelling it in an animal, and we can use such information to design new diagnoses and treatments. But the new technologies might have a more direct impact on human health and disease. For instance, could genome editing be used therapeutically on human beings themselves? To assess this possibility, we need to consider first our current state of understanding of the human genome and its link with human disease; and second, the practical challenges of seeking to manipulate the genome of a living person.

We saw in Chapter 1 how scientific understanding of the link between genes and an organism's characteristics was first shaped by Mendel's discovery of recessive and dominant inheritance patterns in pea plants.[447] Such patterns are also true of single-gene disorders in humans. So the dominant Huntington's disorder makes its presence known in each generation of a family, while the recessive disorder cystic fibrosis only occurs when passed down by two unaffected carriers. A slight variation on this theme is shown by disorders like haemophilia or DMD. These recessive disorders are due to defects in genes on the X chromosome and, since men only have one of this chromosome, these disorders generally only affect men, with women as the carriers. A major step forward in genetics was made from the mid-1980s onwards with the discovery of the genes associated with cystic fibrosis, Huntington's, and DMD. Since then, many more human diseases with a Mendelian pattern of inheritance have been linked to specific genes.[448] The discovery of genetic defects associated with such disorders has particularly accelerated with the development of rapid and economical 'next-generation' DNA sequencing technologies, with a recent study reporting that almost 3,000 gene defects are now associated with disorders showing a Mendelian inheritance pattern.[449] And while such disorders are rare, once added together it's estimated that they affect around 25 million people in the USA alone.[449]

The prospect of being able to correct 'defective' genes has been a dream in medicine ever since the link between genetics and human disease was first identified. Yet, as we saw in Chapter 2, gene therapy via standard transgenic approaches has been far from a success story—even for well-characterized single-gene disorders—because of two main obstacles.[450] One is the difficulty in getting gene constructs into tissues and across the membranes of cells in the body. Viruses can effectively transport gene constructs into cells, but their use also carries risks. The second obstacle has been the lack of a technology to precisely engineer the genomes of treated cells. Instead, traditional gene therapy has meant introducing a foreign DNA construct randomly into the genome of the host cell. Not only can this disrupt the host genome and cause damage, such as activating an oncogene (which can lead to cancer), but it's also only useful for treating recessive disorders in which a gene product is absent, as in cystic fibrosis, not for treating dominant ones like Huntington's where a defective gene product disrupts normal cell function.

Since genome editing is so new, particularly the CRISPR/CAS9 version that's rapidly becoming the dominant approach, its full potential as a therapeutic strategy in humans is still only at the first stages of assessment, yet already there are highly promising signs. Demonstrations of the technology's therapeutic potential have so far mainly involved mouse models of disease, although, encouragingly, there are now ongoing clinical trials and an apparent recent success in treating childhood leukaemia that we'll look at in the section 'New Cancer Cures'. Two main approaches are being pursued. First, genome editing is being used to modify cells outside the body. Although this is much easier to achieve in a controlled fashion, it does limit this approach to only a few cell types, such as those in the bone marrow; consequently, this restricts the types of diseases that can be treated to those affecting the blood or immune system. Second, genome editing is being employed to target cells located within the body. Although this opens up the possibility of using the technology to modify practically any type of genetic disease, technically this second approach is a lot more challenging.

One animal study that has demonstrated the clinical potential of genome editing was carried out by Jinsong Li's team at the Shanghai Institute for

Biological Sciences. They used CRISPR/CAS9 to target a mutant gene in a mouse model of cataract formation.[451] These mice have a naturally occurring mutation in the crystallin gamma C (CRYGC) gene which codes for a protein that is a key component of the lens of the eye. Such mice develop cataracts at an early age, but by using genome editing to correct the CRYGC mutation in fertilized eggs of this mutant, a third of offspring grew up free of cataracts. Li acknowledges that the efficiency of the technique was low, 'and, for clinical purpose, the efficiency should reach 100 per cent'.[451] However, the findings impressed Charles Gersbach, a geneticist at Duke University, enough for him to say that 'what's significant about this is it's taking CRISPR to that next step of what it can be used for, and in this case, it's correcting mutations that cause disease'.[451]

A limitation of Li's study was that the mutation was corrected only in an embryo, not in an adult animal. However, such an employment of genome editing in an adult mouse has been achieved by Daniel Anderson and colleagues at the Massachusetts Institute of Technology; they reported that they had 'cured' mice of a rare liver disorder.[452] The disorder in this mouse model was caused by a mutation in the gene coding for a liver enzyme called fumarylacetoacetate hydrolase (FAH), which breaks down the amino acid tyrosine. In this disorder, which affects about 1 in 100,000 people, the defect causes a build-up of tyrosine in the liver, leading to eventual liver failure. Current treatments in humans involve a low-protein diet and a drug called nitisinone, also known as Orfadin®, which inhibits tyrosine production, but these measures are only partially effective.

To treat the disorder in mice, Anderson and his team injected CRISPR/CAS9 constructs into the animals' blood supply under high pressure. The constructs were subsequently taken up by the liver, leading to a correction of the gene defect in some cells of that organ. In fact, only one in 250 cells were corrected, but it nonetheless allowed the corrected, now healthy liver cells to proliferate and replace diseased liver cells. This was sufficient to cure the disease, allowing the mice to survive after being taken off the nitisinone drug. 'The disease is caused by a single point mutation and we showed that the CRISPR system can be delivered in an adult animal and result in a cure,' says Anderson. 'We think it's an important proof of principle that this technology can be applied to animals to cure disease. The fundamental advantage is that you are repairing the defect, you are actually correcting the DNA itself.'[453] However, he also acknowledges that the efficiency and safety of the approach would need to be significantly improved before the technology could be tested in humans.

Huntington's disease is a well-known single-gene disorder of the brain.[454] This dominant disorder affects a relatively large number of individuals—12 in every 100,000 people in Britain. The condition typically starts with twitching and mood swings, but rapidly progresses to full-scale dementia and death, generally in middle age. Because sufferers generally already have children before they begin to show symptoms, the disease is passed down through the generations. The identification of the genetic defect underlying Huntington's disease in 1993—a decade before the completion of the Human Genome Project—was a triumph of modern genetics.[455] Sufferers of the disorder were found to have a defect in a gene named huntingtin. This gene has a repetitive DNA sequence, CAG, at its start that is repeated around 17 times in normal individuals. Each CAG codes for the amino acid glutamine, so a typical individual will have 17 glutamine amino acid units at the start of the huntingtin protein. However, faulty DNA replication can cause the CAG repeats to expand, and if a person inherits 36 or more, they will succumb to Huntington's disorder. The reason is that the extra glutamines cause the huntingtin protein to form aggregates within cells, making those cells—particularly neurons—dysfunctional.

The identification of the huntingtin gene defect was greeted with hope that this would quickly lead to a cure for this disorder. Unfortunately, all it has meant is that people at risk can now have a genetic test that reveals how many CAG repeats are present in their huntingtin gene. This is obviously welcome news if it shows an individual is free of risk, and can also help those considering having children. However, since a positive result is effectively a death sentence, it's not surprising that most people at risk choose not to take the test. In 2010, journalist Charlotte Raven who took the test and found she was positive described how she initially 'thought taking the test would be like finding out the weather before you go on holiday'.[442] Instead, it was more like finding that 'there was a bomb on the plane when you were already airborne. I felt impotent and envious of the uninformed majority. I wished I didn't know.'[456]

Now though, there has been a glimmer of renewed hope for those with the disorder, following a recent study by Nicole Déglon and colleagues at the University of Lausanne. The researchers infected two groups of healthy adult mice with a virus expressing the mutant huntingtin gene. One of the groups was also injected with a virus expressing CAS9 enzyme and a guide RNA targeting the huntingtin gene. Déglon believes that what her team found is 'remarkably encouraging'.[457] After only three weeks, the two groups of mice showed a striking contrast. While those only treated with the mutant huntingtin showed large amounts of protein aggregation in their brain cells, those with the mutant protein and the CRISPR/CAS9 treatment had almost none—the genome editing had prevented expression of almost 90 per cent of the mutant huntingtin. 'Having reached about 90% [blockage of production], this changes the story [of Huntington's therapy] completely,' says Déglon. 'It opens new treatment strategies that are based in DNA, and so would have a permanent benefit for the rest of someone's life.'[457]

In fact, there are many issues that need addressing before this approach is considered both effective and safe as a way to target Huntington's disease in humans. In Déglon's study the guide RNAs targeted both normal and mutant huntingtin. 'If there is no specificity for mutant huntingtin, that's a concern,' says Abdellatif Benraiss, a neuroscientist at the University of Rochester in New York. 'This is not a treatment for 4 weeks or 4 months, this is going to be permanent.'[457] For although huntingtin's normal role remains unclear, it is thought to be involved in important functions like transporting substances in cells. 'As bad as too much huntingtin is, we still need one copy [of its gene] so it can do its job in our bodies,' says Benraiss.[457] Another issue is that the mutant protein was expressed artificially in the mice. However, Déglon's team are designing guide RNAs that distinguish the normal and mutant huntingtin gene and only target the latter. And they are also planning to test the CRISPR/CAS9 approach in mice engineered to have both normal and mutant copies of the huntingtin gene, as in human sufferers. 'We're just at the beginning of the story,' says Déglon.[457]

A particularly interesting demonstration of the potential power of genome editing has been the successful use of CRISPR/CAS9 to treat DMD in a mouse model. DMD is a recessive single-gene disorder. Because the dystrophin gene that's defective in this disorder is located on the X chromosome, the disorder generally affects males, as they only have a single X chromosome and are more vulnerable to its loss. Loss of dystrophin causes muscle wasting that starts in early childhood. Currently, there is no cure for this devastating disorder, which usually results in death by the late teens. A recent article by the father of a boy with DMD described how he realized his son 'would never play rugby, never make love, never make it to university, never realise his full potential'.[458]

A 2016 study by Charles Gersbach and colleagues at Duke University in the USA showed how CRISPR/CAS9 could be used to partially correct the defect in dystrophin gene expression in mice with a mutation in the gene.[459] Gersbach's team used a type of virus called an adenovirus to deliver the CRISPR/CAS9 tools to the muscles and bloodstream of the mutant mice. Chris Nelson, one of the researchers involved in the study, said: 'We know what genes need to be fixed for certain diseases, but getting the gene editing tools where they need to go is a huge challenge. The best way . . . is to take advantage of viruses, because they have spent billions of years evolving to figure out how to get their own viral genes into cells.'[459] When the team injected the virus direct into the legs of adult mice, it led to improved muscle strength. When injected into the bloodstream, this resulted in improvements in heart and lung function. This is important because failure of these organs is what generally leads to the death of young men that suffer from the disorder.

More recently in 2018, Eric Olson and colleagues at the University of Texas used a similar approach to correct a mutated version of dystrophin in dogs. 'We wanted to put this to the ultimate test and see if we could do it in a large animal,' said Olson.[460] The CRISPR/Cas9 genome editing restored the dystrophin protein in critical body muscles, including the heart, but the study offered little evidence that the dogs regained muscle function. And indeed, while Adrian Thrasher, of London's Institute of Child Health and Great Ormond Street Hospital, sees animal studies like these as 'demonstrating proof of principle of gene editing *in vivo* for neuromuscular disease', he has also pointed out that there is 'still some way to go' before such an approach is 'translatable to human subjects'.[460]

New Cancer Cures

Another exciting aspect of genome editing is the possibility of using it to treat more common human diseases, like cancer. Forms of cancer can either be caused by loss of an inhibitory tumour-suppressor gene or activation of an oncogene, both of which can lead to abnormal cell growth and tumour progression.[461] Recent studies have shown that, in a human population, many different gene mutations seem to be involved in driving the progression of a particular cancer. A recent investigation of women with breast cancer showed that, in 50 patients, over 1,700 mutations were detected in their tumours, most being unique to the individual.[462] 'Cancer genomes are extraordinarily complicated,' says Matthew Ellis of Washington University who led the study. 'This explains our difficulty in predicting outcomes and finding new treatments.'[462] The increasing ability to rapidly sequence the whole genome of a sufferer's cancer and compare that to their normal genome—so-called 'cancer genomics'—means it's becoming possible to pinpoint the exact molecular cause of a cancer in an individual.[463] Such genetic diversity poses problems for standard drug therapy. But with genome editing it may be possible to correct the specific cancer-causing defects in a tumour, whether due to loss of tumour-suppressor genes or activation of oncogenes.

Recently, scientists at the Walter and Eliza Hall Institute of Medical Research in Melbourne targeted and killed human lymphoma cells by using CRISPR/CAS9 to knock out a gene that is essential for the survival of these cells. One of the researchers leading the study, Brandon Aubrey, explains that 'we were able to kill human Burkitt lymphoma cells by deleting myeloid cell leukaemia 1 (MCL1), a gene that has been shown to keep cancer cells alive'.[464] Although this was a 'pre-clinical' study involving only human cells grown in culture, Aubrey, who's also a haematologist at the Royal Melbourne Hospital, believes that, 'as a clinician, it is very exciting to see the prospect of new technology that could in the future provide new treatment options for cancer patients'.[464] Another researcher on the study, Marco Herold, is equally optimistic about the potential of genome editing for cancer therapy as well as research into the molecular basis of tumour formation, saying that, 'in addition to its very exciting potential for disease treatment, we have shown that it has the potential to identify novel mutations in cancer-causing genes and genes that "suppress" cancer development, which will help us to identify how they initiate or accelerate the development of cancer'.[464]

Most exciting of all was the news in 2015 about an apparently successful use of genome editing to treat leukaemia in a British baby. The story had all the elements of a film script—a dying baby, desperate parents, and a team of doctors with a highly experimental new treatment.[465] Yet this was real life, and the parents, Lisa Foley and Ashleigh Richards, only allowed the new therapy to be used to treat their daughter Layla's cancer after all else had failed. Layla was born a healthy 7lb 10oz in June 2014, but three months later she developed a fast heartbeat, went off her milk, and cried more than usual.[466] At first the problem was thought to be nothing more than a stomach bug, but blood tests revealed she had infant acute lymphoblastic leukaemia; indeed, Layla's doctors described it as one of the most aggressive forms of this cancer they'd ever seen. Layla was immediately given chemotherapy and a bone marrow transplant to replace her cancerous blood cells. Yet despite several rounds of the treatment, the leukaemia returned. At that point the doctors told Layla's parents there was nothing more they could offer, apart from palliative care to ease the baby's suffering before she succumbed to the cancer. But Ashleigh and Lisa begged the doctors not to give up. 'We didn't want to accept palliative care and give up on our daughter, so we asked the doctors to try anything for our daughter, even if it hadn't been tried before,' recalls Lisa.[466]

The plea was enough to make the doctors at Great Ormond Street Children's Hospital in London, where Layla was being treated, reconsider their options. And they decided to try a genome-editing approach, despite the fact that it had only been tested on mice. 'The treatment was highly experimental and we had to get special permissions, but she appeared ideally suited for this type of approach,' says Waseem Qasim of University College London's Institute of Child Health, a consultant immunologist at the hospital who led the treatment.[466] This involved taking T-cells—a central component of the immune system—from a donor, and using TALENs to engineer the cells in order to prevent them from attacking the baby's own cells and becoming resistant to the chemotherapy drugs, and also to give them the capacity to attack the leukaemic cells. Renier Brentjens of the Memorial Sloan Kettering Cancer Center in New York explains why this latter modification was important. 'Your own T-cells won't recognize your tumor cells,' he says. 'They think the tumor cell is, in fact, a normal cell. You need to re-educate these T-cells.'[465] At first nothing seemed to be happening, but after two weeks a rash appeared, showing the engineered cells were having an effect. Two months later, Layla was completely clear of the cancer, which allowed the doctors to give her a second bone marrow transplant to replace her entire blood and immune system. After three months she was well enough to go home.

Waseem Qasim believes the approach could have much wider application for other types of childhood leukaemia. 'We have only used this treatment on one very strong little girl, and we have to be cautious about claiming this will be a suitable treatment option for all children,' he says. 'But this is a landmark in the use of new gene engineering technology and the effects for this child have been staggering.'[466] Other experts have cautiously welcomed the news. 'This is a very exciting first effort and the authors imply that they are taking this to wider trial,' says Stephan Grupp of the University of Pennsylvania. 'More patients treated will give us a better idea of what the true impact these genetically engineered T cells will have on leukaemia.'[465] What does seem likely is that this is probably only the first of many such interventions.

Indeed, a number of clinical trials are now underway to test the potential of CRISPR/Cas9 genome editing for the treatment of a variety of human disorders; the first of these to be approved, in 2016, led by Lu You, an oncologist at Sichuan University in Chengdu, China, is currently employing this approach to treat people with lung cancer.[467]

Genes that Protect

If genome editing has such potential for treating cancer, what about other common human disorders, like diabetes, heart disease, stroke, or mental conditions like schizophrenia, bipolar disorder, or depression? The main difficulty is that, in spite of initial hopes following completion of the Human Genome Project that we would soon identify clear genetic differences underlying such conditions, the reality has proved to be rather more complex, as we saw in Chapter 5.[468] To reiterate, there is currently a major debate about whether such disorders are due to many common genetic differences, each with a small effect, or a few rare ones with large effects in specific individuals. However, a recent review of the genetic basis of mental disorder concluded that, 'in bipolar disorder and schizophrenia, increasing evidence supports the role of rare, disease-causing mutations in brain-expressed genes'.[469] Working out the role of these mutations is complicated by the fact that, unlike Mendelian disorders, which tend to be due to mutations that affect the amino acid sequence of a gene's protein product, most mutations associated with common disorders are located in the regulatory elements that control the gene's expression.[470]

This may explain why mental disorders like schizophrenia don't appear to follow a Mendelian pattern of inheritance, because the effects of mutations in regulatory elements (of which there are often many for each gene, all playing a contributing role to its expression) are likely to have much more subtle effects on a gene than those that change its protein-coding sequence.[470] Such a mutation may mean that an individual is more susceptible to schizophrenia, but only if certain environmental triggers are also present. For instance, one recent study found that possession of a mutation that affects expression of protein kinase B alpha (PKB), a gene involved in several important processes in the brain, increases a person's risk of schizophrenia, but only if the individual also takes cannabis as a teenager.[471] Findings like these help to explain previous observations of a link between this disorder and cannabis use, but also why most young people who take this drug do not succumb to schizophrenia.[472]

The complexity of the link between genetics and common disorders poses a challenge for the idea of treating such disorders using conventional drugs. If many genetic differences can contribute to a person succumbing to a disease, then it's difficult to see how drugs could be used to treat such a large number of targets. However, if each person with a particular disease is susceptible because of a rare gene difference only found in a few individuals, then it could make drug treatment of a disorder difficult to justify economically, since a different drug would potentially be needed for each of the hundreds of different molecular causes of the disorder.

While either scenario poses potential problems for conventional drug therapy, this is not necessarily the case for genome editing. So if a large number of genetic differences each only make a small contribution to susceptibility to a disease, it might be possible simultaneously to make multiple corrective modifications at the genetic level, since in theory there is no limit to how many genes can be targeted at once with CRISPR/CAS9. Indeed, as we saw in Chapter 5, George Church's team at Harvard University used this technology to target 62 different retrovirus DNA sequences in the pig genome. In fact, Church himself has said, 'we're not convinced that what we did is generalizable. It doesn't mean that we can now change 62 different genes easily.'[473] But as CRISPR/CAS9 technology continues to evolve, accurate multigene editing may well soon become a practical proposition. And if disease susceptibility is due to a rare genetic difference in a particular individual, this could be treated relatively easily by genome editing, with its easy-to-generate tools specific to any region in the genome, as opposed to the creation of new drugs for each new molecular target identified in an individual.

As well as using genome editing to specifically target mutations associated with common disorders, another possibility suggested by Feng Zhang is that the technology might be used to introduce 'protective' mutations that would insulate a sufferer from the negative effects of a common disorder.[474] Such mutations exist naturally in the human population, and have been shown to protect the individuals in which they occur from both rare single-gene disorders and more common ones. The severity of sickle cell anaemia, a recessive single-gene disorder mentioned in Chapter 2, which is caused by a mutation in the β-globin gene that codes for adult haemoglobin, varies depending on whether the sufferer can produce the foetal form of haemoglobin, known as HbF.[474] This isn't normally produced in adults, but it is in individuals with a mutation in the HbF gene promoter; if they also suffer from sickle cell anaemia, the foetal protein can partially substitute for the loss of normal haemoglobin, protecting them from many effects of the disorder. Genetic analyses of human populations have also identified naturally occurring mutations that protect against more common disorders, including cardiovascular disease and Alzheimer's.[474] It might therefore be possible to introduce such mutations into people as a way to protect them from these conditions.

In addition, genome editing might be used to introduce naturally occurring protective mutations into people as a way to combat infectious diseases, particularly HIV. There were approximately 37.9 million people across the globe with HIV/AIDS in 2018. Of these, 36.2 million were adults and 1.7 million were children under 15 years old.[475] As is well known, deaths due to HIV infection occur because the virus prevents the immune system from functioning properly, leaving sufferers vulnerable to infection by a range of other infectious agents—the so-called acquired immune deficiency syndrome, or AIDS. This remains a major cause of death in some parts of the world, particularly Sub-Saharan Africa, 70 per cent of the people infected with HIV live.[476]

The most successful general route to combating viral infection is vaccination. Unfortunately, HIV is particularly resistant to this approach, since the virus mutates so fast that there is a running battle for the body's immune system to keep up with it. And the fact that HIV can hide away within cells and also disables the very immune system that protects us against infection, are two other reasons why this virus has proven so lethal and hard to combat.[477] A major step forward in AIDS treatment was the development of a range of anti-HIV drugs, particularly those that inhibit the virus's reverse transcriptase enzyme, or that inhibit the protease which HIV uses to mature into infectious viral particles.[478]

The success of anti-HIV drugs means that infection with the virus no longer constitutes a death sentence. People diagnosed sufficiently early and who take a cocktail of such drugs can now expect to live long and fruitful lives.[479] The continuing large number of deaths from AIDS worldwide, which totalled around 770,000 in 2018[475] is mainly due to lack of access to these drugs, and general poverty and lack of proper health care in the developing countries in which most such deaths occur. Yet, despite the success of current anti-HIV drugs, the search continues for more effective treatments for a number of reasons. One is that although they keep HIV in check, and thereby prevent its destructive effects on the immune system, such drugs do not eradicate the virus from the body.[478] In part this is because retroviruses such as HIV integrate their genomes into those of the host cells they infect, as we saw in Chapter 2. If an infected person stops drug treatment, the integrated virus can become reactivated, which means that, currently, people with HIV need to continue taking the drug cocktail for the rest of their lives, which is both expensive from a health provision point of view, but also creates the risk of development of drug resistance and toxic side effects.[479]

Genome editing offers new possibilities for the treatment of HIV infection in a number of important ways. One is to make an infected person resistant to the virus by genetically modifying the cells of the immune system that HIV normally infects. This strategy seeks to mimic a genetic difference found naturally in certain rare human individuals who are resistant to HIV. Such individuals have been identified by the fact that although they have repeatedly come into contact with the virus, for instance as prostitutes or drug abusers who have shared needles, they have nevertheless remained free of HIV infection.[480] And studies have shown that a key way in which

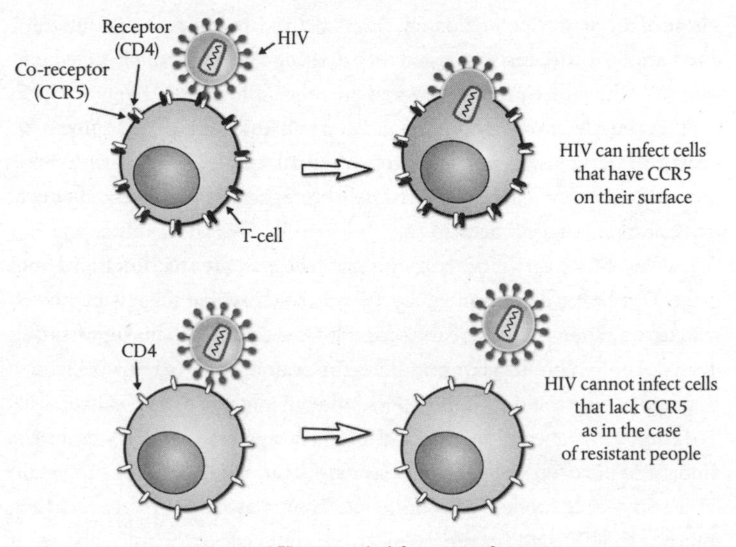

Fig. 25. CCR5 is needed for HIV infection

such natural immunity occurs in these individuals is through loss of function of the C-C chemokine receptor 5 (CCR5) gene, which normally acts as a cooperative partner, or co-receptor, to the cluster of differentiation 4 (CD4) receptor that exists on white blood cells.

HIV normally uses the CCR5 and CD4 proteins as a molecular gateway to gain entry into the immune system's T-cells (see Figure 25). In the rare individuals who don't produce a functional CCR5 protein, the virus can't do this and fails to infect their T-cells or compromise their immune system, which then promptly eliminates the virus. Remarkably, clinicians in Berlin showed that bone marrow from someone with natural resistance to HIV, if transplanted into an infected person, can cure them of the disease. Or at least this appears to be the case for one individual, Timothy Ray Brown, a man infected with HIV and treated this way in 2008, who has reportedly been free of the virus ever since.[481] Brown was lucky in that the MHC proteins on his tissues, whose role in transplant rejection we noted in Chapter 5, matched those of the donor. Unfortunately, this isn't the case for most HIV sufferers, but a study in 2014 first suggested that deleting CCR5 from an infected person's T-cells could eliminate HIV and cure them of AIDS.[481]

The study, led by Chad Cowan and Derrick Rossi at Harvard University, used CRISPR/CAS9 genome editing to knock out CCR5 in human bone marrow cells grown in culture. The cells were then treated with a chemical cocktail that induced them to turn into T-cells. 'We showed that you can knock out CCR5 very efficaciously...that the cells are still functional, and we did very, very deep sequencing analysis to show that there were no unwanted mutations, so it appears to be safe,' said Cowan.[482] This suggests that it may be possible to take an infected person's bone marrow cells, use genome editing to knock out CCR5, and then introduce the modified cells back into the patient, where hopefully they will eradicate HIV from the body. The next step is to test this strategy in an animal model. 'There are excellent mouse models you can give a human immune system and then infect with HIV,' said Cowan. 'We can give our cells to the mice and see if they're protected from HIV.'[482]

More recently, in 2019, a team led by Hongkui Deng of Beijing University used CRISPR/Cas9 to disable the CCR5 gene in human bone marrow stem cells and showed that they could be used to successfully treat HIV in an infected man when transplanted into him.[482] Of course as we saw in Chapter 4, the mutation of CCR5 to generate resistance to HIV was what the renegade scientist Jiankui He did when he engineered the genomes of human embryos that were later born as human twins.[483] The rationale used by He for this intervention was that it was the only way that the twin's father, who has HIV, could have children. But in fact there are ways to remove HIV from sperm so this reasoning was mistaken. And now following a recent study that suggests that people with loss of CCR5 function may have shortened lifespan, concerns have been raised that the twins may have been engineered with this adverse effect.[483] The new findings have also led to questions about the safety of this approach for treating HIV infection more generally. However, since in the intervention by Hongkui Deng's team, the treated patient only received white blood cells that had loss of CCR5 function, this might not have an effect on his lifespan. Clearly though, there now needs to be proper discussion about the pros and cons of disabling a gene such as CCR5 from a living person.

Targeting Killer Viruses

Targeting CCR5 is one way in which genome editing may be used to treat HIV; an alternative is to disable the virus itself. A number of studies have shown that CRISPR/CAS9 can be used to excise the integrated viral DNA from the genomes of infected cells. For instance, in 2014, Kamel Khalili and his team at Temple University in Philadelphia showed that they could use this approach to completely remove the HIV genome from several human cell lines, including one derived from T-cells of the immune system. 'We were extremely happy with the outcome,' said Khalili. 'It was a little bit mind-boggling how this system really can identify a single copy of the virus in a chromosome, which is highly packed DNA, and exactly cleave that region.'[484]

Genome editing offers a potential way to combat many types of viruses that cause human disease, not just HIV. The hepatitis C virus, or HCV, is spread by blood-to-blood contact associated with intravenous drug use, poorly sterilized medical equipment, and transfusions.[485] An estimated 130 to 150 million people worldwide are infected by HCV. Because the virus does not cause any obvious initial symptoms, infected individuals often go undetected. But, over time, HCV can cause chronic scarring of the liver and cirrhosis, which can ultimately lead to liver failure or cancer of the liver. The genome of HCV is composed of RNA, not DNA. In retroviruses like HIV, which also have an RNA genome, the RNA is converted into DNA by the viral enzyme reverse transcriptase, and this DNA can then integrate into the host genome in the nucleus of the host cell. In contrast, the HCV RNA genome remains outside the nucleus, in the host cell's cytoplasm, where it is replicated by a special viral enzyme—RNA-dependent RNA polymerase—to produce more copies of the RNA genome, which are packaged with proteins to form further infectious viral particles.[486]

The fact that HCV is a non-retroviral type of RNA virus (with no DNA stage as part of its lifecycle) suggested that it might be impossible to target it by genome editing. However, recently a team led by David Weiss and Arash Grakoui at Emory University achieved this goal by using a modified form of the CAS9 enzyme. This modified CAS9 uses a guide RNA to find its target in the HCV RNA genome, but, instead of cutting the RNA, it creates a roadblock that prevents the viral genome from being replicated by its RNA-dependent RNA polymerase. Weiss, Grakoui, and their colleagues found that after introducing the modified CAS9 and its guide RNA into human liver cells in culture, the cells became resistant to infection by HCV.[487] The researchers believe this approach might eventually be used to treat chronic HCV infection. And given that other non-retroviral RNA viruses like 'flu or Ebola also have an RNA genome, this strategy may have much wider applicability. Currently, one clinical approach being developed to combat Ebola involves the use of RNA interference or RNAi, the technology mentioned in Chapter 4. But according to Weiss, viruses can develop mechanisms to thwart RNAi. 'Since Cas9 [*sic*] is a bacterial protein and eukaryotic viruses have likely not encountered it, they would not have ways to evade Cas9,' he says. 'Thus, Cas9 could be effective in inhibiting viruses when the RNAi system cannot.'[488]

Possibilities and Problems

Besides viruses, another main source of infectious disease is bacteria. Our current main line of defence against bacteria is antibiotics—drugs that slow the growth of bacteria or kill them outright.[489] These drugs are now so important in our lives that antibiotics like penicillin are household names. While penicillin blocks the process by which bacteria develop protective cell walls, other antibiotics inhibit bacterial gene expression in various ways. For instance, streptomycin and chloramphenicol block the ability of a bacterium to translate genes into their protein products.[489] A cellular structure called a ribosome catalyses this process in both humans and bacteria. The reason why such antibiotics do not also affect translation in human cells is that the ribosome's structure is slightly different in our cells compared to bacteria.

As is becoming all too clear, bacteria can develop resistance to these drugs.[489] This partially occurs through natural selection, by which a mutation that confers resistance in one bacterium can spread rapidly through a bacterial population if such resistance allows survival. Even if the mutation only occurs in one in a million cells, the fact that bacteria can reproduce themselves in less than half an hour makes the spread of resistance a real problem. In addition, bacteria can swap antibiotic resistance genes through a process named horizontal gene transfer. Most dangerous of all, bacteria can develop resistance to multiple antibiotics.[489]

Another negative feature of antibiotics is that, since they affect all bacteria, they can adversely affect those that live in our guts and other parts of our body. This would not be a problem if such bacteria were just useless parasites, but this is far from the case. An increasing number of studies show that many bacteria in our bodies have a beneficial role. A study by Andrés Moya of the University of Valencia on the effects of antibiotic treatment in human patients found that their 'gut bacteria present a lower capacity to produce proteins, as well as deficiencies in key activities, during and after the treatment'.[490] Specifically, the study suggested that, after the treatment, the bacteria showed less capacity to absorb iron, digest certain foods, and produce essential molecules for the patient. Findings like these imply that excessive antibiotic use can lead to health problems, such as those associated with digestion.

Genome editing offers a way to target harmful bacteria without affecting the beneficial ones that help keep us healthy. With its precision, the technique can be used to target one bacterial species while leaving others unharmed. A challenge here, as with targeting of human cells, is to get the genome-editing tools into their bacterial target. Here, viruses may provide a solution. We saw in Chapter 2 how, just as viruses can infect our cells and cause disease, so bacteria have their own viral problem to deal with, namely viruses called bacteriophage, or phage for short. Indeed, the CRISPR/CAS9 process evolved precisely to combat infection by such viruses. And just as disabled retroviruses have been used to introduce gene constructs into human cells, so there is now the possibility of engineering phage to carry gene-editing tools into a target bacterium (see Plate 3). Recently Chase Beisel and colleagues at North Carolina State University showed that this approach could be used to target pathogenic bacteria. The researchers tested the approach with different combinations of bacteria present, and showed they could eliminate just the targeted strain. 'We were able to eliminate *Salmonella* in a culture without affecting good bacteria,' says Beisel.[491] Another benefit of the approach, according to Beisel, is that 'by targeting specific DNA strands through the CRISPR/CAS9 system, we're able to bypass the mechanisms underlying the many examples of antibiotic resistance'.[491]

There is another way in which genome editing might be used to combat infectious disease, and it's based on the fact that some infectious agents are introduced into the body through another organism. An example is the malaria parasite, a pathogenic microorganism carried by mosquitoes. In March 2015, a study in fruit flies by Ethan Bier and colleagues at the University of California, San Diego, showed how the spread of malaria might be tackled not by targeting the microorganism itself, but the mosquito. Bier's team developed a form of genome editing they have named 'mutation chain reaction', or MCR.[492] The researchers tweaked the CRISPR/CAS9 process to make a mutation generated on one copy of a chromosome spread automatically to the other copy (Figure 26). 'MCR is remarkably active in all cells of the body, with one result being that such mutations are transmitted to offspring via the germline with 95 per cent efficiency,' said Valentino Gantz, a researcher involved in the study.[493] If this approach, part of a general strategy known as a 'gene drive', were applied to disease-carrying species in the wild, a single mosquito equipped with a parasite-blocking gene could in theory spread malaria resistance through an entire breeding population in a single season. The approach might be taken even further, by using MCR to spread

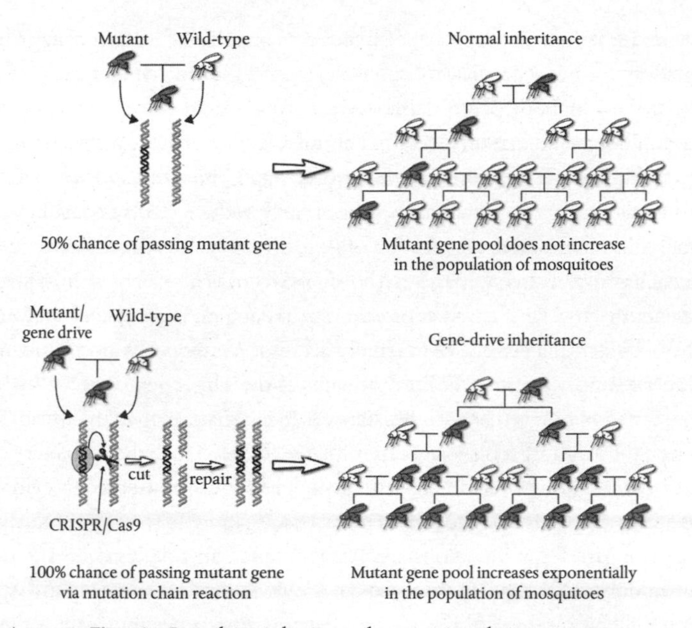

Fig. 26. Gene drive inheritance by mutation chain reaction

a defect that kills the malaria-carrying mosquito species. Showing just how fast research in this area is moving, later in 2015, Anthony James and colleagues at the University of California, Irvine, developed an anti-malarial gene drive strategy in mosquitoes themselves.[494] James and his colleagues engineered *Anopheles stephensi* mosquitoes—the type responsible for more than 10 per cent of malaria cases in India—to express a gene coding for an antibody against the malaria parasite, *Plasmodium falciparum*. And tests showed that the modified mosquitoes passed on the gene to 99.5 per cent of their offspring. James believes the 'technology could have a major role in sustaining malaria control and elimination as part of the eradication agenda'.[494] A type of gene drive called a 'crash drive' can even dramatically reduce numbers of a malarial mosquito species. In another study published in 2015, Andrea Crisanti and Tony Nolan at Imperial College London reported their development of mosquitoes with gene drives that disrupt three genes for female fertility, each of which acts at a different stage of egg formation.[495] Since the female mosquitoes are infertile only when a copy is inherited from both parents, the gene drives would have to be spread throughout a population before having an effect. 'The field has been trying to tackle malaria for more than 100 years,' said Crisanti. 'If successful, this technology has the potential to substantially reduce the transmission of malaria.'[495] One interesting feature of gene drives such as this is how resistant some organisms seem to be to attempts to eradicate them. So in experiments in 2017 with caged mosquitoes, Crisanti and his colleague Tony Nolan watched a gene drive gradually decrease in frequency over multiple generations owing to resistant mutations at the target gene.[496] This raised the question of whether such resistance is likely to develop in the world. In fact if the gene target is chosen carefully, it seems that such the possibility of resistance can be overcome. Some genes are highly conserved, meaning that any change is likely to kill their owners. Picking these genes as a drive target means fewer mutations and less resistance. In 2018, Crisanti and his team crashed a population of caged *Anopheles gambiae* mosquitoes with 100% efficiency by making a drive that disrupts a fertility gene called DOUBLESEX.[496] Because it is crucial for procreation, this gene is resistant to mutations, including those that would confer resistance to a drive construct.

Although the main focus of gene drive studies has been to target disease-carrying mosquitoes, there are also current investigations into the possibility of using this approach against rodents, which in some parts of the world, having been introduced by people, are now wreaking havoc on native ecosystems.[496] But such studies are still at a very early stage compared to those in insects.

Not everyone is happy about the pace of this type of research. A recent letter, published in *Science* by teams in Britain, the USA, Australia, and Japan, stated that while gene drives might save lives and bring other benefits, the accidental release of modified organisms 'could have unpredictable ecological consequences'.[494] For instance, according to molecular biologist and bioethicist Natalie Kofler of Yale University, they might negatively affect

1 Brainbow identifies neurons by colour.

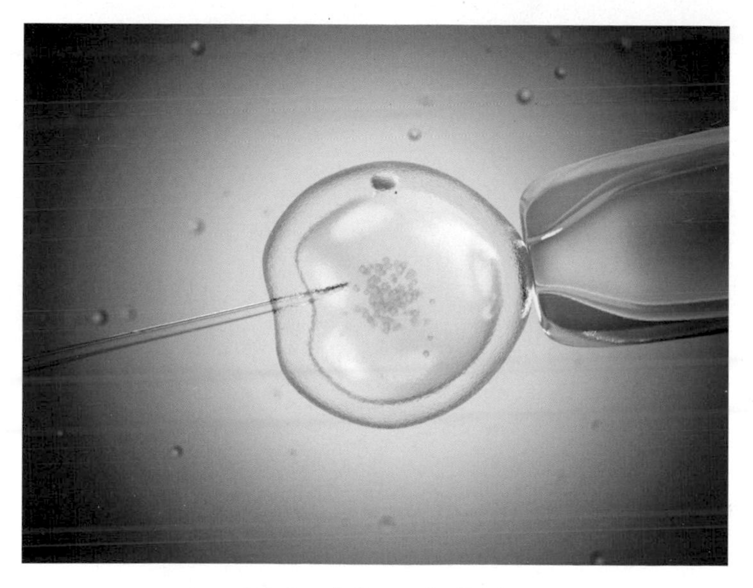

2 Injection of a human fertilized egg.

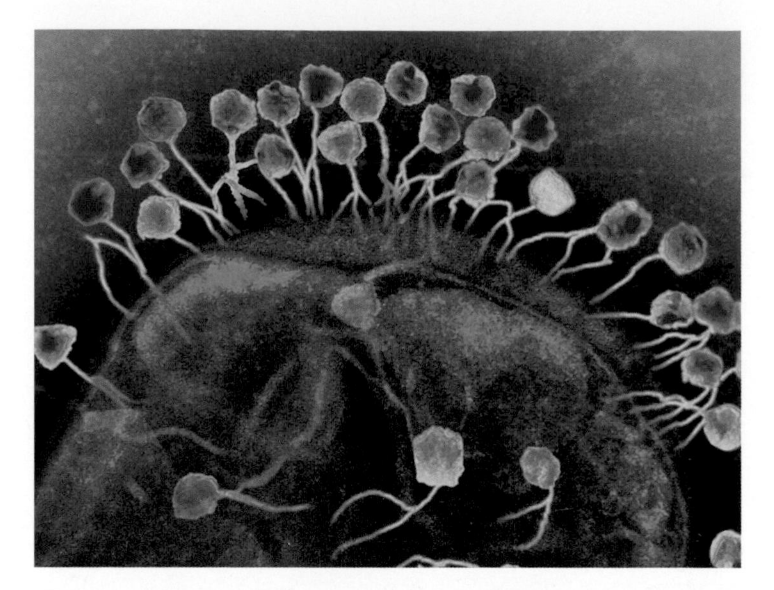

3 Bacteriophage infecting a bacterium.

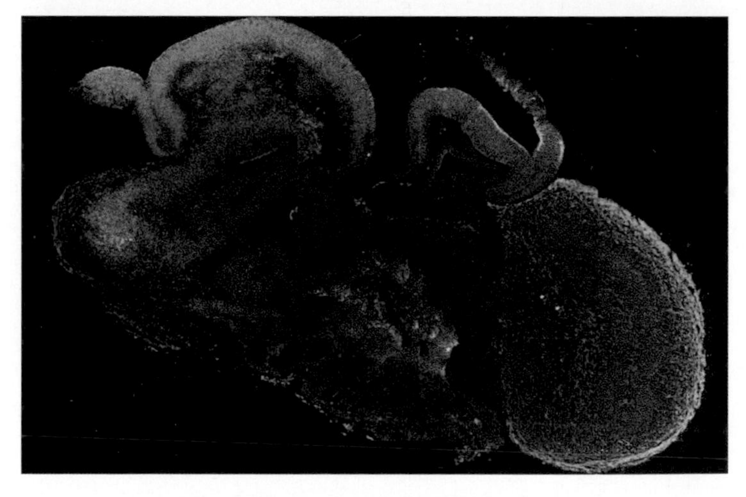

4 Human brain organoid with different structures.

human health by causing the malaria parasite to evolve to be more virulent or be carried by another host.[496] Another fear is that a gene altered in mosquitoes to dramatically reduce their population numbers might somehow be transferred into a beneficial insect species.

What if it jumped, say, into honey bees, whose populations in the wild are already declining? If this happened, farmers might have a hard time pollinating their crops and the world could face food shortages. For this reason, Kevin Esvelt of Harvard University, who is also researching gene drives to eradicate malaria, has called for a wider debate involving scientists, policymakers, and the public about the pros and cons of the strategy. 'There is no societal precedent whatsoever for a widely accessible and inexpensive technology capable of altering the shared environment,' he says.[497]

A Question of Delivery

Such are the possibilities of genome editing for directly treating both the genetic basis of disease and that caused by various infectious agents. Yet some major obstacles remain. One is that the efficiency and accuracy of the technology still needs to be improved considerably before it can be applied routinely to living human beings. However, such is the speed of development of the technology that it's quite likely that such problems will eventually be ironed out. A more fundamental obstacle is getting the genome-editing tools into the cells of the tissues or organs to be treated.[498] This has been a major challenge for previous attempts at gene therapy. A central problem is that cells are enclosed by a protective membrane that acts as a barrier to large molecules like proteins, or nucleic acids like DNA and RNA. An additional obstacle is delivering genome-editing tools to the tissue of interest without their being degraded in the process by the body's natural defence mechanisms.

It is because of this risk of degradation that the genetic or infectious diseases most likely to be treated initially using genome editing will be those affecting tissues that are most accessible or can even be removed from the body and then replaced once a genetic defect is corrected. Diseases that affect the blood cells generally fall into the latter category, whether the red blood cells that carry oxygen around the body or the white ones that provide our immune system, since both cell types originate from stem cells in the bone marrow. A sample of bone marrow can be removed from a sufferer, the cells within it treated, and then the sample replaced, so that cells treated in this way can repopulate the blood.[499] Potential treatable disorders in this category include different forms of leukaemia, but also sickle cell anaemia and thalassaemia, which affect the red blood cells' oxygen-carrying capacity, and severe combined immune disorder. The possibility that such disorders may finally be closer to a real cure is leading to hope but also concerns about how soon such a cure may take to develop amongst people with such disorders. Robert Rosen of Chicago had a rare bone marrow abnormality; these can lead to leukaemia or other life-threatening conditions. He remarked that 'it's like a race. Will the research provide a cure while we're still alive?'[497] To try to hasten a cure for people with these disorders, Rosen co-founded the myeloproliferative neoplasms (MPN) Research Foundation, which is now funding genome-editing studies. Sadly, Rosen himself succumbed to his disorder in January 2018.[500] Because so much is at stake for some desperately ill people, it will be important for scientists to be realistic about the likely timescale of clinical genome-editing applications. For, as bioethicist Alta Charo of the University of Wisconsin Law School has pointed out in relation to this issue, 'As money and excitement enter a field, there is a risk the patient population will incorrectly believe that clinical treatments are—or at least ought to be—available.'[501]

The challenges faced in applying genome editing in gene therapy are similar in some respects to those faced by the traditional transgenic approaches mentioned in Chapter 2, but in other ways they are quite different. In the case of CRISPR/CAS9, the two key components are the CAS9 cutting enzyme and the RNAs that guide this to the genomic region that needs modifying, plus a DNA fragment coding for a replacement part of the genome to be modified. These components can either be delivered as they are or encoded in DNA constructs. The challenge now is to find a route of delivery that is effective, so the target gene becomes modified but doesn't disrupt normal genome function. To achieve these aims there are two likely routes. One is to use a virus to deliver a DNA construct to the target cell.[502] Instead of needing to integrate the DNA construct into the genome of the host cell, with genome editing it's merely enough to deliver the tools to the cell nucleus, so a genome-integrating virus is not necessarily required, merely one that enters the nucleus. This could help minimize the risk of disruption to the host cell genome. The second approach seeks to add something to the CAS9 enzyme, the guide RNA, and replacement DNA fragment that would allow them to cross the cell membrane.[503] Given the potential of gene editing for gene therapy, there are now likely to be concerted efforts to try to solve the issue of delivery.

All the genome-editing approaches so far mentioned in this chapter have been aimed at combating genetically based or infectious disease in children and adults. A much more controversial use of the technology would be to target genetic disease at the point of conception. As we saw in Chapter 4, genome editing is different from previous precision genetic engineering approaches in that it can be applied to a fertilized egg, including that of our own species. Human embryos with a specific gene mutation linked to disease, such as those that cause cystic fibrosis or Huntington's, might be treated in this way. And indeed, as we saw in Chapter 4, there has been much controversy in this area because of the actions of certain scientists. So first Junjiu Huang and his team at Sun Yat-Sen University in Guangzhou, China, used genome editing to correct the gene defect that causes β-thalassaemia, a potentially fatal blood disorder, in (non-viable) human embryos.[504] And then subsequently Jiankui He of Southern University of Science and Technology in Shenzhen, China, announced that he had created genome-edited human babies, despite this being against the law.[505] While He has become a pariah in the scientific community, and indeed has now been jailed for three years for his actions, the debate about whether genome editing of human embryos should be allowed, continues. And indeed, as we have seen, opinions about this amongst scientists have been quite varied, with some calling for an outright ban and others believing that, as long as such studies are purely carried out for research purposes, they should be allowed to go ahead.

The Germline Taboo

The potential use of an approach like CRISPR/CAS9 to modify the genome of a human fertilized egg or early embryo is far more controversial than gene therapy in an adult human, or even a child, because it represents a change to the 'germline'. This is the name given to the egg and sperm cells in a multicellular, sexually reproducing organism like ourselves that gives rise to the next generation.[506] In contrast, all the other cells in our bodies are known as 'somatic'. This distinction dates back to August Weismann, who first developed the concept in opposition to Darwin, who had suggested that the body's cells emit hundreds of 'gemmules', which congregate in the reproductive organs prior to fertilization. In contrast, Weismann argued that the cells that go on to form the eggs and sperm are segregated early on in embryo development.[507] And while somatic cells are subject to ageing and other changes due to environmental influences, Weismann saw the germ cells as immortal and unaffected by such influences. He tested this idea in rather barbaric fashion by cutting off the tails of 68 mice and showing that none of their subsequent offspring lacked a tail.

Recent studies have challenged Weismann's dogma by showing that an organism's life experiences have a more direct effect on its offspring's genomes than previously thought. Such 'epigenetic' effects on the genome involve chemical changes to the DNA and its associated regulatory proteins.[508] In humans, such effects mean that diet, exposure to stress, and perhaps more positive life experiences of one generation may have a profound impact on subsequent generations. None the less, the idea that the germline genome is special and should not be tampered with remains a strong one, not least because there is no way the next generation can give their consent to such an action. So while somatic gene therapy is generally seen as permissible if carried out safely and effectively to treat a genetic disease, germline gene therapy is far more controversial, because even if it were safe and effective, its effects would not be confined to a single individual but would potentially affect generations to come.

In fact, there is another way to modify the genome of the next generation besides genetic engineering of the fertilized egg or embryo, and that is to target the sperm or egg, or alternatively the stem cells that give rise to these in the testicles or ovary. Showing that it's possible to use genome editing to target the latter, Kent Hamra and colleagues at the University of Texas Southwestern cultured rat testicular stem cells and then used CRISPR/CAS9 to knock out certain selected genes in them.[509] The GM stem cells were then reintroduced into the testicles of rats, whose own stem cells had been destroyed by chemotherapy. After allowing the transplanted stem cells to produce sperm in the host testicles, the rats were then mated with females. What Hamra and his colleagues found was that the resultant offspring had been genetically modified in a single step.

Such findings are likely to be of future importance for biomedical science for a number of reasons. One is that although Hamra's team followed quite a complicated route to modify testicular stem cells—which involved isolation and culture of such cells and transplantation into a host testicle—their findings suggest that it might be possible to carry out genome editing on the stem cells of the intact testicles. If so, this could greatly simplify not only production of GM versions of rodents but also of other mammalian species; such treated animals could simply be mated to females to produce GM offspring.

Such an approach in human beings might be used to treat certain types of male infertility, particularly those in which a genetic defect prevents the testicular stem cells giving rise to sperm.[510] Currently this type of infertility is untreatable, for although IVF can be used to treat patients whose sperm are immobile or cannot bind or fuse with the egg, it cannot be employed if no sperm are produced at all. Of course, any use of genome editing in this manner would be controversial, since, while curing infertility, it would also result in the modification of the genome not only in that individual but in future generations. Nevertheless, a case might be made that, if such an approach could be carried out effectively and safely, it should be considered as a way to allow infertile couples to conceive.

Another reason why an ability to precisely modify the genome of a testicular stem cell has important wider implications for medicine is that recent studies have shown that such stem cells have an astonishing capacity to give rise, in certain artificial conditions, to all the different cell types of the body.[511] Quite why this is so, and what general potential pluripotent stem cells offer for the treatment of a variety of human diseases, are issues that we will explore in more detail in Chapter 8.

8

Regenerating Life

Imagine being able to replace human tissues or organs whenever one became faulty, diseased, damaged, or simply old. If this were possible, then disease might become a rarity and the natural human age-span greatly extended, perhaps forever. In fact, such a dream has surprisingly old roots in human society, as shown by the ancient Greek myth of the Titan, Prometheus. As punishment for offending Zeus by stealing fire from Mount Olympus, he was chained to a rock and tortured by an eagle that came each night to devour his liver.[512] Because of Prometheus's immortality, his liver regenerated itself each day, leaving him ready for the same torture for eternity. The choice of organ in this grisly story was well considered, given we now know the liver is the only human organ that can spontaneously regenerate itself after injury. And the ancient Greeks' name for the liver—'hepar', derived from 'hepaomai', which means to repair oneself—suggests that they may have been aware of this property.[513]

While the liver is an organ that can regenerate itself, so can some tissues. One human tissue that has long been known to have a capacity for regeneration is the skin. Skin transplantation in medicine has an ancient history, with the Indian surgeon Sushrutha, who lived on the banks of the Ganges around 600 BC, being its first recorded practitioner. He was an expert in nose jobs, an operation in some demand at that time because of the official practice of disfiguring the nose as a punishment for crimes such as theft and adultery.[514] Sushrutha's method involved cutting a strip of skin from the forehead and using this, still attached to its original location, to reconstruct the damaged nose. After the graft had fully established itself, its connection to the forehead was severed.[514]

The other human tissue that most lends itself to transplantation is the bone marrow. This tissue produces our blood cells—both the red cells that carry oxygen around the body and remove waste carbon dioxide, and the white cells that constitute our immune system. The first person to recognize the potential of bone marrow transplantation to treat leukaemia and other blood cancers was E. Donnall Thomas of the Fred Hutchinson Cancer Research Center in Seattle.[515] In the 1960s, Thomas and his team began developing ways to use radiation and chemotherapy to destroy a cancer patient's diseased bone marrow and then replace it with new marrow from a healthy donor. Owing to this technique, some leukaemias that were once a death sentence now have cure rates of up to 90 per cent. Bone marrow transplantation has also been used to successfully treat non-malignant blood disorders, such as sickle cell anaemia, when an appropriate donor can be found.[516] In recognition of his achievements, Thomas was awarded a Nobel Prize in 1990.

We now recognize that the capacity of skin to regrow when grafted to another part of the body, or transplanted bone marrow to regenerate a whole new blood system, is due to a special type of cell—the stem cell.[517] These cells are distinguished from normal cells by their ability to divide indefinitely and capacity to give rise to more specialized cell types. Stem cells are a reflection of the fact that we all start life as a single cell—the fertilized egg. By the process of embryo development, this single cell will eventually develop into a human being composed of almost 4 trillion cells. These cells can be distinguished by their specific properties, there being over 200 cell types in a human being.[518] Cell types are distinguished by shape, size, and functional properties, but ultimately this reflects the fact that, while all have the same genome, the extent to which different genes are turned on or off varies.

A Very Gifted Cell

In the very early embryo, all cells can give rise to any cell in the body—so-called 'pluripotency'.[519] We know this because, as we saw in Chapter 2, isolation of such cells from an early mouse embryo provides us with ES cells which, when injected into another mouse embryo, can give rise to any cell type in the body, including the eggs and sperm. Following Martin Evans' discovery of such ES cells, which allowed the production of knockout and knockin mice, there was a concerted effort to identify ES cells in other mammalian species. As we saw, this effort has been unsuccessful, except recently in another rodent species, rats. Yet, as we also noted, there is one other mammalian species in which ES cells have been isolated besides these two rodent ones, and that is our own.

James Thomson and colleagues at the University of Wisconsin first achieved this feat in 1998, using 'spare' early embryos donated by IVF patients.[520] Thomson's team generated five immortalized ES cell lines from these embryos; subsequent work by other researchers has isolated hundreds more. For obvious ethical reasons, it's impossible to carry out the ultimate test of human ES cell pluripotency—to inject them into an early human embryo, implant this into a woman, and see whether the ES cells give rise to all the different human cell types in a resulting chimaera human being. However, human ES cells have all the expected properties of such cells. For instance, they form the teratomas mentioned in Chapter 2 if injected into a mouse, which is also a key property of mouse ES cells. Most importantly for therapeutic purposes, human ES cells can give rise to different specialized human cell types in culture, if exposed to chemical agents that induce this change. The reason is that this process mimics events that take place in the embryo.

During normal embryo development, stem cells give rise to more specialized cell types through a process called differentiation.[521] This process, which produces all the different human cell types and the tissues and organs that contain them, is triggered by growth factors, hormones, and other chemical messengers that bind to receptors on the surface of the stem cell. When these receptors are activated, they send signals to the cell nucleus. These signals activate regulatory factors that switch some genes on or others off, in a specific pattern. As a result, proteins specific to a particular cell type are produced. That is why one cell becomes a beating heart cell, while another becomes a neuron that relays electrical impulses in the brain.

In the embryo, differentiation takes place in a highly ordered fashion, with first the basic tissue layers being formed, then the main organs, and finally the specialized cell types within those organs.[522] A major focus of developmental biologists over the past few decades has been to identify the genes involved in regulating this symphony of gene expression. This quest has resulted in the identification of the genes that regulate the basic body plan—which, remarkably, are largely conserved from fruit flies to humans—as well as those that determine the development of specific tissues and organs.[522, 523]

In order to create human cell types, tissues, and organs from ES cells for therapeutic purposes—for instance to provide a new pancreas for a type 1 diabetic or replace the neurons faulty in Parkinson's disease—a major challenge is to mimic the signals that generate these cell types in the embryo. So, information emerging from studies of pancreas or brain development in the living embryo is highly important. However, it is far from straightforward to reproduce a process that normally takes place in the highly structured environment of an embryo in a plastic culture dish.[524]

One success has been the generation, from human ES cells, of the pancreatic beta cells that produce insulin.[525] Over 400 million people worldwide have diabetes. In type 1 diabetes, which accounts for between 5 to 10 per cent of these cases, the beta cells are typically destroyed by an auto-immune disorder.[526] Type 1 diabetics can lead a relatively normal life thanks to the availability of artificial human insulin, generated by genetically engineering a bacterium to express the human insulin gene, as we saw in Chapter 2. But they must also regularly inject such insulin, constantly monitor their blood glucose by pricking a finger to obtain a sample, and be very careful about what and when they eat.[526] The routine is especially hard on children. If not managed properly, high blood glucose levels can lead to nerve and kidney damage, blindness, and a shortened life span.

Doug Melton of Harvard University, who previously identified some key molecular mechanisms underlying the development of frog embryos, began trying to generate human pancreatic beta cells in culture when his own two children were diagnosed with type 1 diabetes. 'We wanted to replace insulin injections with "nature's own solution",' he says.[527] Over the last decade, Melton and his colleagues have devised a method for deriving huge numbers of pancreatic beta cells from human ES cells. 'What we did to solve this problem is study all the genes that come on and go off during the normal development of a beta cell,' he recalls. 'Once we knew which genes come on and go off, we then had to find a way to manipulate their activity.'[527] Melton's team tested hundreds of culture conditions before hitting on a six-step procedure that takes 40 days to produce the beta cells. This generates hundreds of millions of beta cells—the quantities required for transplantation into a type 1 diabetic. To demonstrate that the beta cells were functional, Melton's team showed they could produce insulin when treated with glucose—the normal stimulus that triggers the hormone's secretion from the pancreas—in cell culture.

Elaine Fuchs, of Rockefeller University, has hailed the discovery: 'For decades, researchers have tried to generate human pancreatic beta cells that could be cultured and passaged long term under conditions where they produce insulin. Melton's team have now overcome this hurdle and opened the door for drug discovery and transplantation therapy in diabetes.'[527] However, some obstacles still stand in the way of such cells being used therapeutically. One concern is that since type 1 diabetes is generally caused by an auto-immune disorder, the transplanted beta cells will need protection from an attack by the body's own defence mechanisms. One possibility would be to give the patient drugs that supress the immune system, but as these carry their own risks Melton favours other strategies, for instance encapsulating the transplanted cells in a mesh-like device to protect them from attack by cells of the immune system.

Melton likens this approach to putting the beta cells into a tea bag. 'The insulin and the glucose can go across, but the cells stay in,' he says. 'And importantly the immune cells can't go in and attack it.'[528] Another approach that Melton's team are pursuing involves looking for genes they can edit in the manufactured beta cells to make them invisible to the immune system. By editing such genes and identifying the effect this has, the researchers have identified one that seems to hide the beta cells from the immune system of mice with a version of type 1 diabetes. 'We think that when you mutate that one gene, you protect the cells against the autoimmune attack without really changing the beta cell and without preventing it from being recognized by the immune system,' says Stephan Kissler, who has been working with Melton at Harvard.[528] Eventually, he hopes the team will be able to genome-edit manufactured beta cells before they're delivered to a patient, so they'll work like regular beta cells but won't suffer the same immune attacks.

While the pluripotent properties of human ES cells have led to great excitement about their therapeutic potential, development and use of these cells has its problems. For a start, there are ethical concerns about the derivation of these cells from human embryos. For people who believe an early human embryo has the same rights as a human child or adult, use of such cells is tantamount to murder. Such a viewpoint had a significant impact on public funding for ES cell research in the USA under President George Bush.[529] During this period, all federal funding for such research was withdrawn, and only private funds and initiatives like the California Institute for Regenerative Medicine allowed cutting-edge ES cell work to proceed. In Germany, use of embryos for research is restricted by the 1991 Embryo Protection Act, which makes the derivation of ES cell lines a criminal offence.[530]

Ethical concerns are not the only obstacle faced by scientists seeking to develop ES cells for therapeutic purposes. In Chapter 5, I mentioned how the transplantation of tissues or organs from one human to another usually leads to rejection of the transplanted tissue or organ because of differences in the MHC proteins between the two individuals. The mismatch is detected by the immune system, which then attacks the transplanted tissue or organ as foreign. Because of this, people requiring a new liver, heart, or kidney must be precisely matched with a donor who happens to share a similar MHC profile. MHC mismatch is also a problem for those seeking to use cells derived from ES cells therapeutically.[531] Since they are originally derived from human embryos of a particular genetic make-up, ES cells also have a specific MHC profile, so again a precise match is needed.

The Cloning Controversy

One solution to this problem came from the discovery that it was possible to clone mammals. In 1996, the birth of Dolly the sheep, the first mammal cloned from a differentiated adult cell, shattered the dogma that once a cell becomes differentiated it no longer has the potential to give rise to other cell types. Keith Campbell, Ian Wilmut, and their colleagues at the Roslin Institute took an udder cell from a sheep, removed its nucleus, and transplanted the nucleus into a sheep egg whose own nucleus had been removed. They showed that when a differentiated cell genome is exposed to the environment of the egg cytoplasm it can be 'reprogrammed' to develop into a whole new organism.[532] In fact, Dolly the sheep was not the first demonstration of cloning. In the 1960s, John Gurdon at Oxford University produced cloned frogs from differentiated frog cell nuclei in exactly this manner.[533] However, failure to reproduce this finding in mice led to the idea that this was a peculiarity of amphibians. So Dolly's birth came as a revelation and also acted as a major stimulus to exploring the phenomenon of cloning and pluripotency in greater detail.

The birth of Dolly, and subsequent success in cloning other mammalian species including mice, raised the question of whether it would be possible to clone a human being. Since cloning is an inefficient procedure, with many cloned embryos failing to develop and those that do having various defects, on safety grounds alone it would be highly inadvisable to try to clone a person, never mind the many ethical issues. However, a cloned human embryo could be a valuable source of ES cells, and ultimately of tissues used for therapeutic purposes.[534] So someone requiring a replacement tissue or organ could supply a differentiated cell, say a skin cell, whose nucleus would be removed and implanted into a human egg to create a cloned embryo. This could then be used to create ES cells, and ultimately tissues or organs for transplantation. And since these would be genetically identical to the person requiring the transplant, they would not be rejected. Cloned ES cells could also be important for investigating the molecular basis of disease, since they could be used to generate a tissue or organ defective in a particular individual, to explore how a genetic difference caused the defect. For instance, ES cells cloned from someone with muscular dystrophy could be differentiated into muscle cells, to explore how loss of dystrophin, the gene defective in this disorder, causes a breakdown in muscle integrity and strength.[535]

So there are important reasons for creating cloned human ES cells. However, as well as ethical concerns, the technological path to obtaining such cells has been far from straightforward. In 2004, Woo-Suk Hwang and colleagues at Seoul National University published a study in *Nature* presenting evidence that they had cloned a human embryo and isolated ES cells from it.[536] A subsequent study published in *Science* extended the findings by demonstrating the generation of 11 ES cell lines from skin cells taken from patients. The discovery led to Hwang becoming a celebrity. Korean Airlines offered him free first-class seats and the government gave him the title of 'Supreme Scientist'. The country even released a stamp in his honour; on it was the silhouette of a man standing up out of a wheelchair—a symbol of the potential practical benefits of stem-cell research. Hwang himself said: 'I want to be remembered in history as a pure scientist. I want this technology applied to the whole of mankind.'[537] Outside Korea, there was much acclaim too, with Gerald Schatten, a reproductive biologist at the University of Pittsburgh, saying that Hwang's findings were 'proof that you don't just have to be at a Howard Hughes institute or a first-world country to make such splendid discoveries. It'd be unfair to say that they're just karate kids.'[538]

Yet even as Hwang's fame spread, doubts were beginning to be raised about him. In particular, Young-Joon Ryu, a postdoctoral scientist who had left Hwang's group, grew suspicious about irregularities in the published reports.[537] He tipped off some Korean journalists, and this led to a media investigation that unearthed evidence that Hwang had coerced junior female scientists in his lab to donate eggs—necessary for the cloning work but obtained in a completely unethical manner—and had also fabricated data. Seoul National University opened an investigation into Hwang's work and reached the damning conclusion that 'the data in the 2005 *Science* paper cannot be some error from a simple mistake, but can only be seen as a deliberate fabrication to make it look like 11 stem-cell lines using results from just two'.[537] In 2009, Hwang was 'convicted of fabricating data, misusing research funds, and trading illegally in human eggs' and received a two-year suspended prison term, although he never went to jail. Looking back at the incident in 2014, Hwang said: 'I created an illusion and made it look as if it were real. I was drunk in the bubble I created.'[537]

The exposure of Hwang's fabrication of data led some people to wonder whether it would ever be possible to clone human embryos to provide a source of ES cells, and a lack of positive results in this direction seemed in line with such a conclusion. But in May 2013, Shoukhrat Mitalipov and colleagues at the Oregon Health and Science University produced cloned human ES cells from foetal skin cells and cells taken from an 8-month-old baby with a rare metabolic disorder called Leigh syndrome.[539] Mitalipov's team succeeded where others had failed by pre-testing a variety of different procedures in studies of cloning in monkeys. The researchers also carried out tests to prove that their cloned ES cells could form various cell types, including heart cells that were able to contract spontaneously.

Mitalipov's success still left open the question of whether ES cells could be cloned from an adult human being. This was demonstrated in April 2014 by Young Gie Chung and Dong Ryul Lee and colleagues at the CHA University in Seoul, and independently by Dieter Egli and his team at the New York Stem Cell Foundation Research Institute. In the first study, the cloned ES cells were generated using nuclei from two healthy men, aged 35 and 75, while, in the second, they were cloned from a 32-year-old woman with type 1 diabetes.[540] In this second case the researchers also succeeded in differentiating these ES cells into insulin-producing cells.

Ironically, the news that it had finally proved possible to create human ES cells from embryos cloned from adult humans generated far less excitement than the original claim made by WooSuk Hwang a decade earlier. The reason was that, in the intervening decade, alternative routes had emerged for the generation of pluripotent cells. One such route, mentioned in Chapter 7, was identified by scientists studying the stem cells in testicles that normally give rise to sperm. These studies showed that, with the right combination of growth factors and hormones, such cells could differentiate into many different cell types. A study led by Martin Dym of Georgetown University Medical Center in Washington DC in 2009 showed that human testicular stem cells could be induced to differentiate into cell types like those of the pancreas, heart, or brain. 'Given these advances, and with further validation, it is possible that in the not-too-distant-future, men could be cured of disease with a biopsy of their own testes,' said Dym.[541] This discovery suggested that the stem cells that give rise to the eggs and sperm were not as different from ES cells as had been supposed. In fact, other findings were already challenging the unique status of ES cells in an even more dramatic fashion.

Reprogramming Revolution

In particular, Shinya Yamanaka of Kyoto University began a line of investigation that would lead to a startling conclusion. I mentioned in the section 'The Cloning Controversy' how, when a differentiated cell nucleus is transplanted into an egg with its own nucleus removed, this can lead to a 'reprogramming' of the genetic information in the nucleus so that it becomes capable of generating all the different cell types of the organism. And this must reflect the fact that regulatory factors in the egg cytoplasm modify gene expression to create such a pluripotent state. Yamanaka reasoned that if it were possible to determine the changes taking place in gene expression during the cloning process, it might be possible to create a pluripotent cell equivalent to an ES cell by artificially inducing such changes in a differentiated cell. And indeed, he identified four genes coding for transcription factors that played a crucial role in the reprogramming process. By expressing these genes in a skin cell, he was able to induce a pluripotent state, which he named induced pluripotent stem cells, or iPS cells (see Figure 27).[542] By exposing such iPS cells to different types of environmental conditions, it is possible to induce them to differentiate into a variety of specialized cell

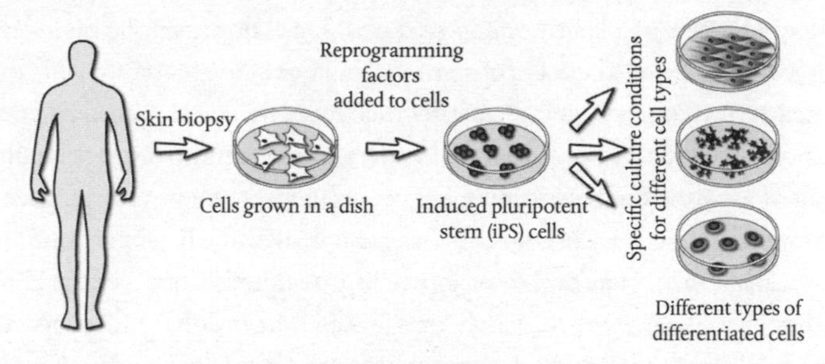

Fig. 27. Induced pluripotent stem cells

types. This discovery led to Yamanaka being awarded a Nobel Prize with John Gurdon in 2012.[543] Yamanaka expressed the inducing genes using retroviral constructs. This poses safety issues for the use of iPS cells for therapeutic purposes, since, as we saw in Chapter 2, expression of genes using such a viral route can lead to cancer. However, it has since proved possible to create iPS cells by introducing the transcription factors as proteins that can enter the cell thanks to special tags which allow them to gain entry across the cell's surface membrane.[544]

Indeed, for a brief period after the publication of two papers in *Nature* in January 2014, it seemed that generating iPS cells might be almost trivial.[545] The studies, by Haruko Obokata of the RIKEN Centre for Developmental Biology in Kobe, Japan, brought the 30-year-old researcher to international prominence. For what Obokata claimed to have discovered was that mouse skin cells only needed to be exposed to a weak citric acid solution for half an hour, or alternatively simply squeezed, and, amazingly, after this treatment they were transformed into iPS cells. Or rather STAP cells, for Obokata coined her own acronym, standing for 'stimulus-triggered acquisition of pluripotency', to describe her creation.[545] The discovery seemed to have wider relevance than simply providing an easier way of generating iPS cells, because, as Obokata pointed out, the STAP mechanism might shed light on the way cells gather wear and tear during our lifetimes. 'By studying the mechanism we might be able to learn more about how the age of cells is also locked in,' she said.[545] Commentating on the study, Dusko Ilic, a stem-cell biologist at King's College London said it was 'a major scientific discovery that will be opening a new era in stem-cell biology'.[545]

Apart from concerns about whether this would mean drinking fruit juice would cause cells in the throat to become pluripotent, from a scientific point of view the findings seemed almost too good to be true. And indeed they were. For even while Obokata was being hailed as a bright new star in the scientific firmament and some journalists wondered if she were heading for her own Nobel Prize to match Yamanaka and Gurdon's, her story was already falling apart. Within days of Obokata's papers being published, allegations began surfacing in science blogs and on Twitter that some of the images looked doctored and that parts of the text had been lifted from other papers.[546] RIKEN began an investigation, and on 1 April 2014 it announced that Obokata was guilty of scientific misconduct. Faced with a media backlash, Obokata's response was to apologize for 'insufficient efforts, ill-preparedness and unskilfulness', which had led to errors of methodology and sloppy data management.[546] But she denied fabricating her results.

Yet attempts to reproduce Obokata's findings by other researchers failed. And then further investigations revealed that the STAP cells didn't genetically match the mice from which they had supposedly been derived. Through her lawyer, Obokata said she couldn't understand how that was possible. But the logical conclusion was that the STAP cells were just standard ES cells that had been relabelled.[546] Obokata's downfall seemed complete, but there was a particularly tragic ending to the story. For, following the exposure of Obokata's fraud, RIKEN began an investigation into how the scandal had been allowed to occur. One scientist particularly criticized for not scrutinizing the findings more carefully was stem-cell biologist Yoshiki Sasai, deputy director of RIKEN and Obokata's line manager. Despite having no direct role in the STAP discovery, in his own words, Sasai was 'overwhelmed with shame'.[546] In early August 2014, after spending a month in a psychiatric hospital being treated for depression, the scientist committed suicide by jumping from a stairwell at a research facility opposite the RIKEN Institute, leaving behind three farewell notes. One was addressed to Obokata with the plea: 'Be sure to reproduce STAP cells.'[546] Yet, in the subsequent period, neither Obokata herself, nor any other researcher, has reproduced her findings.

The scandal surrounding STAP cells should not detract from the very real existence of bona fide iPS cells, for these have now been reproducibly generated in thousands of laboratories across the world. And compared to ES cells, iPS cells have many potential advantages. One is that there are far fewer ethical concerns about them, because they are not derived from a human embryo. And since these cells can be generated from differentiated cells from any human individual, this means it should be possible to generate tissues or organs from iPS cells from an individual and use them to treat the same individual. In addition, the relative ease with which it's possible to generate iPS cells, compared to the complicated cloning and ES cell derivation procedures, is one factor underlying the current excitement about these cells and their research and therapeutic potential. This also distinguishes them from testicular stem cells, which can only be obtained via a testicular biopsy, which carries its own risks and is obviously only possible in men. Both ES cells and iPS cells can give rise to therapeutically important cell types. We've seen in the section 'A Very Gifted Cell' how Doug Melton and his colleagues have identified a way to generate many pancreatic beta cells from human ES cells. Melton also found that he could produce such beta cells from human iPS cells.[547]

Will it ever be possible to grow whole organs in culture using either ES or iPS cells? One potential problem is that an organ is a complicated structure, often consisting of multiple cell types and blood vessels assembled in a precise fashion. But some exciting progress has recently been made. In July 2013, Takanori Takebe and colleagues at Yokohama City University created 'mini-livers' from iPS cells.[548] Takebe's team used iPS cells to make three cell types that normally combine to form the developing liver in a human embryo—hepatic endoderm cells, mesenchymal stem cells, and endothelial cells. When mixed together, the three cell types not only divided in culture but also organized themselves into 3D liver 'buds', complete with blood vessels. Transplanted into a mouse whose immune system had been engineered not to reject human tissues, the human liver buds matured, the human blood vessels connected to the mouse host's blood vessels, and the liver buds began performing many functions of the mature organ, like metabolizing sugars and drugs. When the mouse's own liver was disabled, the human liver buds kept the animal alive for two months. The researchers claimed 'this is the first report demonstrating the generation of a functional human organ from pluripotent stem cells'.[548] Malcolm Allison, a stem-cell expert at Queen Mary University of London, believes the study's findings offer 'the distinct possibility of being able to create mini-livers from the skin cells of a patient dying of liver failure, and when transplanted would not be subjected to immune rejection as happens with conventional liver transplants today'.[549]

Self-Organizing Organs

The ability of cells to self-organize into organs, or parts of organs, is not specific to the liver. Over recent years, researchers around the world have coaxed stem cells to develop into structures resembling tissues from the eye, gut, kidney, pancreas, prostate, lung, stomach, and breast.[550] Named 'organoids' because they mimic some of the structure and function of real organs, these miniature approximations of organs are providing insights into human embryo development, serving as disease models and drug-screening platforms, and may eventually be used to rescue damaged organs. In 2015 Austin Smith, director of the Wellcome Trust/MRC Stem Cell Institute at Cambridge University, stated that this is 'probably the most significant development in the stem-cell field in the last five or six years'.[550]

Still, organoids are far from perfect representations of the organs they partially resemble. Some lack key cell types, while others recapitulate only the initial stages of an organ's embryonic development. But an interesting aspect of these studies is how little encouragement stem cells require to self-assemble into intricate structures. Some scientists are not surprised at this self-organizing feature of stem cells. The developmental biologist Melissa Little at the University of Queensland believes 'the embryo itself is incredibly able to self-organize; it doesn't need a template or a map'.[550] To some extent this has been known since the 1900s, when embryologists showed that sponges which were broken into single cells could reassemble. But many scientists were sceptical about the idea that the organs of more complex animals would show such straightforward self-organizing principles.

In part this was because of results achieved with stem cells cultured in standard conditions. Traditionally, this has involved growing cells in 2D layers on a plastic tissue culture dish. A major step forward came with the realization that cells behave very differently when grown in matrigel—a soft jelly that resembles the extracellular matrix, the 3D mesh of molecules that surrounds cells in the body. Mina Bissell, a cancer researcher at the Lawrence Berkeley National Laboratory in California, helped to changed people's perceptions by showing in the 1990s that breast cells behave very differently in 3D cultures than in conventional ones.[550] Another important step forward occurred in 2011 when Yoshiki Sasai generated much excitement by showing he could grow bodily structures such as part of an eye and a pituitary gland.[550]

As well as being a potential step on the way to growing organs in culture for transplantation purposes, generation of organoids is proving important for biomedical research. Hans Clevers and his colleagues at the Hubrecht Institute in Utrecht have been making important progress in the creation of 'mini-gut' structures. After identifying intestinal stem cells in mice in 2007, Clevers decided to see how such cells would grow in matrigel as opposed to standard cell culture. 'We were just trying things,' he says. 'We hoped that we would make maybe a sphere or a blob of cells.'[550] Yet remarkably, after several months of culture, the stem cells had differentiated into structures that resembled the intestine's nutrient-absorbing villi, as well as the deep valleys between them called crypts. 'The structures, to our total astonishment, looked like real guts,' recalls Clevers.[550]

Clevers' team is now using mini-guts grown from human intestinal stem cells to study the effectiveness of drugs to treat cystic fibrosis. As we saw in Chapter 2, this disorder is caused by the absence of a protein, CFTR, that allows chloride ions out of cells lining the lungs, but also the pancreas and intestine, which is why people with this disorder also have problems with digestion. Clevers' team take rectal biopsies from people with cystic fibrosis, use the cells to create personalized gut organoids, and then apply a potential drug. If the treatment opens the ion channels, then water can flow inwards and the gut organoids swell up. 'It's a black-and-white assay,' says Clevers, and one that is much quicker, cheaper, and safer as a first resort than trying drugs in people.[550] This approach has already been used to assess the effectiveness of a drug called ivacaftor, also known as Kalydeco® and five other cystic-fibrosis drugs in about a hundred patients; at least two are now taking Kalydeco® as a result.

Clevers and his colleagues are also using organoid culture to test therapies for treating cancer. They have grown gut organoids from cells extracted from colorectal tumours and, with David Tuveson, a cancer researcher at Cold Spring Harbor Laboratory in New York, they have also generated pancreas organoids using biopsies taken from people with pancreatic cancer. In both cases, the organoids are now being used to identify drugs that work best on particular tumours. 'What patients are looking for is a logical approach to their cancer,' says Tuveson. 'I'm very excited about what we're learning.'[550]

Undoubtedly the biggest challenge for those seeking to grow organs in culture is the brain. As we mentioned in Chapter 3, the human brain is the most complex structure in our bodies, indeed in the known universe (alien brains may be more complex!). It may be asking a lot to grow a whole human brain in culture. Still, Jürgen Knoblich and his colleagues at the Institute of Molecular Biotechnology in Vienna have recently had some success in growing formations of cells that mimic some of the brain's regions. Knoblich's team created iPS cells from human skin and cultured these with growth factors and other chemicals identified as being important for brain development. The iPS cells first differentiated into neuroectoderm, the layer of cells that eventually gives rise to the embryo's nervous system. This was then suspended in a gel scaffold to help it develop into 3D structures. Remarkably, in less than a month, the stem cells developed into tiny organoids corresponding to most of the regions of the brain. 'If you zoom out and look at the whole, it's not a brain,' says Knoblich. 'But our cultures contain individual brain regions that have a functional relationship with one another.'[551] Besides parts of cortex, which normally forms the outer layer of the brain, the structures also contained regions of forebrain, which makes neurons that connect to the cortex, and the choroid plexus, which generates the spinal fluid (see Plate 4).

Such organoids are useful for studying human brain disorders. People with a condition called microcephaly are born with a head that is much smaller than normal. Most children with microcephaly also have a small brain and intellectual disability. Knoblich and his colleagues created iPS cells from an individual with this condition and used them to create brain organoids.[551] During the initial stages of brain development, stem cells go through a phase in which they divide to make more stem cells, increasing their numbers. After a certain period, some of these cells switch to producing neurons. Knoblich's team found that the period of stem-cell multiplication was reduced in the case of the microcephalic iPS cells. This suggests that one cause of microcephaly is that there aren't enough stem cells available to turn into neurons, leading to a smaller brain. The researchers also found that the reduced number of neurons in the microcephalic brain structures was associated with lack of a protein called centrosomin, which is known to play an important role as a regulator of neuronal growth.[552] When Knoblich's team added this protein to the microcephalic organoids, the number of neurons increased.[551] So one way to treat this condition might be to enhance expression of this protein in the brain.

Recently, Rene Anand of Ohio State University announced that he had made a major breakthrough in the creation of brain organoids by growing structures from human iPS cells that included 98 per cent of the cells that exist in the brain of a 5-week human foetus. Amazingly, the mini-brains contained a spinal cord and even a retina. Anand claims his team's work is different from previous studies, because 'our organoids have most of the brain parts'.[553] This is important, he adds, because 'if you want to study Parkinson's, you need the midbrain. The best I can tell from all published research on organoids is they don't have the midbrain. We have midbrains and we are already moving toward trying to study them.'[553] Anand believes the development of the organoids may be pushed even further. 'If we let it go to 16 or 20 weeks that might complete it, filling in that 1% of missing genes,' he says.[553]

Reactions by other scientists to Anand's claims have been mixed. Zameel Cader, a neurologist at the John Radcliffe Hospital in Oxford, said that although the work sounds very exciting, 'when someone makes such an extraordinary claim as this, you have to be cautious until they are willing to reveal their data'.[554] However, Rudolph Tanzi, an Alzheimer's research pioneer at Harvard University, said 'I think it took all of us by surprise. The results were absolutely astounding… it's an incredible achievement.'[553] Creating a foetal brain that includes so many different types of brain cells amounts to a 'quantum leap forward', he added.[553] Anand has said that using this approach to learn more about Alzheimer's is a 'high priority' for his team. This would involve taking skin cells from Alzheimer sufferers from which to create iPS cells, allowing these to differentiate into brain organoids, and then investigating whether differences can be detected with the normal brain in the development of the organoids in a 3D matrix. These differences might cast light on the molecular and cellular mechanisms that underlie Alzheimer's.

One strategy being pursued in such studies is to focus on individuals with a particularly severe form of Alzheimer's, which shows early onset, affecting people in their 30s and 40s. It is hoped that the behaviour of brain organoids created from such individuals may show more obvious differences in their growth and development. More generally, the strategy of focusing on a specific subset of people with a common brain disorder is already leading to some interesting insights. Flora Vaccarino and colleagues at Yale University selected autistic patients with enlarged heads, a condition that affects about a fifth of people with the disorder.[555] The researchers then created brain organoids from these patients and also from the patients' fathers, who did not have autism. Vaccarino's team found that genes involved with directing the proliferation of cells were overexpressed in the autistic organoids. What's more, according to Vaccarino, the analysis revealed that 'the cells of the patients divide faster than the fathers".[555]

Further investigation showed that this proliferation was particularly associated with overgrowth of 'inhibitory' neurons. When Vaccarino's team used genetic engineering to reduce the expression of a gene they had identified called forkhead box G1 (FOXG1), which is known to be involved in early brain development, they were able to create organoids from the autistic patients without the neuronal imbalance seen previously. Quite how an increase in FOXG1 and inhibitory neurons might lead to autism is unclear, but Vaccarino thinks it's possible that excess inhibition early in development 'is affecting the way neurons connect with each other'.[555] These are preliminary findings, but they suggest that further studies of this type may yield important new insights into disorders like autism and point to possible treatments. Rene Anand has suggested other uses for brain organoids too, like testing the effect of environmental toxins on the brain. 'We can look at the expression of every gene in the human genome at every step of the development process and see how they change with different toxins. Maybe then we'll be able to say "holy cow, this one isn't good for you"', he says.[554]

When Technologies Meet

Developing new approaches for getting ES or iPS cells to develop into complex 3D structures is one important aspect of stem-cell technology. An equally crucial one is the ability to control the differentiation pathway. At the heart of such control is an understanding of the genetic pathways underlying the development of specific cell types, tissues, and organs in the normal developing embryo. Just as important is having the ability to manipulate such pathways in a culture dish. And it's here that a combination of genome editing and stem-cell technology is proving particularly fertile. In particular, the flexibility and efficiency of approaches like CRISPR/CAS9 are taking the genetic manipulation of ES and iPS cells, and the differentiated cells derived from them, to a previously undreamt of level of sophistication.

As I mentioned in Chapter 2, the development of knockout and knockin mice was made possible by the use of homologous recombination to target specific genes in mouse ES cells. However, not only is this approach inefficient, requiring drug selection to identify the one-in-a-million event in which it takes place, but for some reason it has never worked well with human ES cells. In contrast, CRISPR/CAS9 is highly efficient in such cells. Even more important, an adapted version of this method, developed by Su-Chun Zhang of the University of Wisconsin, can now be applied at any stage of differentiation.[556] To do this, Zhang's team developed a version of CRISPR/CAS9 in which the CAS9 enzyme can only be activated by stimulation with a specific chemical. This means that human ES cells can be engineered so that a specific gene is primed to be edited, but this will only happen when the cells, or their differentiated cell progeny, are treated with the chemical.

With this approach, Zhang's team can now 'take out the gene at any given time, in any type of cell'.[556] This is important because shutting a gene off too soon can kill the stem cell or inhibit its development. And, according to Zhang, 'you may want to delete it after the cells have differentiated into heart, brain or liver cells...That precision is one reason I see so much promise in this technology.' Zhang now wants to apply his new approach to the study of brain development. 'You can very quickly pin down exactly what [a] gene does, at the stem cell stage, neural stem cell stage or at the differentiated neuron stage,' he says.[556]

To demonstrate this precision, Zhang's team engineered human ES cells so that they could knock out orthodenticle homeobox 2 (OTX2), a gene known to be involved in formation of the midbrain, at any stage of the differentiation of human ES cells into different types of brain structures. During brain development, the midbrain develops before the forebrain, the site of higher mental functions. By delaying the knock out of the gene, Zhang and his colleagues were able to show that the gene is also essential for the formation of the forebrain. 'If you knock it out, you simply don't have the cerebral cortical cells, and they are essential to what it takes to be human,' says Zhang. 'This is a really definitive way to show what genes are doing.'[556]

While the combination of stem technology and genome editing promises to have a major impact on biomedical research, it also has important implications for the development of new therapies. In particular, genome editing provides a new level of genetic precision that looks set to transform the way stem cells are used to create replacement cells, tissues, and organs. And the ease with which it's now possible to generate iPS cells from particular individuals and use genome editing to alter them, means that this specific combination may be a powerful new type of disease treatment.

Take for instance a study by Linzhao Cheng and colleagues at Johns Hopkins University, which demonstrated how CRISPR/CAS9 could be used to treat the recessive single-gene disorder sickle cell anaemia.[557] We've come across the molecular basis of this disorder in previous chapters: it results from a single amino acid change in the β-globin protein. That alteration makes the haemoglobin molecules form rope-like cables that cause the red blood cells that contain them to bend into rigid, sickle shapes. The sickle cells lodge in narrow capillaries, cutting off local blood supplies and causing great pain in the affected body parts, especially the hands, feet, and intestines. The depletion of the red blood cells also causes the overwhelming fatigue of anaemia. Eventually, this disorder can prove fatal.[557]

Cheng's team took blood cells from people suffering from sickle cell anaemia and induced them to change into iPS cells. They then used CRISPR/CAS9 to correct the mutation in the β-globin gene that causes the disorder. Finally, the researchers coaxed the corrected iPS cells to differentiate into mature red blood cells that did not have the abnormal sickle shape. To become medically useful, the technique of growing blood cells from stem cells will need to be much more efficient and scaled up significantly. The lab-grown stem cells would also need to be tested for safety. But Cheng believes that 'this study shows it may be possible in the not-too-distant future to provide patients with sickle cell disease with an exciting new treatment option'.[558] Cheng's approach may also be useful for treating other disorders of the blood.

In Chapter 7 we saw how a virus was used to deliver genome-editing tools that partially reversed the muscle defects in a mouse model of Duchenne muscular dystrophy, or DMD, raising the possibility that this approach might be used to treat boys with this disorder. But a strategy involving both genome editing and iPS cells might also offer a route to treat DMD in humans. Showing the potential of such a combined approach, Akitsu Hotta and colleagues at Kyoto University have generated iPS cells from boys with DMD.[559] They then used genome editing to correct the defect in the dystrophin gene in these cells. Finally, they showed that the genetically corrected iPS cells could differentiate into skeletal muscle cells and express full-length dystrophin protein. A major goal now is to see whether such corrected muscle cells can be introduced back into DMD sufferers to treat their condition.

Perhaps the biggest hope for the use of combined genome editing and iPS technology for therapeutic purposes is for the treatment of brain disorders like multiple sclerosis, Parkinson's, and Alzheimer's disease. One scientist interested in the treatment of Alzheimer's is Mathew Blurton-Jones of the University of California, Irvine.[560] He plans to use CRISPR/CAS9 to insert potentially therapeutic genes—like the growth factor brain-derived neurotrophic factor (BDNF) or the enzyme neprilysin, which can degrade the plaques that form in the brains of people with Alzheimer's—directly into iPS cells. Blurton-Jones has had some success in treating a mouse model of this disorder by injecting the mice with iPS-cell-derived neurons overexpressing neprilysin. However, he inserted that transgene randomly into the cell genome. 'That's fine for a basic science experiment, but not for the clinic,' he says. 'We would need to target a specific locus that we know is safe, and CRISPR dramatically increases our ability to do that.'[560]

Such is the potential of the approaches mentioned that it's not surprising that a pioneer of them, Su-Chun Zhang, believes 'this marriage between human stem cells and genome editing technology will revolutionize the way we do science'.[561] But amid this excitement some scientists are sounding a note of caution about the safety issues underlying the use of genome editing in biomedical research. One such person is Jennifer Doudna—the CRISPR/CAS9 pioneer. Her concerns began at a meeting in which a postdoctoral researcher presented work in which a virus was used to carry the CRISPR tools into mice. The mice breathed in the virus, allowing the CRISPR system to engineer mutations and create a model for human lung cancer. It occurred to Doudna that a minor mistake in the design of the guide RNA could create a tool that worked in human lungs as well. 'It seemed incredibly scary that you might have students who were working with such a thing,' said Doudna.[562] In fact, Andrea Ventura of the Memorial Sloan Kettering Cancer Center in New York, whose postdoc carried out the work, believes his lab carefully considered the safety implications: the guide RNAs were designed to target genome regions unique to mice and the virus was disabled so it could not replicate. However, he agreed that it's important to anticipate even remote risks. 'The guides are not designed to cut the human genome, but you never know,' he said. 'It's not very likely, but it still needs to be considered.'[562]

Similarly, some scientists are urging caution about the need to consider potential adverse effects of the new technology when used as a therapeutic strategy. In particular, concerns have been raised about the need to make sure this technology does not introduce unwanted changes elsewhere in the genome that have consequences for health. 'These enzymes will cut in places other than the places you have designed them to cut, and that has lots of implications,' says James Haber, a molecular biologist at Brandeis University. 'If you're going to replace somebody's sickle-cell gene in a stem cell, you're going to be asked, "Well, what other damage might you have done at other sites in the genome?"'[562] In fact, much work is being done to eliminate such unwanted 'off-target' effects. However, Haber believes the technology will have to be very precise indeed, since low-frequency events could potentially be dangerous if they accelerate a cell's growth and lead to cancer.

Ironically, the very ease with which CRISPR/CAS9 can be employed poses challenges in ensuring it's used responsibly. So Katrine Bosley, CEO of Editas, a company in Cambridge, Massachusetts, pursuing CRISPR/CAS9-mediated gene therapy, and a veteran of commercializing new technologies, said that while the problem in the past has been convincing others that an approach will work, 'with CRISPR it's almost the opposite. There's so much excitement and support, but we have to be realistic about what it takes to get there.'[562] And, given the controversy about the use of genome editing to modify the human germline, that means there are many technical, but also ethical, issues to discuss relating to technologies like CRISPR/CAS9. This is something we'll look at in the concluding chapter, but before we do so, let's consider, in Chapter 9, an approach to redesigning life that has fewer immediate practical implications, but that, in the long term, may have an even greater impact upon humanity.

9

Life as a Machine

The Rio Tinto in Andalucia, Spain, is one of the strangest rivers on Earth. Near its source deep in the Sierra Morena Mountains, its deep crimson colour does indeed look more like sangria made from the local vino tinto than water. But with their extreme acidity (pH less than 2) and high concentration of dissolved heavy metals, it would be inadvisable to swim in these waters, never mind drink them.[563] Red is just one of the shades of the waterways in these parts; other streams are a violent shade of orange or an emerald green. The strange colouration is due to the huge concentration of metal ores in this region, a feature that has drawn prospectors to this area since ancient times. According to myth, the fabled mines of King Solomon were located here.[563] More reliable historically is that successive waves of Greeks, Carthaginians, and Romans came to this region to mine its iron, copper, and silver.[563]

The Romans in particular mined silver here on an industrial scale, using slaves kept in appalling conditions to work the underground waterwheels that raised water from 100 metres deep to keep the mines dry. Later, in the nineteenth century, the pollution from the copper and sulphur mines run by the British-owned Rio Tinto Mining Company was so bad that locals demonstrated in 1888, in what has been described as the first ecological protest.[564] In a brutal response to this protest, 200 of the unarmed demonstrators were mown down by armed troops, brought in to quell the uprising. Such was the power of the mining company at this time that those responsible for this massacre were never punished; indeed, the events went practically unreported in Spanish newspapers.[563]

Today, mining operations around the Rio Tinto are a shadow of their former self, due to competition from other parts of the world. Visiting this region now, you're more likely to bump into tourists drawn to the alien-looking landscapes or come to learn about the history of the mines than actual miners. But these days there's another type of visitor to the Rio Tinto too: biologists fascinated by its unusual life forms.[565] For this region is also known for its extremophiles, the name given to microorganisms that live in the most inhospitable conditions on Earth. Such is the interest in these microorganisms that the US National Aeronautics and Space Administration (NASA) has set up a project named the Mars Astrobiology Research and Technology Experiment, on the basis that, if life exists on Mars, it may have common features with the life forms that thrive in this region's iron-rich soil which has similarities to that on the red planet. Justifying the interest, Carol Stoker, leader of the project, argues that 'the Rio Tinto area is an important analogue to searching for life in liquid water, deep beneath the subsurface of Mars'.[565]

Life at the Extremes

While the Rio Tinto is a particularly striking example of an extreme environment that supports a wide variety of extremophile microorganisms, it's far from the only one. We now know that a rich diversity of such life forms exist in locations as different as the boiling hot springs of Yellowstone Park or the frozen wastes of Antarctica.[566] Perhaps most surprising is the discovery that, far from life just about hanging on for survival in these environments, these locations seem to be simply teeming with microorganisms. For instance, a recent study led by Steven Chown at Monash University in Melbourne, Australia, identified an unexpected richness of life forms in the icy interior of the Antarctic. 'Most people think of the continent as a vast, icy waste,' says Chown. 'But that's simply not true. There's much biodiversity on land, especially among the microorganisms.'[567]

Meanwhile, a study by Philippa Stoddard, Mark Brandon, and their colleagues at Yale University indicates that the Earth's inhabitable zone may be far deeper than previously suspected.[568] Bacteria make their presence known by a specific mix of lighter molecular weight carbon isotopes, derived from the methane these microorganisms excrete, and the Yale researchers have identified such tell-tale signs in rock samples that were once as much as 19 kilometres below the surface of the Earth. 'These really light signals are only observed when you have biological processes,' says Stoddard.[568] 'Assuming our data are correct, this greatly expands our understanding of the extent of the Earth's biosphere.'[569]

More direct evidence by Mark Lever and colleagues at Aarhus University, Denmark, has come from drilling deep into the crust at the bottom of the ocean and carrying out DNA and metabolic analysis on the samples obtained. This showed that bacteria definitely live as deep as 1.6 kilometres beneath the seabed. 'We found live bacteria in the earth's crust below the sea, regardless of how deep down we drilled,' says Lever. 'We are talking about a huge ecosystem in the rocks below the seabed.'[570] This means the living world we're familiar with on the Earth's surface may represent just a fraction of the total. But how do such subterranean microbes live? Light is central to most life as it fuels photosynthesis, which turns solar energy into organic molecules. Deep in the Earth's crust, there is no light. Instead, bacteria at this depth use a process called chemosynthesis, which allows them to extract energy from the rocks themselves. 'The bacteria feed on chemicals that are released when water seeps down through the rocks,' says Lever. 'The rocks contain iron ions that can react with sea water and produce hydrogen, which the bacteria can use as an energy source for producing their own organic matter.'[570]

The study of extremophile bacteria is of interest not only because it reveals the fascinating diversity of life on Earth but also because of the potential practical usefulness of such microorganisms. One of the most well-known uses of extremophile bacteria is in the so-called 'polymerase chain reaction', or PCR, invented by Kary Mullis in 1984.[571] PCR is a method for amplifying a single copy or a few copies of a segment of DNA by several orders of magnitude, generating thousands to millions of copies of a particular DNA sequence. Its uses range from detection of mutations in human genes, including those in IVF embryos, to forensic analysis of crime scenes and identification of family relationships between Egyptian mummies.[572] Mullis claimed the idea for PCR came to him during a late-night drive through the mountains of California with a friend. 'I was just driving and thinking about ideas and suddenly I saw it,' he said. 'I saw the polymerase chain reaction as clear as if it were up on a blackboard in my head, so I pulled over and started scribbling.'[573] Mullis further recalled that his friend had been asleep but now woke up and 'objected groggily to the delay and the light, but I exclaimed I had discovered something fantastic. Unimpressed, she went back to sleep.'[573]

Mullis also credited his experimentation with the drug LSD with aiding the discovery. When asked whether the drug helped in coming up with the idea for the technique, he replied, 'What if I had not taken LSD ever; would I have still invented PCR? I don't know. I doubt it. I seriously doubt it.'[574] Mullis was working for a small biotech company called Cetus at this time and, after much persuasion, he convinced them of the importance of his discovery. It was lucky for them that they finally listened, for Cetus went on to sell the patent for PCR to Hoffman-La Roche for $300 million—the most money ever paid for a patent. All Mullis got was a $10,000 bonus. However, he did receive the Nobel Prize for his discovery in 1993.[571]

There are three temperature-sensitive steps in PCR (see Figure 28).[575] First a DNA 'template' containing a sequence that requires amplification is incubated at 94°C to break the bonds holding the two strands of the double helix

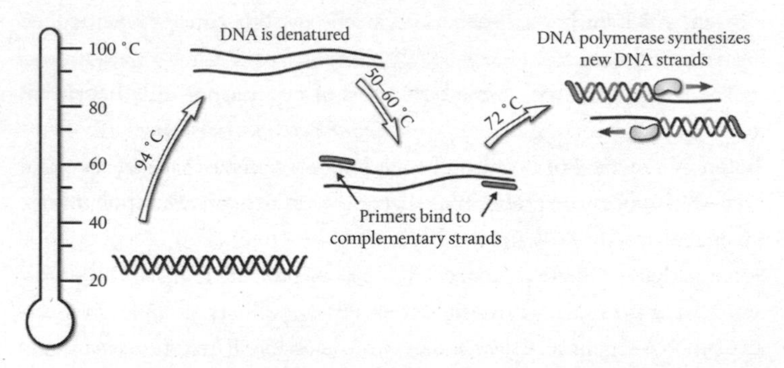

Fig. 28. The three temperature-dependent stages in a PCR cycle

together. Then the sample is cooled to around 50–60°C, which allows two 'primers'—short DNA sequences that match those at the start and end of the region to be amplified—to bind to their complementary sequences in the two separated strands of the DNA template. Next the temperature is raised to 72°C, which allows DNA polymerase, the enzyme that replicates DNA, to synthesize a new strand between the two primers. Finally, the temperature is raised to 94°C and the process begins again. Typically, a PCR reaction will undergo around 30 such cycles, all in the same plastic tube, on a device called a thermal cycler that rapidly varies its temperature. However, an equally crucial aspect of PCR is that the DNA polymerase used is not the usual variety taken from an *E. coli* bacterium, since this becomes rapidly inactive above 37°C, the normal temperature for such bacteria that live in environments like the human gut. Instead, the polymerase comes from a heat-resistant bacterial species called *Thermus aquaticus* that lives in boiling hot springs at Yellowstone Park.[575]

PCR involves one very specific use of a protein from a thermophilic bacterium, but do extremophiles also have a more general practical importance? Certainly Peter Golyshin of Bangor University, who studies such life forms, believes so. 'Chemical synthesis is often conducted in harsh conditions with high temperatures, high pressures and high solvent concentrations,' he says. 'The anticipation is that enzyme catalysts produced by microbes that live in hostile environments could be used in these industrial processes.'[576] Unfortunately, many extremophile microorganisms have proved difficult to culture in the laboratory using tried and tested microbiology techniques. This may be because standard culture media—a nutrient-rich jelly on which bacteria are grown in a lab—does not contain all the other microbes that live in their natural environment. It seems that extremophile bacteria rely partly on the metabolic by-products of other microbial species that thrive in their shared habitat.

According to Golyshin, one way around this problem is 'instead of trying to grow the microbes in the usual way we can just harvest their DNA, express their genes in surrogate hosts such as yeast or *E. coli*, and apply substrates to see if they actively convert it or not'.[576] The greatest diversity of extremophiles is in the sea, for instance near hydrothermal volcanic vents close to Milos Island in Greece. Mohamed Jebbar of the University of Brest, France, has been collecting microorganisms from these vents. 'The average depth of the deep sea floor is 3,800 metres so we need very large and sophisticated technology to explore these remote areas,' he says.[576] However, the potential practical benefits are substantial, as the enzymes obtained from such microbes may be useful for breaking down tough material such as cellulose from plants and agricultural or urban waste to make biofuels. Other potential uses of extremophiles are as biocatalysts that could break up cancerous tumours.

The search for novel extremophile microorganisms is based on the idea that amongst nature's great diversity there will be unusual life forms with biological properties that can be harnessed for industrial or medical purposes. There is, after all, a precedent in the wide variety of antibiotics that are so vital to modern medicine.[577] Typically, such antibiotics are natural products generated by microorganisms. It may seem strange that bacteria themselves produce these deadly anti-bacterial substances, but they do so to maintain an advantage over other bacterial species, with which they compete for resources. With antibiotics we've been able to make war on harmful bacteria, thereby revolutionizing the treatment of infections and providing a key preventative measure to minimize infection during surgery. Unfortunately, as we saw in Chapter 7, this is a war that works both ways, as bacteria have evolved ways to block antibiotic action, with the consequence that many harmful bacteria are increasingly antibiotic resistant. We badly need new antibiotics.

One of the richest potential sources of new antibiotics is common soil: a gram of soil contains more bacteria than there are people on the planet.[578] The difficulties of growing soil bacteria in the lab have hampered their antibiotic potential. But recently Kim Lewis's team at Northeastern University, Boston, developed a 'soil hotel' that recreates soil's chemistry for its bacterial inhabitants.[578] This allowed them to identify 25 new antibiotics. One, named teixobactin, kills some bacteria as quickly as established antibiotics; it also cured laboratory mice suffering from bacterial infections with no toxic side effects.[578] Commentating on the study, Mark Woolhouse of the University of Edinburgh says: 'any report of a new antibiotic is auspicious, but what most excites me about [this research] is the tantalising prospect that this discovery is just the tip of the iceberg... It may be that we will find more, perhaps many more, antibiotics using these latest techniques.'[578]

Such is the potential that lies at the bottom of the ocean or in the ground beneath our feet. Even so, there are an increasing number of scientists who are not content with working with existing life forms but believe the time is now ripe for designing completely new ones. So while genome editing is revolutionizing the genetic engineering of existing species, 'synthetic biology' aims to go even further and redesign life from scratch.

Synthetic Life Code

One pioneer of this approach is Craig Venter, who led one of the teams that first sequenced the human genome. Since then, Venter has headed several biotechnology projects, like the development of genome-edited pigs for xenotransplantation mentioned in Chapter 5. But one of his primary interests is synthetic biology. And, in 2010, Venter announced that he and his colleagues had created the world's first synthetic life form, following a project that cost $40 million and engaged 20 scientists working for more than a decade.[579] Certainly it was an extraordinary feat, for Venter's team took the complete genome sequence of an existing bacterium—*Mycoplasma mycoides*—and resynthesized it from scratch using lab chemicals. They then inserted this genome into a bacterial cell whose own genome had been excised. The researchers finally showed that the artificial genome could propagate itself and the cell through subsequent generations. As a twist, the researchers included extra sequences that served as 'watermarks', proving authorship of the new life form.

Opinions vary as to the novelty of the creation. Julian Savulescu, a bioethicist at Oxford University, believes that: 'Venter is creaking open the most profound door in humanity's history, potentially peeking into its destiny... He is going towards the role of a god: creating artificial life that could never have existed naturally.'[579] Others have been more critical, pointing out that, to have truly created synthetic life, Venter's team would need to do more than just copy an existing natural genome, and instead it would also be necessary to synthesize the bacterial cell's wall, membrane, and cytoplasmic contents.[580]

While the criticism that there is more to a bacterium than its genome is a valid one, it's important to recognize that Venter and colleagues' long-term goal was always about more than artificially synthesizing an existing bacterial genome and showing it could be propagated across generations in another bacterium's vacant shell. This was just supposed to be the first stage in a plan to strip life down to its bare essentials, so that entirely novel genomic elements could be added on. Or, as Venter himself put it, 'once we have a minimal chassis, we can add anything else to it'.[581] Such a minimal genome needs to be identified before it can be constructed—and recently a team led by Bernhard Palsson at the University of California, San Diego, did precisely that.[582]

Palsson's team took various bacterial species with different genomes and modelled their growth in a wide variety of environments, with different nutrient requirements. According to the researchers, this 'forces the cell to use a wide array of its biochemical pathways. By defining...those genes expressed across all simulation conditions, we select those that are used regardless of nutrient availability.'[582] The end result was a set of genes, reactions, and processes universally required by bacteria. This minimal definition will be key to creating useful synthetic bacteria in the future, according to Laurence Yang, a researcher involved in the study, because 'by defining the vital set of genes and functions that need to always be present in a cell to sustain life, we can begin to realize new ways to engineer a cell to optimize production of a desired product without sacrificing the cell's health'.[582] Of course, a key question now is identifying which novel elements should be added to Venter's 'minimal chassis' to provide useful new functions.

It's not only bacteria that have potential uses for industry and medicine. Human cells differ from bacteria in that our genomes are contained within a nucleus, the membrane-bounded structure that separates the DNA from the cytoplasm. Possession of a nucleus is a feature of so-called eukaryotes, in contrast to bacteria, which are prokaryotes. All complex multicellular organisms are eukaryotes, but there are also single-celled eukaryotes, the most famous being yeast. I mentioned in Chapter 1 that a key aspect of the agricultural revolution that began 12,000 years ago was the cultivation, for the first time, of crops in a fixed location and the domestication of wild animals for the production of meat. But the simple yeast also had a big impact on society at this time, by allowing the production of beer and leavened bread.

Surprisingly, given we're used to thinking of bread as a wholesome staple of our diet and beer as a rather guilty pleasure, recent evidence suggests that it was the alcoholic beverage that was developed first by humans, long before leavened bread.[583] It's even possible that leavened bread was only created by accident when some baker of old—perhaps a slightly inebriated one—accidentally spilt some beer into a batch of dough and serendipitously discovered the capacity of yeast to make bread rise. Some scientists have even suggested that beer had a crucial role to play in the development of civilization. Brian Hayden, an archaeologist at Simon Fraser University in Burnaby, Canada, has uncovered evidence that beer-making was practised by the Natufians, who are widely regarded as the inventors of agriculture.[583] Hayden believes that once people began to recognize beer's alcoholic effects, it became a central part of feasts and other social gatherings that forged bonds between people and inspired creativity. Political discussions may also have taken place at these get-togethers, important in developing power structures. 'It's not that drinking and brewing by itself helped start cultivation,' says Hayden. 'It's this context of feasts that links beer and the emergence of complex societies.'[583]

While yeast remains central to brewing and baking today, it also has important uses in biotechnology. This is because, for all the usefulness of bacteria in biomedical processes like the production of human insulin, differences between prokaryotes and eukaryotes in how proteins are modified in the cell following their synthesis mean that sometimes it's only possible to produce biologically active proteins for medical use in eukaryotes. So there is research interest in producing not only synthetic bacteria but also their artificial equivalents in yeast. An important step towards this goal was achieved recently when a team led by Jef Boeke of New York University synthesized a yeast chromosome from lab chemicals.[584] A yeast cell has 16 chromosomes, in contrast to the 23 pairs that exist in a human cell, and Boeke's team synthesized chromosome 9. But unlike Venter's effort, which created an exact replica of the *Mycoplasma mycoides* genome—bar the genetic 'watermarks' mentioned—Boeke decided to streamline the synthetic yeast genome. This was done by leaving out 'junk' parts of the chromosome thought to be unimportant for yeast function, such as repetitive DNA elements and the 'introns' that interrupt eukaryotic genes.[584]

Since there is currently a debate about whether some genomic regions classified as 'junk' are as useless as had been thought, this was an important way to test such elements in yeast. Another innovative feature of the project was the nature of the team assembled to synthesize the yeast chromosome. Boeke initially looked into how much it would it cost to pay a company to synthesize the parts of the chromosome but the price quoted was so exorbitant that he decided to try a more cut-price approach. This involved offering the project as a 'build-a-genome' taught course to undergraduates at Johns Hopkins University.[584] Effectively, each student made their own stretch of the chromosome, which involved stitching together short lengths of DNA created by a DNA-synthesis machine into ever-larger chunks. Eventually, this resulted in an entirely synthetic chromosome. And as a reward for their efforts, many of the undergraduates were co-authors on the paper that appeared describing the synthesis, which was published in *Science* in March 2014.

Having synthesized the streamlined chromosome, Boeke's team inserted it into a yeast cell whose own chromosome 9 had been removed. And, despite the altered nature of the substitute, yeast cells containing the synthetic chromosome grew as well as normal yeast. 'What's amazing about it is that there are over 50,000 base pairs that were either deleted, inserted or changed in that chromosome of 250,000 base pairs, and it works,' says Boeke. 'That's kind of a remarkable effect.'[584] But construction of a single synthetic yeast chromosome was only the beginning of this project. Boeke has subsequently enlisted collaborators—including undergraduate students—from around the world to synthesize streamlined versions of the other 15 chromosomes, and thereby the whole yeast genome.[585] One of the collaborators in this effort is Tom Ellis, of Imperial College London, who used a similar approach to Boeke to synthesize yeast chromosome 11. Ellis sees the project as a riposte to expensive, factory-scale synthetic biology. 'This shows that the academic, open-source reply to what Venter did is, let's set up some labs with undergraduates, and they can do the same,' he says.[584] Ultimately, Boeke views creating a yeast cell with an entirely synthetic genome as a way to test what is essential about this genome and which attractive new features can be added on without disrupting the activity of the cell as a whole. And indeed praise for the project and this goal has also come from Venter and his colleagues, who said 'this work is a prelude to and demonstrates the feasibility of extensive refactoring or streamlining of the other chromosomes'.[584] One of the other scientists working towards such a goal is Matthew Chang of the National University of Singapore. 'The whole premise is to reduce the size of the chromosomes to bare necessities and introduce choice elements which will allow design engineering and evolution of the genome,' he says.[586] An important underlying idea behind such a project is the notion that genomes can be broken down into component parts that have been named 'BioBricks'.[587] And just as Lego® bricks can be joined together to make a variety of complex structures, so such BioBricks, which include protein-coding DNA sequences but also promoters, enhancers, and other regulatory regions, can be joined together to make novel types of genomes.[587] This will allow high-speed, human-driven evolution: millions of new types of yeast with different properties that can be tested in the lab for fitness and function in a diverse number of applications for medicine and industry.

One important question is what the practical applications of artificial yeast cells might be. The possibilities include engineering yeast to make new types of antibiotics or drugs, foodstuffs, or novel materials for creating clothes or for building. But Boeke thinks that many scientists and technologists still have to get their heads around the true potential of how to utilize such artificial life forms. 'It's almost a technology that's ahead of its time,' he said. When he gives seminars, Boeke often asks the audience: 'I can make anything up to a million bases, what should I make and why?' Yet according to him, surprisingly few people want to jump in with proposals. 'It's just outside people's comfort zones to think like that,' he says.[588]

Biohackers in Hackney

While many people may lack the knowledge or confidence to suggest possible novel uses for synthetic biology, that may be starting to change if a small but growing movement of 'biohackers' get their way. Although 'hacker' now often has negative connotations, signifying someone who sabotages, or steals information from, the computer of an individual or organization, the word's original meaning is someone who tinkers with technology to make or repurpose things. And biohackers play with biotechnology in their spare time with the aim of making new life forms as part of a new do-it-yourself biology (DIYbio) movement.[589] Cathal Garvey, a biohacker based in Dublin, believes that 'Biohacking, or DIYbio, has got to be one of the most exciting subcultures active today.'[590] The members of this movement pay a small monthly fee to cover the costs of rent, reagents, and equipment to maintain a shared lab, which provides affordable access to anyone interested in biotechnology.[589]

In 2010 there were only a handful of biohacker groups around the world but by the end of 2018, DIYbio.org, a support organization for such groups, listed 44 biohacking labs in North America, 31 in Europe and 17 across Asia, South America and Oceania.[591] One London biohacker group is based in a laboratory called Biohackspace, founded, appropriately, in Hackney.[592] In March 2015 the UK Health and Safety Executive, or HSE, registered Biohackspace as 'GM Centre 3266'—the first laboratory in Britain that allows anyone to have a go at genetic engineering. Biohackspace has members from diverse backgrounds, but few have scientific training.[592] Projects at Biohackspace range from generating genetically modified bacteria to create new forms of art, to modifying brewer's yeast to make new 'craft beers'.

Another biohacker group, BioCurious, is based in Sunnyvale, California. Like other biohackers, its members are enthused by the possibilities offered by CRISPR/CAS9 technology, partly for its precision but also because it's so quick, cheap, and easy to use, making it ideal for amateur biotechnologists. One BioCurious member, Johan Sosa, an IT consultant, is already using the genome-editing technology. 'Currently we're creating the guide RNA that we're going to use to edit a yeast genome,' he says.[592] One use of the technology will be in the 'Real Vegan Cheese' project, whose aim is to modify baker's yeast so that it produces milk proteins.

Although run by amateurs, biohacker groups often seek advice from professional biologists. For instance, Darren Nesbeth, a synthetic biologist at University College London, has advised the members of Biohackspace about health and safety issues. 'They've got a licence now from the HSE to do genetic modification, which requires they have a safety panel of individuals,' he says. 'There's a framework and guidance there equivalent to what happens at a university.'[592] Despite such safeguards, it's perhaps not surprising that the biohacker movement has caught the attention of the security services. The FBI and US Department of Defense have already been in touch with the BioCurious group and sent agents to visit their lab. 'At the beginning they were coming through quite frequently—at least once a month, formally,' said Maria Chavez, director of community for the group. 'Informally, I'm not sure how many times they may have dropped in.'[592]

The response from professional scientists has been mixed. David Relman, a professor of infectious diseases and co-director of Stanford University's Center for International Security and Cooperation, has said that 'I do not think that we want an unregulated, non-overseen community of freelance practitioners of [CRISPR/CAS9] technology.'[593] In contrast, Darren Nesbeth believes that the biohacker movement could help alter the perception that genetic engineering can only be done by academics in universities. 'I see it as a route to demystifying science for the general public,' he said.[592] Certainly, a key attraction of biohacking for many of its members is the possibility that it might help to make scientific know-how more freely available. So Josiah Zayner, of Burlingame in the San Francisco Bay area, states that: 'I want to democratize science'. Zayner, whose left arm is tattooed with the words 'Build Something Beautiful', has raised more than $46,000 through a crowdfunding campaign to make CRISPR/CAS9 kits available to the public for only $120.[593]

With his dyed hair, ear piercings, and a 'Go Ninja Go' T-shirt, Zayner takes his inspiration from the pioneers of personal computing, who, through groups like the Homebrew Computer Club, shared now-legendary ideas and experiments.[593] And the DIYbio movement is closely allied with projects such as BioBricks, which, as well as helping to create a standard set of 'parts' for synthetic biology, aims to make these freely available to anyone who wants to use them.[594] But while such aims can seem laudable, recently biohacking has come in for criticism because of the activities of some within its community.[595] At conferences and on social networks such as Facebook, some biohackers have shown a flair for the dramatic. So Zayner made the headlines when he injected himself with what he said was CRISPR/Cas9 tools designed to enhance his muscles, at a synthetic biology conference in 2018.[595] Another biohacker dosed himself with an untested gene therapy mix in an attempt to cure his herpes infection. Neither approach seems to have worked, and the potential long-term, adverse effects are unclear.

In response the state of California recently passed legislation intended to discourage DIY gene editing, and state regulators also said they were investigating one biohacker for practicing medicine without a license.[595] But there has also been criticism by some biohackers themselves. For instance, speaking at the annual 'Biohack the Planet' conference in 2019, Gabriel Licina, a chef who once developed his own night vision eyedrops, summed up the views of many when he said: 'I would like to propose that we grow up a little bit. Please, for the love of God, stop stabbing yourself.'[595] Licina himself claims to have developed a form of gene therapy for a rare blood disorder but rather than calling for self-experimentation, he has appealed to peers to help refine and test it. 'I would actually like people to start doing responsible work,' he says. 'That means peer review and outside testing, not just stunts.'[595]

Subverting the Genome

As ambitious as the plan to create artificial chromosomes may seem, not everyone is satisfied with the idea of limiting synthetic biology to creating organisms that utilize the existing genetic code. An even more ambitious goal is to create replicating systems of nucleic acids with a different genetic code than the one shared by all organisms on Earth. As we saw in Chapter 2, the normal genetic code is based on the four DNA letters—the bases A, C, G, and T—which pair in the double helix as A-T and G-C, and code for the 20 different amino acids that make up proteins.[596] RNA is essentially the same, except that the base U replaces the T found in DNA. Remarkably, despite the enormous variety of different species on our planet, every life form, from the tiny 'flu virus to a human being, all share the same genetic code. But this could soon be about to change, thanks to attempts to redesign life in a number of different ways. 'To make proteins with more than 20 amino acids—possibly as many as 30 or 40—requires thinking outside of the box of the standard genetic code,' says molecular biologist Patrick O'Donoghue from Western University, Ontario, Canada.[597] One such way of thinking outside the box being pioneered by teams led by George Church at Harvard University and Farren Isaacs at Yale University involves modifying the genetic code in such a way that it can be used to produce additional amino acids.

To achieve this feat, the researchers have manipulated one of the genetic code's elements, the so-called 'stop codon'. In Chapter 2 we mentioned the fact that which amino acid gets added next in a growing protein chain is specified by three-letter DNA sequences, called codons (see Figure 29). However, as well as codons that specify the different amino acids, the protein translation machinery also needs to know where a protein sequence starts and where it stops. For this purpose, the mRNA sequence that acts as

Second letter

First letter	U		C		A		G		Third letter
U	UUU	Phe	UCU		UAU	Tyr	UGU	Cys	U
	UUC	(F)	UCC	Ser	UAC	(Y)	UGC	(C)	C
	UUA	Leu	UCA	(S)	UAA	STOP	UGA	STOP	A
	UUG	(L)	UCG		UAG		UGG	Trp(W)	G
C	CUU		CCU		CAU	His	CGU		U
	CUC	Leu	CCC	Pro	CAC	(H)	CGC	Arg	C
	CUA	(L)	CCA	(P)	CAA	Gln	CGA	(R)	A
	CUG		CCG		CAG	(Q)	CGG		G
A	AUU		ACU		AAU	Asn	AGU	Ser	U
	AUC	Ile	ACC	Thr	AAC	(N)	AGC	(S)	C
	AUA	(I)	ACA	(T)	AAA	Lys	AGA	Arg	A
	AUG	Met(M)	ACG		AAG	(K)	AGG	(R)	G
G	GUU		GCU		GAU	Asp	GGU		U
	GUC	Val	GCC	Ala	GAC	(D)	GGC	Gly	C
	GUA	(V)	GCA	(A)	GAA	Glu	GGA	(G)	A
	GUG		GCG		GAG	(E)	GGG		G

Initiation codon

Chain terminator codon (STOP)

Amino acids shown as three-letter and single-letter abbreviations

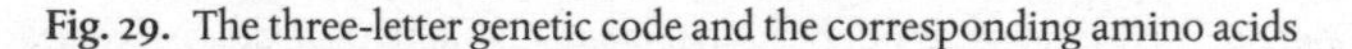

Fig. 29. The three-letter genetic code and the corresponding amino acids

the intermediary between a gene composed of DNA and its protein product also contains special triplet bases called start and stop codons.[596] And while there is only one start codon—AUG in the mRNA intermediary—there are three stop codons, UAG, UGA, and UAA. Known as amber, opal, and ochre, respectively, these stop codons can all act interchangeably of each other.

Church, Isaacs, and their colleagues removed one of these stop codons—the UAG or amber variety—from the whole of the *E. coli* genome and replaced it with the UAA or ochre one. This means that the new codon will no longer play the role of stop signal. Instead it can now be reinserted within a protein coding sequence, together with an altered form of one of the so-called 'transfer RNAs'. These RNA species add amino acids to the growing protein chain by matching each one to a specific codon. By engineering a transfer RNA so that it now adds a novel amino acid not found in nature, it will become possible for a modified bacterial cell to incorporate such a novel amino acid into a protein sequence. 'For the first time, we're showing that you can make genome-wide codon changes,' says Isaacs. 'We're going to be able to start introducing completely new functionality into organisms.'[598]

In fact, this is just the start of the possibilities for reengineering the genome in such a manner, because of natural 'redundancy' in the genetic code (see Figure 29). What this means is that often more than one triplet base codes for a particular amino acid.[599] For instance, the amino acid glutamine can be coded by either CAA or CAG. And such redundant codons could be commandeered in a similar fashion to the amber stop codon, and would thereby be free to code for novel amino acids. Consequently, a bacterial cell could be engineered so that only CAA was used to code for glutamine, freeing up CAG to code for a novel amino acid. Indeed, George Church and his colleagues are already starting to modify different genes in this fashion. Recently, they picked 13 codons and substituted them with alternatives coding for the same amino acid across 42 different *E. coli* genes.[599] Even though the genes' DNA sequence was different, the proteins the cells produced were unchanged. The next step is to endow these freed-up codons with new meaning.

While such studies tamper with the existing genetic code via its codon usage, an even more radical approach seeks to transform the DNA code itself. To do this, some scientists have sought to create a menagerie of exotic letters beyond A, T, C, and G, which can partner up and be copied in similar ways.[600] One such pair is referred to as X-Y, although its true chemical name is more complex. And this extra X-Y pair has now been used to create expanded versions of nucleic acids called XNAs, the X standing for 'xeno'. The development of such XNAs was pioneered by Steven Benner, a biological chemist at the Foundation for Applied Molecular Evolution in Gainesville, Florida. Benner first became interested in this issue as a graduate student in the 1970s. At that time chemists were starting to try to build molecules that could carry out the same functions as natural enzymes or antibodies with different chemical structures. But, according to Benner, DNA was largely ignored. 'Chemists were looking at every other class of molecule from a design perspective except the one at the centre of biology,' he says.[600] Eventually, Benner's interest led to the development of the first XNAs.

A major challenge has been developing a replicating system that can reproduce such XNAs.[600] While XNA, like DNA and RNA, can be synthesized chemically, this is still a relatively inefficient, error-prone, and costly process. In a living cell, DNA and RNA are replicated by enzymes called polymerases, from nucleotides—the units out of which RNA and DNA are made. Such polymerases do not normally synthesize nucleic acids from anything other than the A, C, G, and T varieties used to make DNA, or the U that replaces T in RNA. There is a good reason for this, which is that such enzymes have evolved to precisely recognize these specific nucleotides and no others, in order to maximize the accuracy of the nucleic acid copying process. However, through trial and error, Benner and his colleagues identified XNAs that could be replicated in a test tube by a DNA polymerase.[600]

While this work was done outside the cell, Floyd Romesberg and colleagues at Synthorx, a biotechnology company linked to the Scripps Research Institute in San Diego, recently showed that XNA containing the X-Y addition can be successfully replicated in bacteria over many generations.[601] So aside from curiosity, what is the point of making an organism with such an expanded DNA blueprint? According to Romesberg, there are many reasons: 'People would ask what the big deal is, and I said, "Imagine you had a language with only four letters." It would be clumsy and would really curtail the kinds of stories you could tell. So imagine two more letters. Now you could write more interesting stories.'[601]

In practical terms, such stories might have a number of interesting outcomes. One is that these novel life forms might be a source of important new proteins of biomedical importance. We saw in Chapter 2 how human insulin produced in a bacterium has been important for the treatment of diabetes. Bacteria have also been used to generate other medically important proteins like the red-blood-cell booster erythropoietin or human growth hormone for treating growth disorders. But the available proteins that can be produced this way are limited by the fact that a normal cell builds proteins from just 20 amino acids, which are assembled into long chains. As mentioned earlier in this chapter, the three-letter sequences of DNA called codons specify which amino acid gets added next in a growing

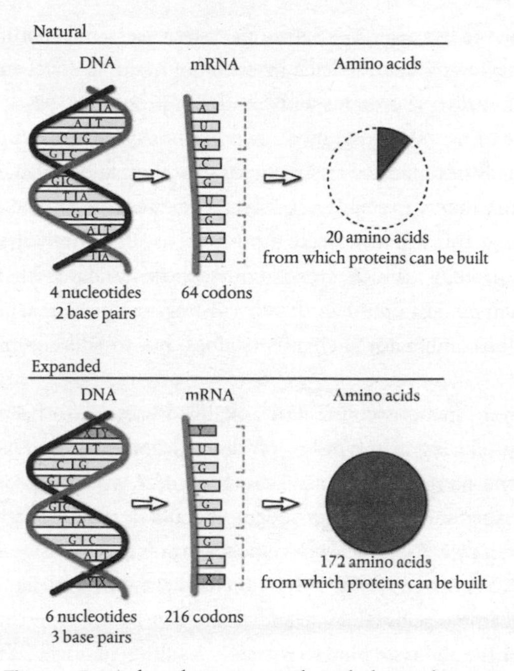

Fig. 30. Expanding the genetic code with the X-Y base pair

protein chain. However, with the extra X-Y base pair, in theory such an expanded genetic code could produce proteins with 172 different amino acids (see Figure 30).[601] This is important because scientists have invented thousands of artificial amino acids. But it's one thing to create such amino acids, quite another to use them to make proteins—for that, a living cell is required. 'When you get to whole proteins, chemists really lose the ability. Protein molecules are too complex, too big,' says Peter Schultz, a biologist at the Scripps Institute.[601]

The goal is to generate bacteria capable of making such proteins. 'To make a billion-dollar business, yes, we need a protein,' says Floyd Romesberg. 'The home run is the ability to produce therapeutic proteins with unnatural amino acids in them.'[601] And recently Romesberg and his colleagues claimed to have achieved this goal. 'The Synthorx system uses a fully natural cellular system and allows for site-specific insertion of multiple novel amino acids to create more diverse proteins with a range of properties,' says Synthorx's CEO, Court Turner. 'We have now shown that our synthetic base pair, X and Y, can not only be replicated *in vivo* but are also compatible with the natural biological machinery to enable novel protein expression.'[602] And, according to Romesberg, this will now make it possible 'to add entirely new types of groups to proteins (which are of course now validated therapeutics) which should not just optimize already existing activities (or arrangements of atoms like combinatorial chemistry does) but to add completely new functions'.[602]

Such novel functions could allow modified bacteria to become living foundries producing new types of products or materials combining both synthetic and natural amino acids, such as new types of protein-based drugs and other important bioproducts. 'It could be the foundation of new types of drug delivery vehicles or nanostructures as well as antimicrobial pesticides,' says Farren Isaacs.[603] The potential range of applications is vast. So artificial amino acids could be added that give proteins unusual properties, such as the ability to bind to metals—resulting in novel adhesives. Or enzymes could be developed that are activated only in the presence of other molecules—which could be useful for drugs. Recoding could also aid biomedical research: novel amino acids, such as those carrying a fluorescent tag, could be inserted and used to track cellular processes.[599]

The potential usefulness of this approach also goes beyond simply making an expanded repertoire of proteins. Scientists, like Philipp Holliger at the MRC Laboratory of Molecular Biology in Cambridge, are interested in how another form of XNAs—this time with a different chemical backbone to the one found in normal DNA—might be an important tool in biomedicine.[604] Importantly, Holliger and his colleagues recently have reported that some XNAs can form 3D structures and catalyse chemical reactions in the same way as protein enzymes. Such XNAzymes, as they've been named, are able to cut RNAs. As regulatory RNAs are increasingly known to play important roles in human health and disease, altering their properties using

XNAzymes might have important therapeutic potential. One advantage of XNAs, according to Holliger, is that they 'are chemically extremely robust and, because they do not occur in nature, they are not recognised by the body's natural degrading enzymes. This might make them an attractive candidate for long-lasting treatments that can disrupt disease-related RNAs.'[604]

As well as acting as catalysts, there's also a hope that XNAs might have an important role to play in the development of nanotechnology, which seeks to create microscopic devices and structures that could have numerous applications.[604] The creation of such devices from DNA makes use of the fact that this molecule can be coaxed into a variety of different shapes by harnessing the same forces that normally hold the double helix together—the attraction between the letters A and T, or G and C. By matching up the letters on a long strand of DNA with those on smaller strands of this molecule, Paul Rothemund of Caltech first showed, in 2006, that the larger strand could be stapled in place to create 3D shapes that Rothemund called 'DNA origami'.[605] This technology has already led to some useful practical applications, such as the creation of a DNA origami ruler, which can be used to measure the space between molecules in high-resolution microscopy.[605] But perhaps the most exciting potential use is for the creation of nanorobots that have flexible joints and sites that could be used to attach antibodies or cancer drugs which could be used to treat cancer or other diseases of the human body.[605]

A particularly important feature of such nanorobots is the presence on their surface of structures that recognize receptors on the surface of target cells, contact with which could lead to the drug or antibody within the nanorobot being delivered in the immediate vicinity of the cell.[605] And by constructing such nanorobots of XNA, it may be possible to ensure that they do not get degraded by the body's defence mechanisms. Such possibilities have led Patrick Maxwell, chair of the British Medical Research Council's Molecular and Cellular Medicine Board, to remark, about Hollinger's findings, that: '[t]his latest advance offers the tantalising prospect of using designer biological parts as a starting point for an entirely new class of therapies and diagnostic tools that are more effective and have a longer shelf-life'.[604]

New Artificial Life Forms

Bacteria with genomes that have been engineered to contain XNAs or which have been dramatically modified might have important uses as organisms in their own right, not just as sources of modified proteins or nucleic acids. One important application could be the creation of bacteria resistant to bacteriophages—the viruses that infect bacteria. By recoding the genes of bacteria used in the industrial production of enzymes, hormones, and food products, researchers could block infection by such viruses, saving tons of material that goes to waste from viral contamination.

The reason that changing the genetic code can lead to such resistance is that viruses replicate by using the host's own replication machinery and the raw molecules—the nucleotides and amino acids—that power this. However, in a bacterium whose genome has been reconfigured so that the amber stop codon no longer signals the end of a protein, viral genes containing this stop codon would no longer be expressed properly, preventing the virus's proteins being produced and thereby halting its spread. Showing the feasibility of this approach, researchers led by Farren Isaacs and George Church have reported that, by reassigning the triplet UAG to UAA, they could heighten the bacterium's ability to resist viral infection from the bacteriophage T7. 'By changing the code, we engineered viral resistance,' says Isaacs.[597] Of course, it is also possible that mutation of viral genomes and natural selection for novel forms that could reproduce in the modified host bacterium might eventually overcome such resistance.

Another potential use for such bacteria that contain the extra X-Y base pair is in the development of new vaccines. It might be possible, for instance, to make a tuberculosis bacterium with unnatural DNA in it. It would be a real, living organism, but without any raw material to copy its genes (that is, no external source of the bases X and Y) it could be injected into a person without having the potential to replicate itself and thereby make them sick. 'If it was TB, but also benign, that would be the perfect vaccine,' says Peter Schultz.[601]

The inability of organisms with a radically altered genetic code to reproduce in the absence of new types of amino acids is also a positive safety feature. For while such organisms may have all sorts of benefits, concerns have been raised about the possibility of them escaping the confines of the laboratory or industrial bioreactor and spreading across the biosphere. This could especially be a problem if such bacteria have a selective advantage over normal bacteria because they are resistant to bacteriophage. However, such a scenario is seen as very unlikely because the altered life forms would be incapable of survival in the wild due to the lack of the altered amino acids they need to make proteins essential for their survival. 'This adds another important safety barrier,' says Farren Isaacs.[603] And, although bacteria have an extraordinary ability to share genetic information in the wild[606]—one cause of the rapid spread of antibiotic resistance—Yizhi 'Patrick' Cai, a biotechnologist at the University of Edinburgh, believes that this is less likely for the modified bacteria. 'They cannot communicate with the wild type,' he says, 'because [the researchers] have engineered a species that speaks a different chemical vocabulary.'[603] However, whether such barriers would be so absolute if these modified bacteria were ever to escape into the wild is a question that would need to be considered very carefully.

While the creation of organisms with completely transformed genetic codes has so far been confined to bacteria, success in this area poses the question of whether it might be possible to apply this approach to more complex plants or animals. For now, the greater complexity of the genomes of these organisms means no one has yet attempted this ambitious goal. But Lei Wang and colleagues at the Salk Institute have introduced novel amino acids into proteins in the brain of a mouse, using a different approach.[599] Wang wanted to modify a protein channel that controls the flow of potassium ions into neurons. He reasoned that if the channel could be engineered to contain an amino acid that changes shape in response to light, the channel could be opened or closed in response to light.

Although similar to the optogenetic approach mentioned in Chapter 3, rather than the light-sensitive channel being a foreign gene from an alga inserted into the mouse genome, in this case the protein would be one of the mouse's own channels normally expressed in neurons. To achieve this goal, Wang and his team injected DNA coding for a modified transfer RNA—which transfers amino acids to the growing protein chain—into the brains of mouse embryos while they were still in the uterus.[599] The modified transfer RNA was designed to attach to an unnatural amino acid that was also injected into the brain. The embryos were then treated with electricity, temporarily rendering the cell membranes of the neurons permeable so they could take up the transfer RNA and amino acid. When the mice were born, some of their neurons contained proteins modified by the unnatural amino acid.

In future, genome editing might be used to modify the genetic code of complex multicellular organisms on a genome-wide scale. If so, one potential use of such an approach could be to create complex organisms resistant to infection by viruses, in the same way that bacteria have been made resistant to bacteriophage. This kind of engineered resistance to viruses in animal cells could be useful for existing biotechnology applications even if applied not to whole organisms but to cells in culture. Animal cells cultured in huge bioreactors play increasingly important roles in industry. For instance, Chinese hamster ovary (CHO) cells are used by biotech company Genzyme to produce the drugs imiglucerase, also known as Cerezyme® and agalsidase beta, known as Fabrazyme®, for treatment of the rare genetic disorders Gaucher disease and Fabry disease, respectively. Recently, the company lost more than $100 million in sales of these drugs after a viral infection impeded the cells' growth and they had to be completely replaced.[603] Thus, the production of modified CHO cells with an altered genetic code that prevents replication of viruses in such cells could have important implications for industry. As Torsten Waldminghaus, a geneticist at Philipps University of Marburg in Germany, puts it, the recoded cell can't play host to viruses because 'it basically speaks another language.'[607]

While the production of cells with modified genomes for cell culture may be an immediate goal, scientists are also talking about the possibility of creating whole animals or plants with radically altered genetic codes. The commercial potential of crops and livestock with a built-in resistance to infection by any type of virus could make such organisms very attractive to farmers. And although more ambitious than engineering a bacterium, genome editing makes such a goal more feasible. Certainly, George Church believes virus-resistant plants and animals could eventually be created. 'It's more of a challenge but it's not out of reach,' he says.[603] Indeed, Church believes that a similar approach might even be used to create virus-resistant humans.[603] If this were the case, then one day we might have the possibility of creating people with a natural built-in immunity to every virus, ranging from the common cold and flu, through to lethal viruses such as HIV or Ebola. Although such a scenario may be scientifically feasible, whether most people would see it as a welcome development remains to be seen. And on this note, it's now time to examine, in the final chapter, exactly how society should seek to deal with the amazing new technologies described so far, and what measures we should be discussing to maximize their potential for human benefit while minimizing the possibility of them being used to cause harm, either accidentally or deliberately, to our own species and all the other life forms with which we co-exist on planet Earth.

10

A Redesigned Planet?

In this book, we have encountered many mutant life forms, whether naturally occurring or produced by treatment with radiation or chemical mutagens. But this concluding chapter begins with a fictional mutant—Spiderman—the comic-book superhero. Spiderman was born when a geeky teenager, Peter Parker, was bitten by a radioactive spider, which turned him into a mutant superhuman gifted with 'the agility and proportionate strength of an arachnid'.[608] Spiderman's story aptly illustrates the fantastical qualities sometimes ascribed to mutants, as well as reflecting the fears and preoccupations of the Cold War era in which he arose. But a quote from this superhero seems particularly relevant to this book: that 'with great power comes great responsibility'.[609]

For there seems no doubt that the new technologies described within these pages—genome editing, optogenetics, stem-cell technology, synthetic biology—provide humanity with unprecedented powers to manipulate the natural world, but also pose crucial questions about how such powers will be used in a responsible fashion. And such is the importance of this issue that the discussion about how to employ such technologies surely cannot be left only to scientists but needs active public involvement. At the same time, the debate is only likely to progress if based on facts and real possibilities, not fears and misconceptions based on a misunderstanding of the science underlying these technologies. With this in mind, I'd like to imagine some possible future scenarios, as one way of assessing the potential, and risks, of these new ways to redesign life. Undoubtedly the most difficult aspect is trying to predict how current discoveries might impact on the future. For not only do science and technology evolve rapidly, but how society handles new scientific and technological breakthroughs is also shaped by the nature of that society.

Ultimately this is a guessing game, but our task is made slightly easier by some talented novelists having imagined possible ways in which new genetic technologies might lead to quite different future scenarios. I'd like to consider two particular visions—one a utopic vision of the potential of science to change society for the better, the other a dystopia in which genetic engineering has created a hell on Earth—and then assess how far such fictional visions match up to likely future developments in science.

Utopias and Dystopias

In the first vision, imagined by Kim Stanley Robinson in his *Mars* trilogy, the quest by future colonizers of the red planet for a fair and equal society requires a bitter revolutionary struggle with echoes of the American War of Independence.[610] Meanwhile, back on Earth, civilization has been rent apart by the disastrous effects of global warming. So it is rebel scientists on Mars who develop a form of genome editing that makes it possible to repair the effects of ageing and thereby grant human beings a greatly expanded lifespan.[611] The ability to live to an extreme age is mainly presented as a liberating new feature of human existence, but Robinson also explores more negative aspects, such as memory loss, mental instability, and existential boredom. Further on in the trilogy, manipulation of the human genome allows people to breathe the thin Martian atmosphere, which is itself transformed by various means, and as people begin to colonize other planets and moons in the solar system besides Mars, they are artificially adapted to low light levels in these outer reaches. While there is plenty of social strife in this vision of the future, as well as disagreements about the extent to which humans should seek to transform the environments of other planets to make them habitable, genetic engineering is mainly portrayed in a positive light.

This is definitely not the case in another futuristic trilogy, by Margaret Atwood. The first novel in this series, *Oryx and Crake*, is told in flashback by the narrator, Jimmy, who may be the last living human being after a deadly plague has devastated humanity.[612] We gradually realize this is no natural disaster, but a deliberate act of terrorism initiated by Jimmy's friend Glenn, nicknamed Crake. We learn that the civilization that has perished was a deeply divided place, dominated by biotechnology multinationals with names like RejoovenEssence, OrganInc, and HelthWyzer. While ostensibly working for RejoovenEssence, Crake's secret plan was to wipe out the human race with an engineered virus. At the same time, he had engineered a new race of people, the Crakers—peaceful vegetarians, with no violent or jealous urges, who are naturally resistant to sunburn or insect bites.[613] Jimmy is now the accidental survivor of the virus, trying to avoid being eaten by escaped pigoons—GM pigs with some disturbingly human qualities—and other genetic creations, while seeking to protect the naive Crakers. In fact, as the trilogy progresses, in *The Year of the Flood* and *MaddAddam*, a glimmer of hope for the future is raised.[613] The overall feeling, though, is of genetic engineering as a catastrophe out of control.

New Gene Therapy

How likely are either of these future scenarios? Beginning with more positive possibilities first, to what extent is genome editing really likely to initiate a revolution in medicine? By providing more sophisticated animal models of human disease, genome editing could help identify new molecular targets for disease treatment and thereby new drugs. Yet a key question, if we are ever to see medical advances of the type envisaged by Kim Stanley Robinson, is how feasible it will be to use genome editing for direct gene therapy in human beings. Genome editing is likely to have the biggest initial impact on genetic disorders of the blood, by making it possible to remove bone marrow, correct a genetic defect in the stem cells that generate the various blood cell types, and then replace the treated tissue. But how feasible will it be to treat genetic defects in other parts of the body?

A major challenge will be finding effective ways to get genome-editing tools into cells within tissues and organs *in situ* in the body. Viruses are particularly powerful delivery vehicles since they have evolved highly sophisticated ways of gaining access to the cell but they also carry risks, as we saw in Chapter 2, when we looked at the use of retroviruses to treat severe combined immune disease (SCID). This treatment was effective, but the insertion of the retrovirus's genetic material into the host cell genome led to the activation of an oncogene, and subsequent leukaemia, in some patients in initial clinical trials. Modified retroviruses have now been developed with much less chance of disrupting the host cell genome in dangerous ways.[614] In addition, there is now a move to develop other types of viruses as delivery vehicles—like the adenovirus, the agent responsible for the common cold.[615] Unlike retroviruses, this virus does not generally integrate its genetic material into the host genome. While seen as a negative characteristic for standard transgenic approaches, this is now viewed as an attractive safety feature, since the adenovirus can deliver genome-editing tools and exit the cell without causing collateral damage.

Another strategy would be to modify the genome-editing tools themselves so they can gain entry to a cell. For instance, the genome-editing enzyme CAS9 could be tagged with a 'cell-penetrating peptide'.[616] These are sequences of amino acids found in proteins that have evolved to naturally cross cell membranes, also known as 'Trojan horse' peptides because of their ability to subvert normal cellular boundaries, just as the Greeks entered Troy in clandestine fashion. The transactivator of transcription (TAT) peptide of HIV is particularly adept at crossing such membranes thanks to its particular chemical properties that allow it to penetrate this normally impervious barrier.[616]

Recently, Gerard Wong and colleagues at the University of California in Los Angeles showed that TAT interacts with the cell's 'cytoskeleton' and specific receptors on the surface of the cell to facilitate its passage across the cell membrane. 'Prior to this, people didn't really know how it all worked, but we found that the HIV TAT peptide is really kind of like a Swiss Army Knife molecule, in that it can interact very strongly with membranes, as well as with the cytoskeletons of cells,' says Wong.[617] By using such information to improve TAT's penetrative capacity, and attaching it to CAS9, the latter may be adapted so that it can penetrate cells in the body as part of a therapeutic strategy. It may also be possible to introduce CRISPR guide RNAs into the cell using this approach.

If delivery of genome-editing tools becomes straightforward, this could mean that single-gene disorders like cystic fibrosis and muscular dystrophy could finally be treated by directly introducing genome-editing tools into the lungs and muscles respectively, or ensuring that they reach their specific destination via the blood through molecular tags that target these organs. Disorders of the brain like Huntington's disease might also be treated in this way, although here an additional obstacle is the 'blood–brain barrier', which protects this vital and sensitive organ from infectious agents. But certain types of adenovirus can cross this barrier, and these may be commandeered to direct genome-editing tools to different brain regions.[618] In fact, we are already beginning to see some exciting progress on these fronts, albeit in the treatment of mouse models of muscular dystrophy and Huntington's, as we saw in Chapter 7. And, given that recent studies suggest there are thousands of single-gene disorders that, while rare in any particular individual, cumulatively affect millions of people,[619] such strategies may, in the future, have a huge impact in reducing pain and suffering in the human population.

How to Conquer Cancer

What about the prospects for treatment of more common diseases like cancer, diabetes, or mental disorders like schizophrenia? Here a complicating factor is identifying which genetic differences underlie such conditions and therefore which might be corrected by gene therapy. For, as we saw in Chapter 7, the more we learn about the genetics of such disorders, the more complex the situation appears. Take, for instance, cancer. We've already mentioned the breast cancer study which showed that, in 50 affected women, over 1,700 mutations were detected in their tumours, most being unique to the individual.[620] And the new field of cancer genomics—in which the whole genome sequence of an individual's tumour is determined and compared to that of a normal cell—indicates that this complexity is characteristic of other cancers too. Making sense of such complexity may seem a daunting prospect but encouraging progress is being made. David Schwartz and colleagues at the University of Wisconsin have even developed what they call a 'Google Map' of the cancer genome.[621]

Schwartz's team subjected cancer samples to fine-detail DNA sequencing, but also a new method called 'optical mapping', which creates an image of the genome as a whole. With this approach it's possible to zoom in for a 'street view' of individual changes in the genetic alphabet of cancer, or out for a more Google Earth-style perspective of whole genome changes. 'Cancer genomes are complicated but we found that, using an approach like this, you can begin to understand them at every level,' says Schwartz.[621] The new approach could make it possible to examine changes in a patient's cancer as it progresses, monitor for signs of drug resistance, and fine-tune treatments. One exciting possibility would be to combine such a diagnostic approach with the use of genome editing to correct the mutations that drive the cancer. If so, then soon treatment of cancer might be carried out in a far more personalized manner. And, as we saw in Chapter 7, the news that, in one patient, an aggressive form of childhood leukaemia appears to have been cured by genome editing T-cells indicates the potential for engineering the immune system so that it itself becomes the anti-cancer therapy.

While cancer is one common disorder of human health, it also has some very specific distinguishing features. Research shows that some people have greater tendencies towards certain types of cancer than others.[622] Such familial predispositions were first recognized in 1866, when the French clinician Paul Broca described a striking history of breast cancer in 15 members of his wife's family.[622] In 1914, German biologist Theodor Boveri suggested that an inherited predisposition to cancer could result from genetic defects that led to a 'weakened resistance against the action of factors that stimulate cell division'.[622] As we saw in Chapter 1, the BRCA mutations that strongly predispose certain women to breast or ovarian cancer do so because they cause a defect in a process used to repair mistakes in DNA. Another type of mutation, in the retinoblastoma (RB1) gene, leads to a predisposition to cancer of the retina.[622] Normally the RB1 gene plays an important role in preventing excessive cell division, in line with Boveri's original suggestion. Other genetic predispositions are less pronounced, but can lead to a greater chance of succumbing to cancers of the bowel, skin, liver, prostate, and lung, to name just some of the tissues in which such predisposition occurs.

Mutations that occur in different cell types during a person's lifetime also increase the chance of that individual succumbing to cancer. Indeed, Boveri recognized this when he said that accumulation of 'particular, incorrect chromosome combinations' underlies cancer.[622] Subsequent research has shown that such changes can be accelerated by environmental insults. We now recognize a strong link between skin cancers like melanoma and excessive exposure, particularly of fair-skinned people, to UV radiation from the Sun or in a tanning salon. And a major breakthrough in our understanding of susceptibility to lung cancer was made by Richard Doll of the British Medical Research Council, who in 1954 showed that smoking is a key risk factor for this type of cancer.[623]

Ian Hall at the University of Nottingham and Martin Tobin at the University of Leicester recently found that the genetic profile of individuals who smoke has an important influence on whether they succumb to lung cancer.[624] People with differences in genes associated with susceptibility to chronic obstructive pulmonary disease (COPD)—a collection of lung disorders including bronchitis and emphysema—were more likely to get lung cancer. These genes play a role in the way lungs grow and respond to injury. However, while providing one explanation for anecdotal tales about individuals who smoked heavily and still lived to a ripe old age, such findings should not be seen as a green light for some to strike up a 40-a-day habit. 'Smoking is the biggest lifestyle risk factor for COPD,' says Tobin. 'Genetics play a big part, as they do in smoking behaviour. Our research helps to tell us why, paving the way for improved prevention and treatment. Stopping smoking is the best way to prevent smoking-related diseases such as COPD, cancers and heart disease.'[624] So while genome editing may eventually be used routinely to treat cancer, preventative measures will continue to be important where there is a clear link with environmental risk factors.

Disorders of the Mind

The environment plays an important role in other common disorders. Obesity is a major risk factor for type 2 diabetes, heart disease, and stroke, with the current rise in obesity leading to predictions of a future 'epidemic' of these conditions.[625] Mental disorders are also affected by environmental factors. Tirril Harris and George Brown of Bedford College in London showed in 1978 that working-class women were far more likely to succumb to depression than more affluent women, which they linked to the impact of 'more severe life events' and greater difficulties with finance and housing.[626] And a recent study showed that black people of Caribbean origin in Britain are nine times as likely to be diagnosed as schizophrenic as white Britons.[627] Arguing against this difference being genetic, reported schizophrenia levels amongst black people in the Caribbean is the same as among whites in Britain. The study concluded that racism was probably a key factor, both in terms of being diagnosed schizophrenic and as a trigger of the condition, but also that other factors specific to particular immigrant communities, such as differences in family structure, may explain why British Afro-Caribbeans are so vulnerable in this respect.[627]

Whether someone succumbs to a disorder of the body, like diabetes, or of the mind, like schizophrenia, is likely to be a complex equation involving both genome and environment. And, the genetic side of this equation is looking far more complicated than some had supposed. For, as we saw in Chapter 7, recent studies show that a mental disorder like schizophrenia is associated with over a hundred different genomic regions, with the majority of these being not in protein-coding genes but in regulatory regions that often have subtle influences on gene expression.[628] In trying to make sense of this complexity, one thing to consider is whether it was a misconception to assume that schizophrenia is a single condition, as opposed to different ones lumped into one diagnosis. This would fit with the wide variety of symptoms used to classify this disorder, which include 'delusions, hallucinations, loosening of associations, disorganized speech and behaviour, illogical thinking, social isolation and cognitive deficit', of which one diagnosed schizophrenic may display one set of such symptoms while another has a completely different set.[629]

Such complexity also seems true for depression. This disorder affects 350 million people worldwide, and as many as two-thirds of people who commit suicide have the condition.[630] The symptoms and severity of depression also vary widely from one person to the next, and between men and women. Recently a genetic link to depression was identified by Jonathan Flint and colleagues at Oxford University, but only by focusing on people with the most serious form of the disorder. This type of depression, called melancholia, robs people of the ability to feel joy. According to Douglas Levinson, a psychiatrist at Stanford University, if you have this severe form of depression 'you can be a doting grandparent and your favourite grandchildren can show up at your door, and you can't feel anything'.[630]

Flint's study found links between melancholia and two genes.[631] One is called sirtuin 1 (SIRT1) and plays an important role in mitochondria, the structures in the cell that produce most of its energy. 'It's an appealing bit of biology for a disorder that makes people tired and unmotivated', says Levinson.[632] The other gene, phospholysine phosphohistidine pyrophosphate phosphatase (LHPP), regulates the action of the thyroid gland, which might also make sense in light of the listlessness associated with severe depression. For now, the significance of these genetic links remains to be established, but what has caught the interest of psychiatrists is how the study focused on people with a very specific form of the disorder.[632] Such an approach might help reveal clearer links to the genetics of other mental disorders. It may also mean that psychiatrists need to reconsider whether terms like 'depression', 'schizophrenia', and 'bipolar disorder' obscure a situation in which there are many partially overlapping conditions, each with a specific molecular basis.

In terms of a gene-therapy approach to treating common disorders, it may be that some disorders may be more treatable than others precisely because they have a clearer genetic basis. In many ways this is similar to the situation with breast and ovarian cancer, in which admittedly drastic measures—a double mastectomy and removal of the ovaries—can greatly reduce the risk of these cancers once a person is shown to have a defect in the BRCA genes.[633] In the future it may be possible to correct the BRCA gene defect using genome editing and thereby avoid such a need for surgical intervention. But it's important to recognize that defects in BRCA1 and BRCA2 only account for about 5 per cent of breast cancers and 10 to 15 per cent of ovarian cancers. Inherited genetic links to the majority of cases of breast and ovarian cancer remain much less clear.[634] And even with discoveries such as the new genetic links to severe depression mentioned, it may be that other forms of this condition may turn out to be far less easy to define in genetic terms. As such, predictions about the possibility of genome-editing treatments as a general strategy for treating common disorders may be rather premature, at least until the genetic links themselves become clearer.

Yet a lack of clarity about the genetic basis of a mental disorder like depression might not necessarily stand in the way of using genetics to treat such a disorder in the future. For, as we saw in Chapter 3, scientists have used optogenetic light activation of neurons associated with feel-good activities to reverse the depressive state in a mouse model of this disorder.[635] Could such an approach ever be used in humans? To do so, it would be necessary to both genetically engineer a person's neurons so they responded to light and find a way to shine light upon these neurons. In the future, such engineering of the human brain may be possible using a virus or cell-permeable editing tools. As for stimulating neurons, we saw in Chapter 3 how scientists are experimenting with long-range methods of stimulation, either by a light device on the surface of the skull or a magnetic field. So one day we might see a situation where optogenetics in humans is used to treat Parkinson's disease and epilepsy, but also depression. And it's also possible that a combination of optogenetics and genome editing might be used to manipulate gene expression in the brain as a form of therapy.

Showing the pace of research in this area, Karl Deisseroth has formed a company to pursue optogenetics trials in human patients.[636] The company, named Circuit Therapeutics, plans to initially focus on the treatment of chronic pain. Neurons affected by chronic pain are located in and outside the spinal cord, making them a more accessible target than the brain. 'In animal models it works incredibly well,' says Scott Delp, a neuroscientist at Stanford, who works closely with Deisseroth.[636] Meanwhile, another company, RetroSense Therapeutics, which is based in Michigan, is soon to begin human trials of optogenetics to treat a genetic condition that causes blindness, which will involve stimulating neurons in the retina to bypass the defect.[636] In both these cases, the accessibility of the spinal cord and eye make these logical starting points for therapy, but the fact that Circuit Therapeutics is also planning to develop treatments for Parkinson's and other neurological disorders of the brain suggests that clinical trials in these areas may not be far behind.[636]

Would such technological solutions to psychiatric disorders risk detracting from other ways of tackling such disorders that focus on the social causes of mental illness? This is particularly important given a study that showed a strong link between economic crisis and stress, anxiety, and depression. The study, by researchers from Roehampton University in London and the children's charity Elizabeth Finn Care, found that the incidence of depression jumped between four- and fivefold during the economic crisis of 2009.[637] Commenting on the findings Steve Field, chairman of the Royal College of General Practitioners, says: 'GPs across the country have been seeing a definite increase in the last year in the number of patients coming to see them with mental health and physical issues. These appeared to be related to either losing their job or fearing their job and livelihood are threatened.'[637]

So even if optogenetics could one day be used to treat depressive people by triggering 'happy' memories, one concern is that, without also tackling the social factors that trigger depression, we might end up with a situation similar to Aldous Huxley's *Brave New World*. In Huxley's novel the government provides its populace with the drug soma, which subdues all 'malice and bad tempers', thus avoiding any need to discover and potentially tackle the true source of a person's sorrow.[638] And, while non-addictive and without negative side effects, soma is used by the government to control people and stifle dissent. The fact that some psychiatrists believe that, even now, antidepressants are often given out not to combat serious depression but merely to 'get rid of unhappiness' shows the dangers of only pursuing a technological solution to mental disorders.[638] The development of new scientific ways to treat depression should not detract from the fact that tackling social issues like unemployment and lack of job security that enhance this disorder should also be a priority, as should the proper provision of counselling and other ways to treat depression.

New Organs for Old

A different strategy for tackling human disease would involve the replacement of diseased, damaged, or aged tissues with replacement organs. As we saw in Chapter 5, one source of such replacements might be pig organs modified by genome editing so they are not rejected by the human host.[639] A further alternative, though, would be to use human tissues and organs developed using stem-cell technology. And the remarkable self-organizing ability of stem cells in 3D cultures mentioned in Chapter 8 means this is not such a far-fetched idea as it would once have seemed.[640] Clearly, much more needs to be done if the organoids created so far can be developed into real replacement organs. None the less, particularly with the precision that can now be achieved in the manipulation of gene expression in living cells via genome editing, it's not implausible to imagine a future in which a person's own cells are reprogrammed into pluripotent stem cells and used to generate replacement organs.[640]

It is one thing to imagine replacing a heart, pancreas, or liver, but the brain of someone can't simply be replaced. For, as Tufts University philosopher Daniel Dennett once put it, a brain transplant would be the one type of transplant surgery where it would be better 'to be the donor not the recipient'.[641] However, in a number of different areas of brain study there have been important recent developments that could aid strategies to repair or regenerate human brains while retaining their basic integrity. One is the recognition that new neurons are produced naturally in some regions of the adult brain.[642] According to neuroscientists Maya Opendak and Elizabeth Gould of Princeton University, such 'neurogenesis' may help animals, including humans, adapt to their current environment and circumstances in a complex and changing world. 'New neurons may serve as a means to fine-tune the hippocampus to the predicted environment,' says Opendak. 'In particular, seeking out rewarding experiences or avoiding stressful experiences may help each individual optimize his or her own brain.'[643] Importantly, stressful experiences like restraint, predator smells, and sleep deprivation decrease the number of new neurons in the mouse hippocampus. In contrast, rewarding experiences like physical exercise and mating increase the production of new neurons in this brain region.[643]

The importance of neurogenesis for normal brain function is indicated by studies demonstrating that Alzheimer's disease—the commonest cause of dementia, the symptoms of which include memory loss and difficulties with problem solving or language—can involve a disruption of this process. Alzheimer's, which affects over half a million people in Britain, and 5 million people in the US,[644] is associated with the presence in the brain of 'plaques' of a protein called amyloid-β.[645] Cell signals regulated by the wingless type MMTV integration site (WNT) proteins, which play important roles in neurogenesis, have been shown to be disrupted by such plaques. And indicating possibilities for therapy, the introduction of neural stem cells into the hippocampus of a mouse model of Alzheimer's by Phil Hyu Lee at the University College of Medicine in Seoul alleviated some symptoms of the disorder.[646] This demonstrates one way in which stem-cell therapy might be employed in humans. However, there are many issues to consider before this approach becomes a therapeutic reality. So transplants must not only have the desired physiological effect but also shouldn't trigger brain tumours. This is not a trivial consideration given that stem cells share many properties with cancer cells, like an ability to divide indefinitely; indeed, one theory of tumour formation sees it as driven by stem cells.[647]

Artificial Sex Cells

A characteristic feature of pluripotent stem cells—whether ES cells or iPS cells—is their potential ability to give rise to any cell type in the body. But identifying the precise culture conditions that will allow an ES or iPS cell to generate a specific type of cell is far from trivial. As we saw in Chapter 8, it took many years to identify the conditions required to generate pancreatic beta cells from ES or iPS cells for the treatment of type 1 diabetes. Nevertheless, we've also seen how remarkable recent progress has been in the generation not just of specific cell types but even 3D structures with many similarities to particular human organs. Sometimes the results of stem-cell transformations can be quite startling, like the use of human iPS cells to create a beating heart. The study, led by Lei Yang of the University of Pittsburgh, took human skin cells, transformed them into iPS cells, and then used these to create heart precursor cells.[648] These cells were then implanted into a mouse heart 'scaffold', a network of non-living tissue composed of proteins and carbohydrates to which cells can attach and grow. The precursor cells grew and developed into heart muscle on the scaffold, and eventually 'began contracting again at the rate of 40 to 50 beats per minute', according to the researchers.[648] 'It is still far from making a whole human heart,' says Yang. 'However, we provide a novel resource of cells . . . for future heart tissue engineering.'[648]

Recent studies show that pluripotent stem cells can be also used to generate artificial eggs and sperm. In 2013, Mitinori Saitou and colleagues at Kyoto University showed that cultured mice iPS or ES cells could be induced to give rise to so-called primordial germ cells (PGCs).[649] These specialized cells are normally formed during embryo development and later give rise to either sperm or eggs. And although the artificial PGCs couldn't develop beyond this stage in the culture dish, Saito and his team showed that, if implanted in mouse testes or ovaries, they could mature into sperm or eggs, respectively.[649] Both the artificial eggs and sperm were capable of fertilization and producing offspring.

A team led by Azim Surani of Cambridge University and Jacob Hanna of the Weizmann Institute of Science in Israel has now reproduced the 'first half' of Saito's study with human cells.[650] A major hurdle that confounded the first attempts to produce human artificial PGCs was the fact that while mouse ES cells are 'naive'—easy to coax into any differentiation path—the human versions are much less adaptable. But by chemically tweaking the human stem cells, Hanna made them naive like the mouse version. 'The first time we used those cells with the Saitou protocol—boom! We got PGCs with high efficiency,' said Hanna.[650] And working with Surani, an expert on PGC biology, he was able to use ES cells and iPS cells, from both males and females, to make human PGCs with high efficiency. 'It's remarkably fast. We can now take any embryonic stem cell line and once we have them in the proper conditions, we can make these primordial cells in five to six days,' said Surani.[650]

By studying this process, the researchers now hope to gain important insights into the molecular mechanisms that regulate sperm and egg formation in the testicles and ovaries and how defects in these might cause infertility. More generally, studying this phenomenon could further understanding and potential treatment of certain age-related disorders. As people age, they accumulate not only mutations in their DNA but also chemical changes that alter gene expression. Such 'epigenetic' changes can be caused by exposure to chemicals in the environment, diet, and even stress, and have been linked to disease and ageing. The DNA in embryonic PGC cells is wiped clean of these epigenetic changes. So studying this process in culture could, according to Surani, 'tell us how to erase these epigenetic mutations'.[650] The process by which pluripotent stem cells become PGCs in culture could also be used to screen drugs used in cancer chemotherapy that cause infertility as a side effect, in order to identify ones that are less damaging to sperm or egg formation.

Such are the potential benefits of studying the transformation of stem cells into sperm or egg precursors in culture. But they also raise a more controversial issue, which is whether this route might be used to create sex cells for making human babies (see Figure 31). This could be great news for

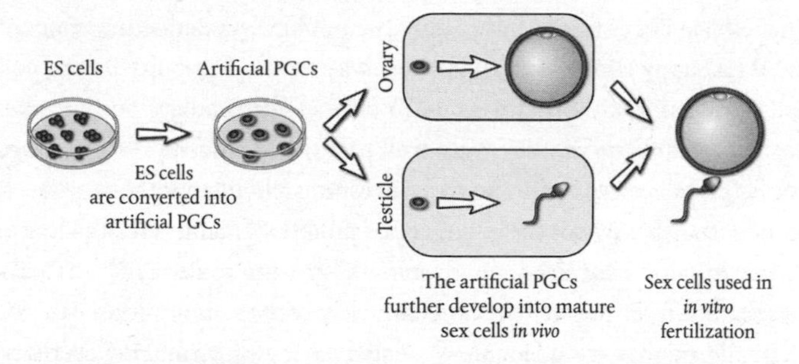
ES cells
Artificial PGCs
ES cells
are converted into
artificial PGCs
Ovary
Testicle
The artificial PGCs
further develop into mature
sex cells *in vivo*
Sex cells used in
in vitro
fertilization

Fig. 31. Use of artificial primordial germ cells to make eggs and sperm

infertile individuals who fail to produce sperm or eggs in the normal manner, for instance due to an early menopause, accident or injury, or exposure to chemicals like the ones used for cancer chemotherapy. And indeed, although the study by Saitou's team was only in mice, soon after reporting their findings the lab began getting emails from infertile couples desperate to have a baby.[649]

There are still many technical challenges before this approach can even be considered for clinical application. In their mice studies, Saitou's team found that their artificial PGCs only produced offspring at a third of the rate for normal IVF. In addition, Yi Zhang, who studies epigenetic mechanisms at Harvard University, found that PGCs produced using Saitou's method do not erase their epigenetic programming as well as naturally occurring ones. 'We have to be aware that these are PGC-like cells and not PGCs,' he says.[649] The difference raises questions about potential health risks for babies created using this approach, because such epigenetic differences might lead to adverse effects in later life. In addition, both iPS and ES cells frequently collect chromosomal abnormalities, genetic mutations, and epigenetic irregularities during culture. According to Harry Moore, a stem-cell biologist at the University of Sheffield, 'there could be potentially far-reaching, multi-generational consequences if something went wrong in a subtle way'.[649]

Let's assume, though, that a way were found to ensure that eggs and sperm produced in this manner were safe to use as a treatment for infertility.

What ethical issues might this raise? One might assume that many people would be happy about such an approach being used to help infertile couples. But what if it allowed women to have children at any point in their lives, for instance when they were well past the normal reproductive age? Would this be seen as liberating or an irresponsible extension of a woman's reproductive age? And if the latter, is that double standards given that actor Anthony Quinn conceived a child naturally at the age of 81?[651] Let's also consider a further possibility. Currently, gay and lesbian couples can only have children that are biologically related to one of them. However, conceivably, with this approach, an egg could be produced from a skin cell of one gay man and fertilized with the sperm of his partner, then implanted into a surrogate mother.[652] And a skin cell from a lesbian might be used to make sperm to inseminate her female partner.

In fact, it may be difficult or even impossible to generate eggs from male cells containing an X and Y chromosome, or sperm from female cells with two X chromosomes, because of the role of the Y chromosome in generating the males of our species.[652] It's possible, though, that genome editing might be used to genetically engineer PGCs to get around such problems. And if this were the case, while some people might welcome same-sex couples being able to reproduce together, would others see this as a step too far away from the 'natural' order of things? Finally, let's suppose that an individual decided they'd like to have both sperm and eggs created from their skin cells. Imagine then that a woman had sperm generated in this way and used them to inseminate herself, or a man pursued this goal using a surrogate mother. This would allow such a person to have a child with themselves.[652] Now while many people might be happy with a same-sex couple procreating using this approach, I imagine far more would have a problem with this final scenario, not only because of the many ethical issues raised by what's been termed 'the ultimate incest',[652] but also since such an extreme form of inbreeding would be very inadvisable from a health point of view. Yet it's important to consider this final scenario, as well as the others, if only to illustrate some potential dilemmas that advances in stem-cell technology are raising for the future.

The New Eugenics?

The use of stem-cell technology to generate artificial eggs and sperm raises exciting future prospects for the treatment of infertility, if also some highly controversial issues, but it seems likely that it will be some time before anyone considers this a safe approach for the treatment of this disorder.

The question of whether genome editing of human embryos might be used for clinical purposes requires more immediate consideration. As we saw in Chapter 4, the news that scientists in China had used the CRISPR/CAS9 approach to correct a gene defect in a human embryo—albeit one that could never have developed into a person—first showed how rapidly research is progressing in this direction. And the news also generated a storm of controversy.[653]

However, the argument by some researchers that there should be a ban on such research has been far from universally taken up by scientists. For instance, Debra Mathews, of the Berman Institute of Bioethics at Johns Hopkins University, has pointed out that: 'while there is controversy and deep moral disagreement about human germline genetic modification, what is needed is not to stop all discussion, debate and research'.[654]

Such a debate will involve discussing how genome editing might be used in research into the mechanisms underlying human embryo development. 'Much of our knowledge of early development comes from studies of mouse embryos, yet it is becoming clear that gene activity and even some cell types are very different in human embryos,' says Robin Lovell-Badge of the Francis Crick Institute in London. 'Genome editing techniques could be used to ask how cell types are specified in the early embryo and the nature and importance of the genes involved.'[654]

In fact, as we have seen, Kathy Niakan and her team at the Crick Institute, are using genome editing precisely to address such questions, in work that has been approved by the UK Human Fertilisation and Embryology Authority (HFEA).[655] Yet other scientists fear such research may prepare the ground for future use of human embryo genome editing for therapeutic purposes. And the news in 2018 that gene-edited babies had been born in China following the illegal use of CRISPR/Cas9 by Jiankui He to target the CCR5 gene in human embryos, showed that such fears were not groundless. Far from being applauded for this, Jiankui has become a pariah in the scientific community, and indeed has now been jailed for three years for his actions.[656] But one could argue that he has potentially opened the door to future attempts at germline gene editing.

It has to be said that the arguments for wanting to modify human embryos for clinical purposes have not been entirely convincing, since it's already possible to analyse human embryos that may have a genetic disorder and distinguish ones that lack the genetic defect underlying the disorder.[657]

This involves taking a single cell from an IVF embryo when it's only a ball of cells and carrying out DNA analysis on the cell. If an embryo is identified as not possessing the defect, then it can be implanted into the mother. This approach is now used to select embryos without defects in the CFTR gene which causes cystic fibrosis; the huntingtin gene that leads to Huntington's disease, an early-onset dementia;[657] and the BRCA genes, associated with susceptibility to breast and ovarian cancer.[658] And despite the method destroying the specific cell that is analysed, the embryo can compensate for this loss so that resulting human individuals appear normal. For this reason, Edward Lanphier, who, as we saw in Chapter 4, is opposed to human embryo genome editing, believes that there are no convincing arguments to justify the technology as a form of germline therapy. 'You can do it,' he said. 'But there really isn't a medical reason.'[659] Yet advocates of genome editing in human embryos for therapeutic purposes point to a growing number of disorders associated with defects in numerous genes.[659] It would be very difficult to select embryos with non-defective versions of multiple genes but feasible to correct them with genome editing. But the genome-editing method carries its own problems, because the more genes that need to be targeted, the greater the chance of incomplete targeting and 'off-target' effects that could cause unwanted changes in other parts of the genome.

Even if germline genome editing in humans was made legal, this could surely be only done on the basis that there was absolute convincing evidence that this wouldn't result in incomplete targeting or other adverse effects.

And if screening of IVF embryos and selection of those without a specific genetic defect nearly always provides an alternative route, this also would be an argument about germline gene editing for clinical purposes.[660] However, as we discussed in Chapter 7, one possible area where a case might be made for germline genome editing is correction of a gene defect associated with male infertility. These days, many types of infertility where sperm are produced but fail to swim or can't bind and fuse with the egg can be treated using a technique called intracytoplasmic sperm injection (ICSI), which involves injecting the sperm into the egg through a glass needle.[660] This technique makes conception possible for men whose sperm would fail to fertilize an egg in standard IVF. But there are other types of infertility in which the testicle is incapable of making sperm at all.[661]

Given that genome analysis is increasingly identifying the genetic defects responsible for such a failure, perhaps genome editing could also be used to correct it? It would be possible to monitor the efficiency of such an approach and confirm the lack of off-target effects by analysing a sperm sample following treatment, before any decision was made to use sperm formed after correction of the defect for IVF treatment. This kind of correction of an infertility defect would be germline genome editing. But given the pressure from infertile couples desperate for their own biological children, it may perhaps be seen as an acceptable use of the technology, if shown to be safe.

Roots of Intelligence

Concern about safety is only one of the issues that trouble people opposed to germline genome editing. So, according to Edward Lanphier: 'People say, well, we don't want children born with this, or born with that—but it's a completely false argument and a slippery slope toward much more unacceptable uses.'[659] A major worry is that using such an approach for therapeutic purposes would lead ultimately to 'designer babies'—human individuals engineered at birth to have beautiful looks, high intelligence, or exceptional ability at sport or music. Such fears run deep among scientists. So Eric Lander recently warned that: 'It has been only about a decade since we first read the human genome. We should exercise great caution before we begin to rewrite it.'[662] And at an international conference that took place in Washington in November 2015 to discuss the science and ethics of genome editing, over 150 biologists submitted a statement calling for a worldwide ban on the genetic editing of embryos, claiming the practice would 'irrevocably alter the human species'.[663] One particular concern was that such technology would only be available to rich people, leading to a world where inequality and discrimination were 'inscribed onto the human genome'.[663]

But before we get too carried away with the idea that genome editing could be used to create such enhanced individuals, it's worth considering the complexity of the interaction between environment and genetics, and between genes themselves, that combines to shape a human being. For, as we saw in Chapter 7, if there's one central message emerging from studies of the link between the human genome and susceptibility to common disorders, it's that this link is far more complex than had been supposed. And this is also turning out to be the case for many other human characteristics.

Take, for instance, intelligence. Comparisons of IQ scores between identical versus non-identical twins have supported the idea that intelligence has a strong genetic component. Yet such studies have a chequered history; Cyril Burt of University College London, who produced the largest body of findings of this sort in the mid-20th century, was identified after his death in 1971 as a fraud who invented data and even a research assistant.[664] More generally, there have been problems with some studies that investigated identical twins separated at birth. These cases are sought after, since identical twins brought up together experience the same environment and may be treated much more similarly by relatives, friends, and acquaintances than non-identical ones. For obvious reasons though, it's difficult to find identical twins separated at birth, and such twins are often self-selecting in making themselves known to researchers.[665] Moreover, few studies of twins, whether reared apart or reared together, have included twins from extremely different backgrounds. And some identical twins said to have been separated at birth have often had some contact; for instance, one set were separated but lived in the same village.[666]

None the less, positive findings from such studies have encouraged efforts to identify the genomic regions associated with IQ and other measures of intelligence. Unfortunately, a recent article in *Behavior Genetics* summarizing such studies concluded that 'it now seems likely that many of the published findings of the last decade are wrong or misleading and have not contributed to real advances in knowledge'.[667] Some critics put this lack of consistency down to 'wishful thinking and shoddy statistics'.[667] Yet the biggest study to date, which investigated over 100,000 people and was meant to introduce a rigour lacking in past studies, also generated inconclusive findings. The study, led by Daniel Benjamin of Cornell University in Ithaca, New York, found three gene variants associated with both educational attainment and higher IQ scores. However, the effect of these variants was, according to an article in *Nature*, 'maddeningly small', with each variant associated with 'roughly one additional month of schooling in people who had it compared with people who did not'.[667]

Talent Born or Made?

Such negative findings do not bode well for anyone seeking to use genome editing to produce the next Einstein. As for creating an artistic genius like Mozart, a study led by Kári Stefánsson, central executive officer at deCODE, a biotech company in Reykjavik, Iceland, points to some potential problems. The study found that genetic factors associated with an increased risk of bipolar disorder and schizophrenia are found more often in writers, painters, and musicians. 'I think these results support the old concept of the mad genius,' says Stefánsson. 'Creativity is a quality that has given us Mozart, Bach, van Gogh. It's a quality that is very important for our society. But it comes at a risk to the individual, and 1% of the population pays the price for it.'[668] In fact, David Cutler, a geneticist at Emory University in Atlanta, believes the study's identification of the genetic factors that raise the risk of mental problems explain only about 0.25 per cent of the variation in artistic ability among individuals. 'If the distance between me, the least artistic person you are going to meet, and an actual artist is one mile, these variants appear to collectively explain 13 feet of the distance,' he says.[669] However, the findings do suggest that tampering with the human genome in order to create a great artist might have an opposite effect to that predicted.

Mozart's ability to compose sonatas, symphonies, and operas before he hit puberty, and masterpieces for the rest of his short life, suggests a high degree of innate talent. However, we shouldn't neglect the important role of environment in creating one of the greatest musical geniuses of all time. So not only was Mozart intensively coached from a young age by his father—a composer, violinist, and author of a popular textbook on violin playing—but his later work was highly influenced by the Enlightenment.[670] This 18th-century movement promoted science and art as a means of individual expression and a challenge to autocratic rule. Such ideas may have inspired Mozart to turn his back on his patron, the Archbishop of Salzburg, and pursue a career as a freelance musician, something almost unheard of at the time. It was a precarious existence, and one factor in Mozart's tragically early death at the age of 35, yet it allowed him to live independently and create his masterpieces.[671] In fact, the Enlightenment's influence upon Mozart went far further than allowing him to practise his art, for its themes of individualism also influenced his music. For instance, Mozart's experience as a member of the Freemasons—at that time a society of radical thinkers—guided his opera *The Magic Flute*, a thinly veiled celebration of Enlightenment ideals.[671]

Of course, only focusing on such social influences is as mistaken as assuming that biology alone can explain Mozart's creativity, as a leading geneticist once did when he told me that we would 'soon know the gene that made Mozart a musical genius'. In fact, it's likely that both innate talent and a very specific set of circumstances helped create such a unique body of music. So attempts to use genome editing to produce the next Mozart are far from likely to succeed. As it is, the possibility of exhuming this particular genius and analysing his genome is impossible, given that, for all his talent, Mozart's precarious financial status at his death means that his remains lie unmarked somewhere in Vienna's St Marx Cemetery.[672]

Nature and Nurture

How about manipulating the human genome to create a great sportsperson? Perhaps the most basic sporting ability is running, and a gene implicated in this ability is actinin alpha 3 (ACTN3).[673] This gene is expressed in the 'fast-twitch' muscle fibres associated with rapid and powerful contractions. ACTN3 varies at position 577 of its protein sequence, and the variants are referred to either as R to denote the amino acid arginine occurring at this position, or X to denote a so-called 'stop codon' which brings the protein sequence to an abrupt, premature halt. Some studies indicate that great sprinters generally have two copies of the R variant, while marathon runners are more likely to have two copies of the shorter X variant. Yet the findings vary depending on whether Africans or Europeans are being studied, and even studies indicating a positive association of the respective variants with sprinting or endurance running ability show that ACTN3 makes only a modest contribution to elite athleticism.[673] Other genes associated with athletic ability include peroxisome proliferator-activated receptor delta (PPARδ), which regulates muscle growth; insulin-like growth factor 1 (IGF-1), which repairs and builds up muscles; and genes that regulate erythropoietin, a hormone controlling the production of red blood cells, which would act to increase blood oxygen levels.[674] So creating a super-sprinter like Usain Bolt is unlikely to be as simple as tweaking a single gene.

The role of genetics may be even more complicated for team sports. And overly focusing on genetics neglects the important role played by environment in the creation of a great sportsperson. Two examples from football show how intertwined the roles of nature and nurture can be. Cristiano Ronaldo is often touted as the greatest goal scorer on the planet. And the recent discovery that he has an extra bone in his ankle has led to claims that this is one factor that allows him to put extra spin on the ball and deceive goalkeepers.[675] Yet to focus only on physical attributes underestimates another crucial factor in Ronaldo's success, namely the 'blood, sweat and tears' shed in pursuit of his goal. So Mike Clegg—Manchester United's power development coach between 2000 and 2011—remembers that, as an 18-year-old newcomer to the club, 'Ronaldo was a natural talent,' but also that 'he crammed in thousands and thousands of hours of graft to turn himself into the perfect player.'[676]

If anyone has a claim to be an even better footballer than Ronaldo it's Lionel Messi. Yet Messi was born with a physical defect that should have cut short any hope of a footballing career—a growth hormone abnormality that would have resulted in him becoming no taller than 4′ 7″ in adulthood if left untreated.[677] But while growing up in Argentina Messi found ways to compensate for his small stature: unable to 'muscle' his way through a team's defence, he learned to glide through it, and in the process became a dribbling maestro. When Barcelona's sporting director Carles Rexach spotted his potential, and the club provided him with growth hormone treatment that allowed him to reach 5′6″, the result was a multiple winner of the Ballon d'Or—the trophy awarded annually to the world's best footballer.[677] Now Messi is clearly gifted with great endurance and speed, which may be rooted in specific genetic qualities. But the fact that what should have been a handicap on the route to sporting success turned out to be a crucial feature along such a route shows the complexities of individual history. What's more, it's hard to imagine any parent choosing genome editing to endow their child with a growth hormone defect so that after a childhood spent battling to overcome this handicap, they might emerge as a world-class football player.

It's not just in sport that genius may require both nature and nurture. For instance, who would have predicted that a youth judged a failure by his teachers and who said 'school failed me, and I failed the school' would become one of the greatest scientists of all time? Yet such was the experience of Albert Einstein.[678] The young Charles Darwin was accused by his father of caring 'for nothing but shooting, dogs and rat-catching', and of being 'a disgrace to yourself and all your family'.[679] In fact, with hindsight, we can recognize qualities shown even at an early age by these two scientists that would later contribute to their great discoveries. So, as a youth, Darwin's interest in cataloguing the wildlife he caught bordered on obsession. Later in his life, he played backgammon each evening with his wife, the results of which he carefully recorded, once reporting to a friend that 'the tally with my wife...stands thus: she, poor creature, has won only 2490 games, whilst I have won, hurrah, hurrah, 2795 games'.[680] Yet his obsessive attention to detail proved of great importance when Darwin was collecting examples from the natural world to back up his theory of evolution by natural selection. And while Einstein didn't impress his school teachers, at the age of 16 he wrote an essay about the physical world that prefigured some of his later insights about relativity.

But would Darwin have come up with his revolutionary theory if he had missed out on a trip around the world on the HMS *Beagle*, which he almost did, being only second choice for the position of a companion for Robert Fitzroy, the ship's captain?[681] And would Einstein have made his amazing discoveries if he had been successful in his applications for university teaching positions? For his failure in these led to him accepting a job at the Swiss Patent Office, which, although routine, provided a 'worldly cloister' that allowed him the time and space to develop ideas that the pressures of an academic career might have denied.[682]

The Case for Regulation

The complexity of the link between biology and life experience, plus the potential risks of off-target effects, should hopefully persuade anyone thinking of trying to create a designer baby that this would not be a good idea. But IVF practice also does not exist in a social vacuum. In Britain all work on human embryos, whether for research or clinical purposes, requires a licence from the Human Fertilisation and Embryology Authority, or HFEA.[683] As already mentioned, the first HFEA application to carry out genome editing on human embryos was made by Kathy Niakan of the

Francis Crick Institute in London; she is using CRISPR/CAS9 to knock out or otherwise manipulate different genes to see what effect this has on embryogenesis. 'The knowledge we acquire will be very important for understanding how a healthy human embryo develops, and this will inform our understanding of the causes of miscarriage,' she says. 'It is not a slippery slope [towards designer babies] because the UK has very tight regulation in this area.'[684] And Robin Lovell-Badge, head of stem-cell biology at the institute, agrees: 'There is clearly lots of interesting and important research you can do with these techniques which has nothing to do with clinical applications.'[684]

In contrast to the situation in Britain, in the USA public funds for research on human embryos have been harder to obtain. We saw in Chapter 8 how no federal government funds were available for research on human ES cells when George Bush was US president. This ban was relaxed under the presidency of Barack Obama.[685] However, demonstrating the generally more conservative stance of the US government on human embryo research, recently the NIH stated that it will not fund any genome-editing research on human embryos.[686] Justifying the ban, the NIH director Francis Collins said that genome editing of embryos is 'viewed almost universally as a line that should not be crossed'.[686] Yet a curious situation exists in the USA whereby, although public funding for human embryo work has waxed and waned, private funds none the less sustain this research at a high level.[686] And ironically, because in the USA there is no organization like the HFEA that regulates human embryo research, there are effectively no legal limits on this type of research, or even on its clinical application, as long as private money funds it. This raises the question of whether a body like the HFEA that allows valuable research on human embryos to proceed, but excludes attempts to pursue such research in an unethical fashion or apply it clinically, may be something other countries, including the USA, should consider.

Despite the controversies about the use of CRISPR/Cas technology to edit human embryos, there is far less pressure on those using this approach to manipulate other species. And, as we observed in Chapters 5 and 6, such manipulation will probably proceed in two main directions—first, to develop GM organisms as models of human health and disease, and second, to create new varieties of animals and plants with commercial value for agriculture. On the positive side, genome editing offers the possibility of dramatically expanding the range of organisms that can be created in these two different areas, as well as greatly increasing the precision of the genome modification, with highly beneficial outcomes for medicine and farming. Yet there are also potentially negative aspects of such a strategy that we should now consider.

A Question of Safety

Let's consider, for example, some possibilities raised by Margaret Atwood in her *Oryx and Crake* trilogy. In this world of extreme division between haves and have-nots, one reason why the better-off live in gated communities is because of the threats posed by genetically engineered viruses let loose in the wider world by bioterrorists.[687] To complicate matters, most of these viruses seem to have started life in the labs of the giant biotech companies that now dominate the world—acts of rebellion by alienated scientists who work for these companies. And given that the final catastrophic act of rebellion is by the talented scientist Crake, who rose to a position of authority in one of the companies but then used it create a super-lethal virus that wipes out civilization, this poses the question of how worried we should be about new genetic technologies being used to create such agents of bioterrorism.

Certainly, the fear of biological weapons being used by terrorists concerns many people. For as Wil Hylton recently wrote in an article entitled 'How Ready Are We for Bioterrorism?' in the *New York Times*: 'The specter of a biological attack is difficult for almost anyone to imagine... Like a nuclear bomb, the biological weapon threatens such a spectacle of horror—skin boiling with smallpox pustules, eyes blackened with anthrax lesions, the rotting bodies of bubonic plagues—that it can seem the province of fantasy or nightmare.'[688] Such are the fears, but what about the reality, and how much more of a threat do such weapons represent now that it's possible to manipulate the genomes of life forms, including harmful bacteria, with an unprecedented level of precision?

In fact, biological weapons are not as new as might be imagined. The ancient Hittites sent plague victims into the camps of their enemies and Herodotus, a Greek historian writing in the 5th century BC, described archers firing arrows tipped with manure to contaminate the wounds of their victims.[688] In 1763, as the British fought the French and their Native American allies for possession of what's now Canada, Sir Jeffrey Amherst, the British commander-in-chief in North America, wrote to Colonel Henry Bouquet: 'Could it not be contrived to send smallpox among these disaffected tribes of Indians'?[689] The colonel replied: 'I will try to inoculate the [Native American tribe] with some blankets that may fall in their hands, and take care not to get the disease myself.'[689] Smallpox decimated the Native Americans, who had never been exposed to the disease and had no immunity. During World War II, British and American scientists investigated the possibility of using smallpox as a biological weapon; however, because of the availability of a vaccine, it was not felt likely to be very effective.[689] But in 1989, Vladimir Pasechnik, a Soviet scientist who defected to Britain, claimed that the Soviet pharmaceutical company Biopreparat was a front for a massive bio-weapons programme, and another defector, Ken Alibek, said that a goal of the programme was to create deadlier forms of smallpox against which current vaccines would be useless.[689]

Yet for all the fears about biological weapons, and the willingness of individuals and governments to develop and even employ them, overall they haven't been a particularly effective weapon in history. For, although use of such agents may tap into human fears about contamination by other life forms that reach deep into our evolutionary past, shooting someone or blowing them up is generally still far more effective than trying to infect them with a biological agent. And we already have many vaccines and drugs to combat known pathogens. But could this situation change if genome editing made it possible to create new, super-lethal forms of known bacterial or viral species, or even invent brand new pathogens?

Bear in mind first that a bioterrorist would find it difficult to compete with some natural viruses already out there in the world.[690] Take HIV for instance—a virus that spreads via sex or exchange of contaminated blood, yet lies dormant in the body for many years so infected people show no symptoms and continue with their lives, infecting others in the process.[691] Then, when the virus does finally reveal itself, it does so by repressing the very immune system that normally protects against viral infection. Or consider Ebola—a virus that causes extensive haemorrhaging from multiple sites in the body of an infected person, resulting in the release of blood that is highly infectious to anyone who comes into contact with it.[692]

An ingenious and dedicated bioterrorist might still try and outdo nature. For instance, a particularly lethal combination might be a virus that causes extensive haemorrhaging like Ebola, but which could be spread by coughing and sneezing, like 'flu. But while it's possible to imagine such a combination of characteristics, it would be difficult, and perhaps even impossible, to create such a combination in practice, because viruses have evolved over millions of years to infect a very specific part of the body and spread from individual to individual in a very precise manner.[693] It may in practice be no easier to create a virus with the characteristics of 'flu and of Ebola, than it would be to make a pig fly. On the other hand, it's worth recalling the study we mentioned in Chapter 4, in which CRISPR/CAS9 was used by Andrea Ventura's group at the Memorial Sloan Kettering Cancer Center in New York to engineer a respiratory virus that could cause lung cancer in mice. For Stanford's David Relman, Ventura's group 'ended up creating, in my mind, a very dangerous virus and showed others how they too could make similar kinds of dangerous viruses'.[694]

Another reason why it could be difficult for terrorists to use genome editing to create a lethal super-virus is the serious amount of expertise, facilities, and funding required for such a feat. For despite what we've said previously about techniques like CRISPR/CAS9 being far easier to carry out than previous approaches, and particularly with individuals like biohacker Josiah Zayner, mentioned in Chapter 9, wanting to 'democratize' this technology by making it freely available for an affordable price, it's important not to overstate the point. To engineer a deadly virus using genome editing it would still be necessary to have access, not only to molecular biology reagents and equipment but also specialized containment facilities so that the virus being created did not pose more of a risk to the terrorists creating it than to those against whom it was aimed.[695]

Perhaps it is for this reason that Margaret Atwood deliberately chose to locate her bioterrorists not on the fringes of her future society, but as alienated individuals working within biotech companies themselves. But how real is such a possibility? In today's academic or commercial biomedical laboratories, work on potential pathogens is generally intensely monitored, as part of the guidelines established after the Asilomar conference that we mentioned in Chapter 2. So if someone really did want to develop a biological weapon within such a lab, they would have to evade the strict scrutiny.

Having said that, one incident illustrates that risks might be present in the most unexpected places. Coming just a month after the 9/11 attacks, the arrival in the USA in October 2011 of letters containing small amounts of powdered anthrax which killed five people and left 17 seriously ill, had huge impact in a country already feeling under siege. One contaminated letter sent to the leader of the Democrat majority in the US Senate, Tom Daschle, read: 'We have this anthrax. You died now. Are you afraid? Death to America.'[696] Who was responsible for such an atrocity? The finger was quickly pointed at Iraq, with former CIA director James Woolsey, Jr arguing that Iraq was the state 'most likely' to support an anthrax attack against the USA. 'Saddam has a festering sense of revenge for his humiliation in the Gulf War,' he said.[696] Indeed, the attacks were used to justify the second Gulf War. So when US Secretary of State Colin Powell made the case for the invasion of Iraq at the UN Security Council, he held up a vial of white powder equivalent to a teaspoon of anthrax, and then stated that Iraq 'could have produced 25,000 litres' of the deadly agent.[696]

Yet years later, the individual the FBI finally named as responsible for the anthrax letters had no link to Iraq. He was Bruce Ivins, a disaffected scientist at the Army's biodefence labs at Fort Detrick, Maryland, who committed suicide in 2008 before he could be brought to trial.[696] This case shows the danger of uncritically believing war propaganda. But it also raises questions about work being carried out in secret military laboratories in the name of defence, and whether this shouldn't also be subject to the sort of guidelines established at Asilomar. The case further poses questions about the need for transparency in work involving GM organisms, no matter what the sector. This is surely important if we are to prevent potentially lethal agents falling into the wrong hands, as well as to safeguard against the potential misuse of such agents by governments. And such precautions may become increasingly important as the genome-editing revolution progresses.

Pigoons and Other Oddities

Moving away from engineered viruses and bacteria, how worried should we be about the possibility of genome editing being used to create new—and potentially dangerous—larger organisms? In *Oryx and Crake*, perhaps the most dangerous aspect of the world that Jimmy must now inhabit is the presence of numerous bizarre animals. Particularly sinister are the pigoons. In the previous society, the 'goal of the pigoon project was to grow an assortment of foolproof human tissue organs in a transgenic knockout pig host—organs that would transplant smoothly and avoid rejection, but would also be able to fend off attacks by opportunistic microbes and viruses'.[697] However, the feral pigoons now roaming the ruins of civilization seem to have acquired a measure of human-like intelligence, so now it is they who hunt Jimmy. Other creations, like the snats—a cross between snakes and rats—and glowing green rabbits, fit the general theme of a society in which genetic engineering had become an anarchic pursuit in which anything went. In one of Jimmy's flashbacks, we learn that, in the biotech companies, 'there'd been a lot of fooling around . . . create-an-animal was so much fun, said the guys doing it; it made you feel like God'.[698]

Back in the real world, it's worth asking just how far genome editing could go in modifying animals used for medical research, or indeed as a source of replacement organs. Here it might help if terms like 'humanized pigs' were avoided by scientists in reference to animals that have been engineered to provide a source of organs for transplants, for in reality such GM porkers are merely animals with changes in proteins involved in the immune response, not the sinister creations of Atwood's imagination.[699] In fact, as we saw in Chapter 5, there are other good reasons for creating genome-edited pigs; for instance, for studies of heart function and disorders of this organ. And as pigs are already kept for food production, it may be that such research will not raise many issues for the general public.

Having said this, the news in 2017 that US researchers had created pigs that contain living human cells within them shows how rapidly science fiction can become fact.[700] A team led by Juan Carlos Izpisúa Belmonte of the Salk Institute in California reported that they had produced a dozen pig embryos containing human cells. By adding human stem cells to pig embryos engineered to lack a particular organ, the hope is that these cells will take over the task of forming the missing organ in an adult pig. A justification for such research is that it could not only provide a new source of organs for transplants but also lead to important new insights into the molecular mechanisms underlying human organ formation.[700] In fact Belmonte had already shown previously that such an approach could be used to grow rat organs in adult mice. But growing human organs in pigs looks set to be far more challenging. One challenge is that the pig pregnancy lasts about 112 days, compared to nine months in humans, meaning that the embryonic cells are developing at completely different rates.[700]

However, genome editing could conceivably be used to synchronise the growth of the human organ with that of its pig host. Today, Belmonte's team is slogging through a process of trial and error, testing how different animal and human cells interact when combined, in hopes that they can apply what they learn to pig-human chimeras. But even that slog is, by the research standards of just a decade ago, proceeding at great speed. With conventional methods, Belmonte says, 'it would take hundreds of years. But thanks to Crispr, we can move quickly to many, many genes and modify them.'[701]

Such are some of the potential technical challenges. But concerns are also being raised about the pace and direction of the research, and the NIH has said that it will not support further studies involving such 'human–animal chimeras' until it has reviewed the scientific and social implications more closely. John De Vos, of the Department of Cell and Tissue Engineering at Montpellier University Hospital and Medical School in France, has envisioned worst-case scenarios involving pig chimeras.[701] He believes that if human cells entered a pig's brain, for instance, the animal might develop new kinds of awareness and intelligence. 'It would be horrible to imagine a form of human consciousness locked in the body of an animal,' he says.[701] While acknowledging such concerns, Belmonte believes there could be safeguards to prevent such a scenario. For instance, the human cells could be edited to prevent them contributing to the developing pig brain.[701]

Meddling with Monkeys

Another area of research that holds great potential promise for the future but also raises many ethical issues is the use of genome editing to manipulate the primate genome. Currently only a small proportion of medical research is carried out on primates. In Britain in 2013, rodents were used in 82 per cent of studies but primates only in 0.05 per cent.[702] Yet the placards held by animal rights protestors often have photos of monkeys on them because these are more likely to strike a chord with the public than pictures of rats and mice. For reasons of cost alone it's likely that primate studies will always represent a tiny proportion of total research. However, our newfound ability to precisely modify the genomes of primates may lead to greater use of this animal group. For, as we saw in Chapter 5, despite widespread use of rodents to model human brain function and disorders, there are fundamental differences between human and rodent brains, not just in size but also in structure, that mean there are limitations to what we can learn from such studies.[703]

For instance, how much can we really learn from rodent studies about complex human disorders like autism, schizophrenia, or bipolar disorder, which have a strong social and language basis? Those who carry out such studies point to the fact that while they may not reproduce the complexity of these human disorders, they can identify cellular mechanisms that reveal important insights into the underlying basis of these psychiatric conditions. However, to model these disorders in a more sophisticated fashion there seems no doubt that we could learn a lot from exploring the role of altered gene expression in species whose brains, and ways of interacting with the world, are much closer to our own, namely other primates.[703] Clearly though, if such studies are to become more commonplace, then we need to consider the ethical questions they may raise.

One strategy would be to engineer monkeys with genetic differences identified in humans with psychiatric disorders. As we already saw in Chapter 5, such studies are already underway, particularly in China. We've already mentioned the large number of different genomic regions now linked to such disorders.[704] Given this complexity, it might make sense to focus on genes linked to more extreme versions of psychiatric disorders, for, as we mentioned in the section 'Disorders of the Mind' with regard to depression, this may help identify a more specific genetic link to the disorder. But if such engineering results in a monkey with features of autism, depression, or schizophrenia, we ought to consider whether creating a serious mental disorder in an animal similar to our own species is ethically acceptable.[703]

A few points are worth noting here. One is that a key aspect of human psychiatric disorders is the dislocation they cause in terms of social interactions. The French philosopher Michel Foucault, and Andrew Scull, a sociologist at the University of California, San Diego, have both argued that while mental disorders were recognized in pre-industrial society, they were not necessarily viewed negatively, with sufferers regarded as part of the spectrum of normal human behaviour, or, even more positively, as visionaries.[705] Foucault, with his notion of a 'great confinement', proposed that mental illness came to be regarded as an affliction at the time of the Industrial Revolution and the rise of a regimented workforce, in which deviations from the norm were seen as a threat to the social order.[705]

The effects of a psychiatric disorder in a laboratory primate model may be quite subtle. This could pose problems in choosing how to assess such changes in a primate, but also means the effects of the disorder may be relatively benign. But what if such genetic changes result in psychological distress in a primate? Thought needs to be given to how such distress might be alleviated while still allowing valuable information to be gained in terms of new ways to understand and treat mental disorders in humans.

A Question of Language

A different way of studying human brain function would be to genetically engineer a primate to be more human-like in its general thought processes. Given the ethical implications, it's perhaps not surprising that no scientists are currently proposing to use such an approach. Yet, as an obvious way to investigate the biological basis of human consciousness,[706] it's something that at least needs to be considered in principle. In Chapter 5 we saw how the FOXP2 gene was identified as a potentially critical link in language capacity in humans.[707] But while introduction of the human version of the FOXP2 gene into mice led to some interesting changes in their behaviour, such as increased vocalization and enhanced ability to negotiate a maze, the major differences between a mouse and a human mean there are limits to interpreting what such changes signify for our understanding of human brain function.[708]

It would, therefore, potentially be valuable to introduce the humanized FOXP2 into a monkey, and assess the effect on learning, memory, and other cognitive functions. And indeed, as was mentioned in Chapter 5, Bing Su, of the Kunming Institute of Zoology in China, has said he plans to use gene editing to make such a change to FOXP2 in macaque monkeys.[709] An even more interesting experiment would be to introduce this genetic difference into a chimpanzee, given that our closest biological cousin already has a capacity for associating abstract symbols with objects or actions. Indeed, the symbol learning feats of some chimpanzees in studies carried out in the 1970s were used to argue that chimps can learn language in sophisticated fashion like humans.[710] In line with such claims, some chimpanzees, like Nim Chimpsky (a pun on the name of the famous linguist Noam Chomsky), could be taught to associate over 100 symbols, and it was even claimed that they could construct primitive sentences.[710]

More recent scrutiny of these studies suggests that even chimps taught intensively from an early age never progress past a simple association of symbols and objects, and while they do show some ability to lump together objects and actions, this is very far removed from the human capacity for forming complex grammatical sentences.[710] But what would be the effect of introducing the humanized version of FOXP2, or other genes now being shown to have a functional relationship to FOXP2, on a chimp's language capacity?[710] Such studies could be important for uncovering the functional basis of human language in an animal model, and could also lead to new insights into language defects and mental disorders like autism or schizophrenia, which are characterized by problems in social interaction.[711] Yet they would also raise some serious ethical issues.

For instance, if introducing genetic changes into a chimpanzee enhanced its language capacity, could this create a chimp with self-conscious awareness? If so, it could raise all sorts of questions about the effect on an animal in a captive state. And of course it might also create fears that this kind of change could lead to a *Planet of the Apes* situation, in which self-consciously aware talking apes take over the world. In the 2011 film *Rise of the Planet of the Apes*, the apes gain consciousness and the ability to talk through a genetically engineered virus that also proves deadly to people, thereby wiping out most of humanity and with it the chance of our species preventing the apes' rise to power.[712] It's hard to imagine a single virus having such an effect, but genome editing, coupled with a better understanding of the genetic basis of human language, means that the creation of a self-conscious chimp capable of conversing via sign language (for a talking ape would also require genetic engineering of the vocal chords) is not such a far-fetched idea as it might once have appeared.

A simple way of ensuring that we do not get into situations that create ethical dilemmas relating both to animal welfare and potential threats to human safety would be to severely restrict the sort of genetic modifications that can be carried out on a monkey or ape. Yet if sophisticated primate models of human brain function and dysfunction prove necessary to understand our brain function and the genetic basis of mental disorders, we may in the future face difficult choices between furthering scientific understanding and developing potential new treatments for psychiatric disorders, and the ethical dilemmas that tampering with the genomes of other primate species may generate.

Fears about Food

The prospect of genetically engineered, self-consciously aware apes probably lies some distance in the future, should it arise at all, but the impact of genome editing on food production may have a more immediate effect on our lives. This too could raise important ethical issues, particularly as GM foodstuffs are already highly controversial. In fact, resistance to novel foodstuffs is not a new phenomenon. When Tsar Peter the Great introduced potatoes into Russia in the 18th century, long after they had been accepted in the rest of Europe, this led to riots amongst the peasants, who saw it as a plan to do away with their traditional source of carbohydrate—black bread. One of the Tsar's secret police reported that 'ignorant allegations... that the potato is a cursed fruit whose cultivation brought about God's refusal to bless the Russian land with fertility, were the cause of disobedience of peasants of the Moscow gubernia, who, in some villages, destroyed entire potato fields'.[713]

The government's own guidelines about how to eat the new plant hardly helped. The peasants were initially told that the edible part of the plant was the fruit on the potato bush, rather than the root growing underground. So deep was this misconception that even the Tsarina Catherine had the fruit of the potato plant served to her husband after he gave her the 'earthly apples' as a gift.[713] In fact, there was a deeper logic at work in the resistance to the potato, for many peasants viewed the new plant as a plot to reduce even further their limited rights and autonomy. Eventually, the potato's value as a source of nutrition won over; ironically, the centres of greatest resistance to the plant are now Russia's main potato-growing regions.[713]

Recent opposition to GM foodstuffs has been focused around three main areas of concern: potential toxicity to human health; detrimental effects on the environment; and further concentration of food production in the hands of the giant agribusiness companies, as we saw in Chapter 6. Currently, there is no convincing evidence that GM crops are toxic to human health.[714] However, important points have been made about the possibility of antibiotic resistance genes spreading from standard GM plants to harmful bacteria, thereby creating antibiotic-resistant pathogens, or genes for herbicide resistance becoming incorporated into weeds, leading to the growth of weeds that cannot be killed by such a herbicide. But these concerns may soon become a thing of the past, given that, with genome editing, it's no longer necessary to use antibiotic selection to create GM crops or animals. And the very subtlety of this approach, which can create the sort of genetic changes associated with mutations that occur naturally, like those that produce hornless cows or disease-resistant potatoes or pigs, may mean GM food is viewed more benevolently by consumers.

Or again, it might not, since genetic modification of food species is still heavily bound up in many people's minds with the nightmarish and the strange. Such concerns surface in the future vision of food production conjured up by Margaret Atwood in *Oryx and Crake*. So we learn in one of Jimmy's flashbacks how the giant agribusiness companies had mutated food into an unrecognizable state in order to fit corporate needs.[715] In this society, chickens were now ChickieNobs, monstrous creations that are all breast, without a head, wings, or feathers. Other foodstuffs like 'SoyOBoyburgers', represent a surfeit of heavily processed foods of debatable nutritious value. We also get hints that humanized pigoons bred for organ replacements, or even disposable human beings from the 'Pleebland' zones outside the walled enclosures where the richer people live, may be ending up in some food products.[715]

Since all detailed food labelling has been abolished in this fictional future, there is no way to prove such adulteration is taking place. It's also clear that food production is now completely outside the control of ordinary farmers. For instance, 'the Happicuppa coffee bush was designed so that all of its beans would ripen simultaneously, and coffee could be grown on huge plantations and harvested with machines. This threw the small growers out of business and reduced both them and their labourers to starvation-level poverty.'[715] All of which gives us a sense of a society in which ordinary people have become disempowered, with no control over what's available to eat or drink—nor what they can produce.

Food for the People

Back in the present, an important question is to what extent new genetic technologies might be leading us to such a nightmarish future vision of food production, or whether we can look to more positive uses for such technologies in agriculture. Here it's important to consider how food production relates to the structure of society as a whole. In the capitalist system that dominates the planet, the driving force is ultimately profit, which leads to a continual drive to revolutionize and expand production. One positive aspect of this for food production is that technological innovations like the original Industrial Revolution, and more recently the Green Revolution, have allowed food production to keep pace with a rising global population. This isn't only true of crops. So Hugh Pennington, emeritus professor of bacteriology at Aberdeen University, has argued that, 'prior to the 1950s, large numbers of people died because of tuberculosis due to a simple lack of nourishment. The wide availability of cheap animal proteins, both chicken and fish, has put an end to that.'[716]

Yet despite the capacity of intensive farming to provide a cheap source of animal protein for poor people, another way of looking at this issue is to examine the quality of such food, and the sustainability of current food production methods. To tackle quality first, it's clear that many people are increasingly eating 'junk' food. This has huge implications for future global health, with a recent report by the United Nations predicting that almost 1 billion people will be obese by 2025.[717] So although fewer people in the developed world now die of a simple lack of nourishment and associated diseases, dramatically rising obesity is starting to pose an equal threat to the lives of the poor. And although not directly a product of intensive farming, one could argue that the latter practice is part of a general cheapening of the quality of food products along with their price.

What's more, accompanying this increase in diet-induced ill health, there is evidence that intensive livestock farming methods are offloading problems on to the rest of society. The massive overuse of antibiotics in farming in order to contain infections such as *Salmonella* has led to the spread of antibiotic-resistant bacteria that threaten modern medicine itself. And a recent study showed that use of antibiotics in agriculture is growing worldwide, particularly in China, used over 84,000 tons of antibiotics for this purpose in 2013[718] and is projected to double its consumption by 2030.[719] Intensive farming also has a negative impact upon animal welfare, both in terms of the physical condition of the livestock and their psychological health. An important question, then, is what genome editing is likely to add to the equation, and whether its impact is likely to be predominantly negative or positive.

The ability to apply genome editing to practically any farmed species, and to use it with a precision completely lacking in past forms of genetic engineering, ought surely to be a positive thing. For instance, we saw in Chapter 6 how genetic differences associated with characteristics found in naturally occurring plant or animal species—like resistance to blight in wild potatoes or a benign response to African swine fever in warthogs—can now be introduced into domestic crops or animals quickly and cheaply. And this ought in theory to lead to much reduced requirement for pesticides and antibiotics. Similarly, it should now be possible to introduce changes that add to the quality of a food product, whether leaner pork, more flavoursome strawberries, riper tomatoes, or reduced amounts of carcinogens after a potato is fried. More fundamentally, the capacity of genome editing for altering multiple genes simultaneously offers the prospect of being able to introduce wholesale, but highly precise, changes. And this might make it possible to radically transform plant or animal species so that they can withstand the extremes of temperature, drought or flood, or changes in the acidity or salinity of our oceans that will occur due to accelerating global warming.

Yet concerns have also been voiced about the use of genome editing in agriculture. One worry is that this technology may put increasing power in the hands of giant agribusiness companies primarily concerned with maximizing short-term profit and not about long-term effects on human health, animal welfare, or the environment. However, genome editing could equally be viewed as an empowering technology available to small producers, not only giant companies, in a way that has not been possible with previous forms of genetic engineering. This is important, because one criticism of current agriculture and food production methods has been the recognition of the need for sustainability and the value of locally sourced produce. Local sourcing is seen as important because it minimizes unnecessary use of fuel for transport, and is also part of a general increased recognition of the fact that different communities have different resources, skills, and requirements.[720]

Far from locally sourced foods being merely a fad in countries like Britain and the USA, there has been a revival of interest in indigenous vegetables in the developing world. A recent report described how, at a popular restaurant in Nairobi, 'the waiting staff run back and forth from the kitchen, bringing out steaming plates of deep-green African nightshade, vibrant amaranth stew and the sautéed leaves of cowpeas'.[721] This is in contrast to several years ago, when European vegetables like kale were the main greens on the menu. According to leading African nutritionists, such indigenous vegetables are not just tasty but also often richer in protein, vitamins, iron, and other nutrients than non-native crops, and they are better able to endure droughts and pests. As Mary Abukutsa-Onyango, a horticultural researcher at Jomo Kenyatta University in Juja, Kenya, has said, 'in Africa, malnutrition is such a problem. We want to see indigenous vegetables play a role.'[721]

Scientists like Abukutsa-Onyango in Africa, as well as those in other parts of the developing world, are interested in studying indigenous vegetables to further tap their health benefits and improve them through selective breeding. An obvious question is whether genome editing could be used to refine such plants, as an alternative to a focus on cash crops that make profits for big companies but do little to help feed ordinary people in far too many developing countries. Meanwhile, the potential genetic value of these crops hasn't escaped the notice of researchers in the developed world. Calestous Juma, director of the Science, Technology, and Globalization Project at Harvard University, believes that as well as being amenable to improvement by genome editing, such native crops may have valuable 'traits that may be useful for other crops',[721] if introduced by genome editing. But if such moves are to help ordinary people in the developing world, the latter need to be properly involved in decisions about the development and use of indigenous crops, and not merely treated as accessories by big companies keen to identify attractive genetic characteristics, but who leave nothing for local people in return.

Editing as Aesthetics

From food to frivolity. Genome editing could be used for purely aesthetic purposes. For instance, how about using it to create designer pets? Recently, researchers at the Beijing Genomics Institute announced that they had created micro-pigs by TALEN genome editing and would be selling them as pets.[722] The micro-pigs were created from a small breed of pig known as Bama, by disabling one of its copies of the gene coding for growth hormone receptor. The animals weigh around 15 kilograms when mature, or about the same as a medium-sized dog, as compared to 100 kilograms for a normal adult pig. Each micro-pig is being offered for 10,000 yuan—about £1,000. The animals are being developed to raise cash to fund stem-cell experiments and other research that take place at the institute. 'We plan to take orders now and see what the scale of the demand is,' says the institute's director, Yong Li.[722] Customers will also be able to select their pet pig's colour and coat pattern.

The project has horrified animal welfare groups. 'The idea is completely unacceptable,' says Penny Hawkins, head of the British Royal Society for the

Prevention of Cruelty to Animals (RSPCA)'s research animals department. 'In the past, pets have been bred by selecting animals, generation by generation, to produce a desired trait. Inducing a massive change in one go risks creating animals that suffer all sorts of horrific impairments.'[722] Some scientists have also expressed caution about the project. 'It's questionable whether we should impact the life, health and well-being of other animal species on this planet light-heartedly,' says Jens Boch of the Martin Luther University of Halle-Wittenberg in Germany, one of the original pioneers of TALENs.[723] But reproductive biologist Willard Eyestone, of Virginia State University, says 'If the micro-pig is carefully evaluated and found to be equal in health compared to a normal pig and differs only in terms of size, there would be little scientific reason to block it from being offered as a pet.'[722] In contrast to the RSPCA, he adds, 'we must bear in mind that we have been altering the genetic make-up of pets for millennia, using the comparatively imprecise method of…selective breeding, which sometimes results in less than healthy traits for the animal. In principle, gene editing should offer a far more predictable and humane alternative.'[722]

However, concerns about GM pets don't only relate to welfare issues. Some scientists fear that using genome editing in this fashion trivializes the technology in the minds of the general public and could lead to an eventual backlash. Max Rothschild, of Iowa State University, explains: 'The micro-pigs produced by gene-editing are "cute" for some people. But they are still pigs and require that their owners know how to raise them properly…More to the point, this more trivial use of gene editing takes away from its important uses to improve livestock welfare, disease resistance and productivity.'[722] Daniel Voytas, who, as we saw in Chapter 6, is pioneering genome editing in agriculture, is similarly concerned. 'I just hope we establish a regulatory framework—guidelines for the safe and ethical use of this technology—that allows the potential to be realized,' he says. 'I worry that pet mini pigs distract and add confusion to efforts to achieve this goal.'[723] Such sentiments may mean little, though, to consumers keen to purchase an exotic pet. When the micro-pigs were first exhibited at the Shenzhen International Biotech Leaders' Summit in China, they stole the show. 'We had a bigger crowd than anyone,' said Lars Bolund, a medical geneticist at Aarhus University in Denmark, who helped develop the pigs with the Beijing Genomics Institute researchers. 'People were attached to them. Everyone wanted to hold them.'[723]

This isn't the only way that biotechnology is being used to create pets. Edgar and Nina Otto from Florida were so attached to their Labrador Sir Lancelot, that, when he died, they sent his frozen remains to Sooam Biotech in South Korea, which created a clone that the couple named Lancelot Encore.[724] Junichi Fukuda, who runs a TV commercial production company in Tokyo, is another customer. He paid Sooam to create a clone of Momoko, his deceased pug, whom he says saw him through a divorce and gave him much love in her 16 years of life. 'She was the best pet in the world for me,' he says. 'The reason I was able to work hard and become more successful was because I was together with Momoko—that was how much I loved her.'[725] However, it's unlikely that many bereaved pet owners will be rushing down this route given that Sooam charges $100,000 to clone a pet. Incidentally, Sooam is the brainchild of Woo-Suk Hwang. As we saw in Chapter 8, Hwang claimed to have isolated stem cells from cloned human embryos, only to be exposed for fraudulent and unethical practices. Yet around the same time as his human embryo claims, Hwang also announced that his team had cloned the first dog, an Afghan called Snuppy, and in this case the claim was real.[725] In fact, even at the height of his disgrace Hwang was already setting up Sooam, backed by private donors. And its current success shows that Hwang seems to have been as good at resurrecting his career as he is at creating cloned copies of people's pets.

Engineering a Winner

So much for using cloning to resurrect a dead pet. But what about using new genetic technologies to create an animal that can generate lots of money by enhancing its sporting performance, like a genome-edited racehorse? Horse racing is the second most popular spectator sport in Britain, with over 6 million race-goers each year; it employs about 90,000 people and generates £3.7 billion a year for the economy.[726] Breeding racehorses is big business. Take Frankel—the greatest and most successful thoroughbred in modern racing history. Not only is Frankel himself valued at more than £100 million, but for a mare to have a quick roll in the hay with him in the hope of producing a similar winner costs £125,000. As with human athletes, both nature and nurture help to create a great racehorse, with the training regime being highly important. Yet the role of genetics is demonstrated by the fact that almost all the world's half a million thoroughbreds derive from only 28 ancestors born in the 18th and 19th centuries. And up to 95 per cent can be traced to just one stallion—the Darley Arabian, born in 1700.[726]

Surprisingly, it's still rare for genetic analysis to be used to identify great potential winning racehorses; instead breeding decisions are usually made by studying pedigree—the records of bloodlines and race results going back generations. There's a problem though in relying on pedigree alone as a guide to quality, for an ancestor five generations back only contributes 3 per cent of an animal's DNA. The Green Monkey is an example of the dangers of relying on bloodlines. In 2006 this colt with an impeccable pedigree sold for $16 million. Yet it ran just four times and failed to win once. To try to improve matters, British scientist Stephen Harrison set up Canterbury-based company Thoroughbred Genetics Limited in 2000. This company was the first to offer DNA screening for racehorse performance. Using such methods to identify the best combination of stallion and mare, Harrison's biggest success is Sacred Choice, with nine wins from 37 starts. Yet judged by traditional criteria, its mother, Sacred Habit, didn't seem up to much. 'Sacred Habit was sold because it was a rusty animal,' says Harrison. 'And yet it bred this multiple group-one winner.'[726]

Until recently, the type of genetic analysis available only provided a low-resolution estimate of a horse's sporting potential. However, the complete sequencing of the horse genome in 2009 has made it possible to begin pinpointing specific genetic differences that identify a great racehorse. In 2010, Emmeline Hill of University College Dublin discovered that variations in the myostatin gene, which codes for a protein that regulates muscle development and muscle fibre type, determines the type of race that most suits an animal and whether or not it will be an early developer. Her company, Equinome, offers three tests to owners and trainers, including one based on this single 'speed' gene. 'It was the first time that anyone had identified a single gene to an athletic trait in a thoroughbred,' says Donal Ryan, managing director of Equinome. 'It's quite astonishing that a single gene has such significance, but it does.'[726]

Yet despite the value of such tests for identifying the best breeding partners, creating a great winner still remains a lottery, for the reason mentioned in Chapter 2—the mixing and matching of the paternal and maternal genomes that occurs during egg and sperm formation often leads to unpredictable results in offspring. Simon Marsh, a top horse breeder, remains sceptical about some claims being made about the value of genetic analysis in selecting the best breeding matches: 'You can predict, if you are on the front row of the grid in a Formula One race that you are probably going to win the grand prix,' he says. 'But if you have the greatest stallion in the world and the best mare in the world, there's no reason why their progeny won't be beaten by something that cost just 10% of the price.'[726]

Such pessimism might be misplaced if we introduce another factor into the equation—the technologies described in this book. One obvious strategy would be to clone a great racer. Currently, the Jockey Club bans the use of cloned horses, although an indication of changing views on this matter was the decision in 2012 of the International Federation for Equestrian Sports to allow cloned horses to compete in future Olympic Games.[727] In the USA, the governing body of quarter horse racing—sprinting contests over short distances between small animals—lost a legal challenge to their ruling against clones, setting a legal precedent that may also affect thoroughbred racing.[727] But cloning is still a relatively inefficient process and can lead to problems of ill health in some offspring. It also can't be applied to dead horses, unless they have been cryopreserved.

In contrast, as further insights emerge about the contribution of specific genetic differences to racehorse biology, genome editing might in the future be used to fine-tune a horse's performance, or even recreate the genetic differences that made a deceased horse a winner. For instance, DNA might be taken from the corpse of a famous horse and used as a guide to genome editing. One of the most famous horses in British racing history is Red Rum, who won the Grand National steeplechase race three times. This sporting legend died in 1995, and is buried at the finishing line at the Aintree ground in Liverpool where the race takes place, with his head facing the winning post.[728] So maybe a DNA sample could be obtained from Red Rum's remains, and his genome sequenced and used as a guide to genome editing another great winner. But would this be considered sacrilegious to the memory of this sporting legend or acceptable practice in the pursuit of creating another winner? And what would an ability to genetically stack the odds in a horse's favour do to the nature of the sport?

Manufacturing Mammoths

If watching horses run around a track is not your cup of tea, you may be interested in much larger animals—like woolly mammoths. Some laboratories are intent in trying to resurrect this icon of the Ice Age using the latest technologies. In particular, as well as raising funds through cloning dead pets, Woo-Suk Hwang's Sooam Biotech company is pursuing this goal. The company recently cloned the endangered coyote, and plans to use this technology to repopulate endangered species such as Ethiopian wolves, the American red wolf, and the Lycaon or African wild dog.[725] However, it's Sooam's particular focus on the woolly mammoth that has excited most commentators' attention.

Recently, scientists from the company established a collaboration with researchers at Russia's North-Eastern Federal University in Yakutsk, the capital of the Sakha Republic in Siberia, to clone this long-extinct mammal.[725] With no live mammoths, success will depend on finding a well-preserved dead mammoth in the frozen tundra, extracting one of its cells, and implanting the cell's nucleus into an elephant egg from which the nucleus has been removed. Finally, the resulting cloned embryo will be implanted into a female elephant. To pursue this goal, Sooam scientists travel every summer to Siberia, deeper and deeper into the Arctic Circle, looking for a mammoth sample suitable for use in cloning. 'The point is to find something that's better than anything we've found before,' says Sooam researcher Insung Hwang. 'That's why we go on expeditions every year. That's why we try to improve our techniques of preserving the tissues during transportation—we even built a lab in Yakutsk to really shorten that time of transporting samples from Russia to Korea.'[725] And the discovery in 2015 of preserved pieces of frozen mammoth skin on the Lyakhovsky Islands of the Siberian north coast have boosted hopes that it may be possible to isolate viable nuclei from such remains.[729] More recently, in 2019, scientists at Kindai University in Japan reported that nuclei from thawed mammoth cells showed biological activity when transferred into elephant eggs.[730] However, this did not result in the cell division that must occur for an egg to develop into an embryo.

Not everyone is convinced it will be possible to recreate a mammoth in this way. It's not just that the DNA in a mammoth cell frozen for thousands of years may be too fragmented for successful cloning. It's also unclear whether a mammoth embryo and foetus will be compatible with an elephant mother. But there may be other ways to achieve the goal. George Church is trying to recreate a mammoth by a different route at Harvard University. Based on information from the sequenced mammoth genome, Church recently used CRISPR/CAS9 to introduce genetic differences that gave this species its small ears, subcutaneous fat, and long hair into an elephant genome.[731] If the hybrid creatures survive, Church's team will next engineer an elephant that can survive in the cold. Church believes that expanding the elephant's range into colder climates could help keep it away from the human conflicts threatening to make Asian and African elephants extinct. Later, after the engineered elephants gain a foothold, Church's team will try to revive the mammoths by integrating higher amounts of mammoth DNA into the hybrids.[731]

A legitimate question, even if such a project is scientifically possible, is why anyone would want to recreate a mammoth. The justification provided by Church is an ecological one. '4,000 years ago the tundras of Russia and Canada consisted of a richer grass- and ice-based ecosystem,' he says. 'Today they are melting, and if that process continues, they could release more greenhouse gas than all the world's forests would if they burned to the ground. Returning mammoths to the tundras could stave off some effects of warming.'[732] Church believes mammoths could keep the region colder by eating dead grass; enabling the Sun to reach spring grass, whose deep roots prevent erosion; increasing reflected light by felling trees, which absorb sunlight; and punching through snow so freezing air penetrates the soil. But Stuart Pimm, an ecologist at Duke University, believes the attempt to resurrect this species 'totally ignores the very practical realities of what conservation is about'.[731] And despite the rationale provided by Church, it's hard not to feel that an unspoken reason for recreating mammoths is because of the thrill value of seeing these great animals in the flesh.

Dino-Chickens and Unicorns

Which brings us to the question of whether it would ever be possible to bring back the dinosaurs, as in *Jurassic Park*. An important premise of the first film in the series was that dinosaurs could be recreated by analysing, and using information from, the genomes of 80-million-year-old dinosaur species present in samples of ancient DNA. One idea was that such DNA could be found in insects preserved in amber, these insects having sucked the blood of dinosaurs.[733] Having identified the full genomic DNA sequence, a dinosaur could be recreated by reconstructing its genome using an existing reptile genome as a template. In fact, in 1993, when the first *Jurassic Park* film was released, such an idea seemed not so far-fetched. Two days before the film's release, Raúl Cano and colleagues at California Polytechnic State University announced they had sequenced the DNA of a

120–135-million-year-old amber-encased weevil.[734] A year later, Scott Woodward at Brigham Young University in Utah announced that he and his team had sequenced DNA from a dinosaur bone. 'I am confident that we have a DNA sequence that belongs to a Cretaceous period bone fragment,' said Woodward. 'Based on the circumstantial evidence, we believe it is a dinosaur.'[733] But as researchers refined the techniques for ancient DNA analysis, they began to realize that these earlier announcements were too good to be true. Geneticists can now retrieve and study the DNA of relatively recently extinct forms—moas, cave bears, Neanderthals—from well-preserved remains, but genetic material seems too fragile to last for tens of millions of years.[734] The more ancient sequences now appear to be contamination from modern sources.

However, scientists may have been looking in the wrong place, for dinosaur descendants turn out to be all around us, in the form of birds. Although a starling, or even a golden eagle, wouldn't have the same effect on your pulse rate as a *Tyrannosaurus rex*, birds, including the domestic chicken, are closer genetically to dinosaurs than previously imagined. The idea that birds evolved from dinosaurs has been around since 1860, when German scientists discovered the fossil of a creature they named *Archaeopteryx*, after the ancient Greek words: archaīos, meaning 'ancient', and ptéryx, meaning 'wing'.[735] *Archaeopteryx* had wings and feathers, yet looked remarkably like a dinosaur. But it was particularly comparisons of the genomes of birds and reptiles that confirmed that our feathered friends are genetically quite close to their dinosaur ancestors. And now some scientists want to create a dinosaur-like organism by modifying a bird genome. One such scientist is Jack Horner of Montana State University.[736] As a child he had two dreams: one was to be a palaeontologist, the other was to have a pet dinosaur. His first dream came true when he was only 8, for Horner found a dinosaur bone near his home in Montana. Since then he's unearthed many more dinosaur remains—including foetuses within eggs. One of Horner's major discoveries was that some dinosaurs built nests, lived in colonies, and cared for their young.[736]

But it's the second part of Horner's dream that's proving to be the most controversial. Since birds are the evolutionary descendants of dinosaurs, Horner believes they have dormant DNA that, if activated, could allow them to develop some of the traits dinosaurs had, such as teeth, three-fingered hands, and tails. 'For me, creating a dinosaur is the biggest project we have,' says Horner. 'It's like the moon project. We know we can do it—it just takes time and money. And we will get it done. We will make a dino-chicken-like animal pretty soon.'[736]

Demonstrating that such ideas are not totally off the wall, recently scientists engineered a chicken embryo so that its beak became a dinosaur-like snout and palate, similar to that of small feathered dinosaurs like *Velociraptor*. The study was led by Bhart-Anjan Bhullar of Yale University in New Haven and Arkhat Abzhanov of Harvard University, who said they did not set out to create a 'dino-chicken'.[737] Rather, they were interested in understanding the molecular processes that led to the development of the beak, a key aspect of bird anatomy. It is also the part of the avian skeleton that has 'diversified most extensively and most radically,' says Bhullar.[737] Yet despite this diversity—ranging from flamingos to pelicans—very little is known about 'what the heck a beak actually is,' he says. 'I wanted to know what the beak was skeletally, functionally and when this major transformation occurred from a normal vertebrate snout to the very unique structures used in birds.'[737]

To study this question, the researchers trawled through the genomes of organisms ranging from mice, emus, alligators, lizards, and turtles. They found that birds have a unique cluster of genes related to facial development, which the non-beaked creatures lack. When they silenced these genes, the beak structure reverted to its ancestral state. So too did the palatal bone in the roof of the mouth. For now Bhullar has no plans, or ethical approval, to hatch the snouted chickens. But he believes they would have been able to survive 'just fine'. 'These weren't drastic modifications,' says Bhullar. 'They are far less weird than many breeds of chicken developed by chicken hobbyists and breeders.'[737]

What other genetic changes would be required to make a chicken more dinosaur-like? Jack Horner believes that, besides the beak change, other modifications are required to make a 'chickenosaurus'. So scientists would have to give it teeth and a long tail, and revert its wings back into arms and hands. Horner likens this to breeding a wolf into a Chihuahua, except on an accelerated timescale.[736] Yet not everyone thinks the task will be so straightforward. Bhart-Anjan Bhullar believes that if dinosaur-like features were to be restored, it's still possible that they wouldn't function correctly: 'You could perhaps give a chicken fingers, but if the fingers don't have the right muscles on them, or if the nervous system and the brain are not properly wired to deal with a hand that has separate digits, then you may have to do a considerable amount of additional engineering.'[738]

Of course, there are other reasons why many people would have concerns about scientists trying to recreate extinct dinosaur species, not least if they've watched any of the *Jurassic Park* films, where things generally end up going badly wrong. Such fears don't seem to deter Horner though. Indeed, he thinks that genome editing might be used not only to resurrect extinct animals but even create mythical creatures. 'As absurd and wild as it sounds, I honestly believe that even before we make a dino-chicken, we could make a unicorn,' he says. 'Wouldn't it be fun to have a unicorn? I mean, just think of the possibilities of making mythical creatures—mixing and matching different characteristics!'[738] All of which gives a flavour of some of the ethical conundrums that genome-editing technology may throw up in the future. But the greatest dilemmas are likely to relate to the use of biotechnology to transform the human species or create human-like organisms.

The Remaking of Humankind

In the section 'Talent Born or Made?', we discussed why the use of genome editing to generate 'designer babies' with desirable characteristics like great intelligence or musical or sporting ability may be far from a trivial matter, if it is possible at all. None the less, it's also important to consider how a greater awareness of the role of genetics in forming a human individual as well as also the technologies discussed in this book—genome editing, but also optogenetics, stem-cell technology, and synthetic biology—might transform humanity in the future. For instance, what if it really does prove feasible to create genome-edited pigs whose hearts, pancreases, lungs, and livers can be transplanted into people requiring these organs? Or what about pig–human chimaeras as a source of actual human organs? Will this mean that human lives will be greatly extended, because any failure of a person's vital organs will be fixed by another transplant, acquired for little more than the price of buying some pork steaks at the butchers? And if such a strategy becomes commonplace in medicine, will this alter our perception of what it means to be human, or will acquiring spare organs in this manner simply be seen as equivalent to having a hearing aid or heart pacemaker?

Of course, there is still the question of what happens when the most distinctive human organ of all fails—a person's brain. For even if human lifespans could be greatly extended with a succession of genetically engineered heart, liver, kidney, or lung transplants, this might be of little use without the means to rejuvenate the brain. We're all increasingly familiar with the spectacle of degenerative diseases like Alzheimer's robbing elderly people of their mental faculties and much of their individuality. So it seems likely that such problems would only increase as people's lifespans are extended, unless we find ways to better understand and treat such forms of dementia. Here, a solution will rely on scientists gaining a clearer idea about the molecular and cellular changes that underlie dementia and thereby identifying new drug targets. But there's also the possibility that creation of human brain structures from ES cells or personalized iPS cells could one day be used to treat degenerative disorders like Alzheimer's and Parkinson's, and even those that affect personality, like depression, schizophrenia or bipolar disorder.

Apart from the safety aspects of introducing artificially generated neurons into the brain—we'd need to be sure such cells wouldn't form tumours—there is also the question of whether such infusions might alter personality. But then, in the future, might stem-cell technology even be used to engineer people with a permanently upbeat mood? And if not through stem-cell technology, could a similar outcome be achieved using optogenetics or other technologies that manipulate neuronal activity, or even the expression of genes, using magnetic fields or radio signals? For this to be possible, individuals' brains would need to be genetically engineered to respond in the correct fashion, but, with future advances in genome editing, it's not inconceivable that this could become routine. If it did, this raises a more worrying possibility, which is that such a technology might be used to brainwash people into accepting a repressive political system or to erase certain memories and plant false ones. There would need to be safeguards against their misuse.

The replacement or rejuvenation of human organs is one way in which an individual human life could be radically altered in the future. But could genome editing also one day allow human beings to acquire completely novel characteristics, borrowed from the rest of the animal kingdom? For instance, could a human gain the ability to detect sensitive odours like a dog, the night vision of a cat, or even a capacity to remain underwater for long periods of time like a dolphin? A potential problem is that such properties have evolved over millions of years in tandem with the other changes that combine to give such organisms their unique characteristics. So it's by no means clear that such characteristics could be engineered into a person in a functional manner, without having detrimental effects on the rest of the human body.

There is another possibility: electronic gadgetry might allow a human to acquire such characteristics. Such a strategy could well involve a combination of electronic devices and tissue implants, perhaps derived from personalized iPS cells. Either way, does this mean that, in the future, humanity might undergo a diversification into individuals with quite different characteristics, depending on which animal quality attracted them? And might people of the future decide to transform their unborn children in this fashion, by engineering IVF embryos, in an even more radical take on designer babies?

The possible fusion of electronics and bioengineering leads us to another possibility—that a human brain grown in culture might be linked to sensory inputs that would allow it to detect and possibly learn. Such a brain might then be allowed to become the controlling device within a computer or robot. While such a scenario may sound like the plot for a bad horror movie, recent advances in the culture of human brains from iPS cells of the sort described in Chapter 8 mean that this possibility can no longer be disregarded as pure fantasy. Of course, there would be many ethical issues involved in treating a human brain in this fashion, but let's imagine that such an experiment went ahead. What would be the nature of the interaction of such a brain with the outside world? Would it see itself as human? And how would it feel about being contained inside a machine in this manner?

Taking our thought experiment a step further, given recent advances in the creation of 3D structures representing diverse human tissues and organs, is it possible that one day different organs could be combined to make an artificial human being? This is the main premise of the classic 1980s science-fiction film *Blade Runner*. In this future dystopia, stem-cell technology is used to create 'replicants', who are visibly indistinguishable from adult humans but engineered to have characteristics like superhuman strength or a beautiful physique.[739] The replicants are slaves, created to take part in military combat or perform dangerous jobs, or who are 'basic pleasure models' that service human sexual needs. To keep them under control, the replicants have been engineered to have a greatly shortened lifespan. This leads to conflict when some replicants rebel, something they have been programmed not to do, showing the ways in which genetic engineering might not always have the desired result, or perhaps just providing an illustration of the freedom of the human spirit, engineered or not. All of which raises the question of what it means to be human, but also what kind of society would allow a division to arise between 'real' humans and those 'replicated' from human stem cells.

Prospects for the Future

It's time to bring this book to a close. Let's return to society as it stands in the first part of the 21st century and assess the relationship of the biotechnologies discussed in these pages to that society. At the start of this book,

I made a point of locating the development of new genetic technologies as part of two unique human characteristics: our ability to make and use tools; and our self-conscious awareness, which allows us to plan how to employ such tools. It is these attributes that have allowed humankind to manipulate the world around us—both living and non-living—in such remarkable fashion. And as a consequence, our species has been able to grow to almost 8 billion in number and also to progress, in less than 50,000 years, from living in caves to sending people into space and robots to explore the surface of Mars.

Yet despite such amazing progress, how much are we in control of our destiny and how sustainable is our society? For, currently, a number of major challenges face humanity and it remains to be seen how well we will stand up to the test. Undoubtedly, the biggest problem of our time is global warming. It is difficult nowadays to find any serious scientist who does not believe that phenomena caused by human activities—primarily rising CO_2 emissions but also other greenhouse gases—are causing a rapid warming of the planet. And numerous studies now point to the dire future consequences of such warming. The NASA glaciologist Eric Rignot of the University of California, Irvine, recently concluded that 'a large sector of the western Antarctic ice sheet has gone into a state of irreversible retreat... This retreat will have major consequences for sea level rise worldwide.'[740] If both the western and eastern ice sheets—which cover an area the size of the USA and Mexico combined—were to melt completely, this would raise global sea levels by an astounding 60 metres or 196 feet.[741]

What remains unclear is how rapidly the process will occur. But even a rise of 7 metres, which some scientists are now predicting by the end of the century, would flood London, New York, and many other major cities.[742] An even more worrying long-term prediction has been made by another NASA scientist, James Hansen, of Columbia University, who has been called the 'father of climate change awareness'.[743] He believes that once it reaches a certain point, global warming may enter a 'runaway' phase, leading eventually to conditions like that on the planet Venus.[744] Given that Venus has a surface temperature of 482°C, this would clearly mean the end not just of human civilization on Earth but probably all other life. However, long before we reach that point, human populations, and the animals and plants we rely on for food, will have to face the effects of rising global temperatures and sea levels.[745] Yet, despite the seriousness of the threat, a succession of conferences involving world leaders have failed even to reduce the rise in CO_2 emissions, never mind reverse them.[746] So Hansen's verdict on the climate conference that took place in Paris in December 2015 was that it was a 'fraud', with 'no action, just promises'.[743] And unfortunately, the largest greenhouse gas emitters in the world—China, the United States, and India—offered either nothing or very little about their commitments to curb emissions at a more recent global summit in 2019.[747]

Genome editing might offer humanity a way of engineering crops and livestock so that they can cope with an increasingly stressful climate. The technology might also be used to create livestock that contribute less to global methane emissions. Methane is a powerful greenhouse gas, being 25 times as efficient as trapping the Sun's heat as CO_2.[748] And a surprising 26 per cent of all methane emissions in the USA come from the burps and farts of cows and other ruminants that produce the gas as a by-product of their digestive process. Yet individual animals can vary considerably in the amount of methane they produce, and a new European Union-funded project named RuminOmics aims to use cutting-edge technologies, including genome editing, to try and produce a milk-producing breed that produces less methane.[749] Such a breed could be attractive to farmers because a herd that emits less methane is likely to be more productive. According to Lorenzo Morelli, director of the faculty of agriculture at the Catholic University of Sacred Heart in Piacenza, Italy, and a partner in the project, 'The methane is lost energy that could go into producing milk. So if we can find the right genetic mix, we can find cattle that are less polluting, more productive, and more profitable for the famer.'[749]

At the same time, genome editing could provide us with ways to keep ahead of microorganisms that are rapidly becoming resistant to current antibiotics. But it seems astonishing that our society is capable of generating such amazing new technologies as genome editing, optogenetics, and organoids created from stem cells, yet lacks the political will to stop global warming, which, if unchecked, will ultimately threaten humanity itself. Because of this, it's worth asking some searching questions—and such questions are increasingly being asked in some very interesting places.

Bill Gates, co-founder of Microsoft and the world's richest man, recently pledged $2 billion to counter climate change, while encouraging other wealthy individuals to do the same.[750] Not so unexpected you might say, given that Gates has a long history of philanthropy, particularly in terms of funding new technologies to help people in the developing world, who will be worst hit by global warming. More surprising was Gates' rejection of claims typically made by free market advocates that the only way to reduce climate change is to leave it to private enterprise.

According to Gates, the problem with such a strategy is that 'there's no fortune to be made. Even if you have a new energy source that costs the same as today's and emits no CO_2, it will be uncertain compared with what's tried-and-true and already operating at unbelievable scale and has gotten through all the regulatory problems.'[751] Instead, he believes that global warming will only be prevented by a 'push-and-pull' strategy—the former provided by 'a substantial carbon tax', the latter by massively increased state investment in technologies that shift energy production from the current oil- and coal-based sources to renewable ones.[751] But this also poses the question of whether such intervention could be useful in other spheres.

Why, for instance, is something as fundamental as the development of new antibiotics only left to the big pharmaceutical companies, which have traditionally not focused much attention on this area of medicine because of its low profit margins? One solution proposed by economist Jim O'Neill is for big pharma to pay into a $2 billion global 'innovation fund' that would fund 'blue-sky' research into antibiotics—with much of the money going to universities and small biotech companies.[752] 'We need to kick-start drug development to make sure the world has the drugs it needs, to treat infections and to enable modern medicine and surgery to continue as we know it,' he says.[752] Against those who see such a contribution as too costly, O'Neill points to the fact that antibiotic-resistant bacteria could kill 10 million people a year worldwide by 2050 and cost $100 trillion in lost economic output. However, such a fund may require higher taxation of pharmaceutical companies if they can't be persuaded to act with what O'Neill calls 'enlightened self-interest'.[752]

At the same time it seems pertinent to ask why agribusiness companies are allowed to use antibiotics in farming in a manner that threatens human health, and whether tighter restrictions are required. An important move forward in this direction was recently made by California Governor Jerry Brown, who signed the USA's most stringent law regulating the use of antibiotics in livestock, banning antibiotics used in human medicine and those used solely for growth promotion. According to Brown: 'The science is clear that the overuse of antibiotics in livestock has contributed to the spread of antibiotic resistance and the undermining of decades of life-saving advances in medicine.'[753] However, while welcoming the new law, Rebecca Spector, director of the Center for Food Safety, wants even further 'restrictions on the use of other antibiotics and other drugs in animal production, and also…stronger sanitary requirements and more space for animals so that they can exhibit natural behaviors', which is 'critical to preserving these drugs for human use, and also to encourage producers to improve the living conditions of these animals so that drugs are less necessary'.[753] In fact, despite such sentiments, the quantities of antibiotics sold for use in US food-producing animals increased 9 percent in 2018, according to the country's Food and Drug Administration.[754]

It is clear that genome editing offers many possibilities for creating animals and plants that can adapt to, or even help offset, a changing climate. And engineering disease resistance into crops and livestock could help reduce the use of pesticides and antibiotics in agriculture. Yet already we are seeing uses of the technology that some might consider frivolous, like apple slices that don't brown, or sinister, such as sterile plants or animals that never reach reproductive age, leading to even more control of production falling into the hands of giant companies. In Chapter 7 we saw how gene drives are a technology that could be used to tackle mosquito-borne diseases like malaria. But they also have other potential uses. A group of Harvard biologists recently wrote in the journal *eLIFE* that gene drives could 'support agriculture by reversing pesticide and herbicide resistance in insects and weeds, and control damaging invasive species'.[755] But, while they might have many benefits, there could be also many negative consequences of tampering with natural ecosystems in this fashion. A reasonable question to ask, therefore, might be whether we need a wider informed public debate about the direction of the new technology in agriculture. Some people may remain hostile to GM technology in this sphere; others will see the potential for human benefit but want evidence that genome editing is really going to be used for making a significant contribution to feeding the world's population, not just increasing the bank balances of giant multinationals. And, at the same time, we surely need a proper debate about the quality of our food, and whether governments should be doing more to restrict the spread of junk food and to subsidize healthier alternatives. For there is a risk of only focusing on 'technological fixes' to health problems, when what are really required are societal measures.

Finally, there is the question of how far new biotechnologies should be allowed to proceed in modifying human cells or modelling human health and disease in animals. One potentially exciting possibility that we've considered in this book is that such technologies might lead to better treatments for mental disorders. Currently, the drugs available to treat such disorders leave much to be desired. Indeed, Simon Schulz, director of the Centre of Neurotechnology at Imperial College London, believes there is a fundamental problem with the present strategy, 'because it is becoming exponentially more difficult to find new compounds that can deal with the disease in question but which don't affect other things'.[756] Instead, Schulz points to technologies like optogenetics, which might be applied in humans after genome editing of cells in the brain. 'This method doesn't need to be used with light,' he says. 'You could put a similar sensitivity in the neuron to make it drug-activated instead. I see that—in particular—as being a really powerful approach in the future.'[756] In fact, such a strategy may have great potential for medicine, but also for misuse. 'We'll have the technology at some point to place invasive implants into people's heads to give them new senses,' says Schulz. 'I'd imagine the military would like to be able to do that. We might consider it unethical for them to do certain things. But on the other hand, if you had someone who is tetraplegic and has to have these implants then you have a strong ethical case to give them control of their body to be able to live an independent life. And why would you stop with normal body functionality? Why would you stop?'[756]

Now I'd be very surprised if there weren't many people who would want to stop not just at this point, but considerably sooner, when it comes to the prospect of genetically engineering human brain cells. But Schulz makes a valid argument when he asks us to consider whether what's considered acceptable in a few decades' time might not be very different from today. For, as he says, 'Look at privacy as a social concept. Think about how our great grandparents would have dealt with the level of exposure we now have online. It's something they would never have considered. What's acceptable now and what will be acceptable in 2035 may be quite different. Are we developing technology for the ethics as they are now or the ethics of 2035?'[756] Clearly, though, in any democratic society, if perspectives are to change, this needs to be on the basis of maximum open public discussion. And while the prospects offered by genome editing or stem-cell technology for medicine are very exciting, evidently there needs to be much thought about the safety and ethics of treating human patients with such radically different forms of therapy, and also about the welfare of the animals employed to model disease using the new technologies.

As for the question of whether genome editing should ever be used to modify the human germline, there are many different viewpoints, as we've seen at various points in this book, with some violently opposed and others who argue that, in some circumstances, it might be justified as a treatment for disease. As we've also seen, some people even believe that, if proved to be safe, then it might be quite valid to use genome editing of the human germline to 'enhance' the species. At the moment it would be hard to find anyone proposing to make practical use of the technology in this way. Yet who knows what the situation might be in five or ten years' time, if genome editing becomes increasingly fail-safe and our ability to interpret the genome itself more refined. At the same time, will genetic modification of our nearest primate relatives begin to blur the boundaries between what is human and what is not? These are as much social questions as scientific ones, and as they concern us all, they should be the subject of a far-reaching, scientifically informed public debate. Hopefully, this book will prove to be a useful starting point.

GLOSSARY

CRISPR/CAS9 Genome editing technology that uses CRISPRs (clustered regularly interspaced short palindromic repeats) and the CAS9 enzyme.

Embryonic stem (ES) cells Pluripotent stem cells isolated from early embryos that have the potential to give rise to any cell type in the body.

Enzyme A biological molecule that acts as a catalyst. Most enzymes are proteins, but certain RNAs, called ribozymes, also have catalytic activity.

Extremophile Micro-organisms with the ability to thrive in extreme environments such as in hot springs, icy wastes, chemically polluted springs, or under high pressure.

Gene expression Overall process by which the information encoded in a gene is converted into an observable phenotype (most commonly production of a protein).

Genome editing A type of genetic engineering in which DNA is inserted, replaced, or removed from a genome of a living cell using artificially engineered nucleases, or 'molecular scissors'.

Genome-wide association study Investigation of many common genetic variants in different individuals to see if any variant is associated with a characteristic.

Induced pluripotent stem (iPS) cell Normal cell that has been altered so that it can give rise to any cell in the body by modifying its pattern of gene expression.

Knockout and knockin Organisms in which a gene has either been totally disabled (knockouts) or subtly altered, for instance by a mutation, or addition of a fluorescent tag (knockins).

Mutation A permanent, heritable change in the DNA sequence of a chromosome, usually in a single gene; commonly leads to a change in or loss of the normal function of the gene product.

Optogenetics A technique that involves the use of light to control cells in a living tissue, typically neurons, that have been genetically modified to express light-sensitive protein pores.

Organoid A three-dimensional organ-like structure grown in a culture dish, using iPS or ES cells that have been allowed to give rise to different cell types in a 3D matrix.

Promoter DNA sequence that determines the site of transcription initiation for RNA polymerase.

Restriction enzyme Proteins found naturally in bacteria that cut at or near specific DNA sequences and that can be used as a tool in molecular biology to make gene constructs.

RNA interference The phenomenon of gene silencing mediated by the interaction of a double-stranded RNA, with a corresponding target messenger RNA.

TALENs Transcription activator-like effector nucleases are artificial restriction enzymes generated by fusing a TALE DNA-binding protein to a DNA cutting protein.

Transcription Process whereby one strand of a DNA molecule is used as a template for synthesis of a complementary RNA by RNA polymerase.

Transcription factor General term for any protein, other than RNA polymerase, required to initiate or regulate transcription in eukaryotic cells.

Translation The ribosome-mediated production of a protein whose amino acid sequence is specified by the nucleotide sequence in an mRNA.

Xeno-DNA (XNA) Artificial forms of DNA that have unusual letters, or bases, or a different chemical backbone, to naturally occurring DNA.

ZFNs Artificial restriction enzymes generated by fusing a zinc finger DNA-binding domain to a DNA cutting protein.

ENDNOTES

Introduction: The Gene Revolution

1. Genes and human disease, *World Health Organization*, <http://www.who.int/genomics/public/geneticdiseases/en/index2.html> (2016).
2. Genome editing, *Science Media Centre*, <http://www.sciencemediacentre.org/genome-editing/> (2015).
3. Quaglia, D., Synthetic biology: the dawn of a new era, *Huffington Post*, <http://www.huffingtonpost.com/daniela-quaglia/synthetic-biology-the-daw_b_7990020.html> (2015).
4. The biotech revolution, *ABC Science*, <http://www.abc.net.au/science/features/biotech/1970.htm> (2004).
5. 1982: the transgenic mouse, *University of Washington*, <http://www.washington.edu/research/pathbreakers/1982b.html> (1996).
6. Loria, K., The genetic technology that's going to change everything is at a critical turning point, *Business Insider*, <http://uk.businessinsider.com/how-crispr-could-change-the-world-2015-8> (2015).
7. Baker, M., Gene editing at CRISPR speed. *Nature Biotechnology* 32: 309–12 (2014).
8. McNutt, M., Breakthrough to genome editing. *Science* 350: 1445 (2015).
9. Lewis, T., Scientists may soon be able to 'cut and paste' DNA to cure deadly diseases and design perfect babies, *Business Insider*, <http://uk.businessinsider.com/how-crispr-will-revolutionize-biology-2015-10?r=US&IR=T> (2015).
10. Macrae, F., Our little miracle! Baby girl battling leukaemia saved by 'revolutionary' cell treatment, *Daily Mail*, <http://www.dailymail.co.uk/health/article-3305603/World-baby-girl-battling-leukaemia-saved-miracle-treatment-Genetically-modified-cells-hunt-kill-disease-transform-cancer-care.html> (2015).
11. Normile, D., Chinese scientist who produced genetically altered babies sentenced to 3 years in jail, Science, <https://www.sciencemag.org/news/2019/12/chinese-scientist-who-produced-genetically-altered-babies-sentenced-3-years-jail> (2019).
12. Johnston, M. and Loria, K., This is the game-changing technology that was used to genetically modify a human embryo, *Business Insider*, <http://uk.businessinsider.com/how-to-genetically-edit-a-human-embryo-2015-8> (2015).
13. Deisseroth, K., Optogenetics: controlling the brain with light [extended version], *Scientific American*, <http://www.scientificamerican.com/article/optogenetics-controlling/> (2010).

14. Sutherland, S., Revolutionary neuroscience technique slated for human clinical trials, *Scientific American*, <http://www.scientificamerican.com/article/revolutionary-neuroscience-technique-slated-for-human-clinical-trials/> (2016).
15. Gorman, C., What's next for stem cells and regenerative medicine?, *Scientific American*, <http://www.scientificamerican.com/article/regenerative-medicine-whats-next-stem-cells/> (2013).
16. Yee, J., Turning somatic cells into pluripotent stem cells. *Nature Education* 3: 25 (2010).
17. McGowan, K., Scientists make progress in growing organs from stem cells, *Discover Magazine*, <http://discovermagazine.com/2014/jan-feb/05-stem-cell-future> (2014).
18. Yong, E., Synthetic yeast chromosome, *The Scientist*, <http://www.the-scientist.com/?articles.view/articleNo/39573/title/Synthetic-Yeast-Chromosome/> (2014).
19. Fecht, S., XNA: synthetic DNA that can evolve, *Popular Mechanics*, <http://www.popularmechanics.com/science/health/a7636/xna-synthetic-dna-that-can-evolve-8210483/> (2012).

Chapter 1: Natural Born Mutants

20. Moss, L., 12 bizarre examples of genetic engineering, *Mother Nature Network*, <http://www.mnn.com/green-tech/research-innovations/photos/12-bizarre-examples-of-genetic-engineering/mad-science> (2015).
21. Parrington, J., *The Deeper Genome* (Oxford University Press, 2015), pp. 166–80.
22. The development of agriculture, *National Geographic*, <https://genographic.nationalgeographic.com/development-of-agriculture/> (2016).
23. We were wolves, once, *Pin it*, <https://www.pinterest.com/pin/476466835552028679/> (2015).
24. Melina, R., How did dogs get to be dogs?, *Live Science*, <http://www.livescience.com/8405-dogs-dogs.html> (2010).
25. University of Cambridge, New research confirms 'out of Africa' theory of human evolution, *Science Daily*, <http://www.sciencedaily.com/releases/2007/05/070509161829.htm> (2007).
26. Yong, E., Origin of domestic dogs, *The Scientist*, <http://www.the-scientist.com/?articles.view/articleNo/38279/title/Origin-of-Domestic-Dogs/> (2013).
27. Dog has been man's best friend for 33,000 years, DNA study finds, *The Telegraph*, <http://www.telegraph.co.uk/news/science/science-news/12052798/Dog-has-been-mans-best-friend-for-33000-years-DNA-study-finds.html> (2015).
28. Ghosh, P., DNA hints at earlier dog evolution, *BBC News*, <http://www.bbc.co.uk/news/science-environment-32691843> (2015).
29. Underwood, F. A. and Radcliffe, J., Mowgli's brothers, *Kipling Society*, <http://www.kiplingsociety.co.uk/rg_mowglibros1.htm> (2008).
30. Hare, B. and Woods, V., Opinion: we didn't domesticate dogs. They domesticated us, *National Geographic*, <http://news.nationalgeographic.com/news/2013/03/130302-dog-domestic-evolution-science-wolf-wolves-human/> (2013).

31. Chaika, E. O., Evolution of wolf to dog, *Elaine Ostrach Chaika*, <http://elainechaika.com/2013/01/mainstream-scholars-specializing-in.html> (2013).
32. McKie, R., How hunting with wolves helped humans outsmart the Neanderthals, *The Guardian*, <https://www.theguardian.com/science/2015/mar/01/hunting-with-wolves-humans-conquered-the-world-neanderthal-evolution> (2015).
33. Choi, C. Q., Being more infantile may have led to bigger brains, *Scientific American*, <https://www.scientificamerican.com/article/being-more-infantile/> (2009).
34. Bradshaw, J. W., Pullen, A. J. and Rooney, N., Why do adult dogs 'play'? *Behavioural Processes* 110: 82–7 (2015).
35. Callaway, E., Dog's dinner was key to domestication, *Nature News*, <http://www.nature.com/news/dog-s-dinner-was-key-to-domestication-1.12280> (2013).
36. Teh, B., Scientists discover the ancestor of modern dogs and wolves, *Regal Tribune*, <http://www.regaltribune.com/scientists-discover-the-ancestor-of-modern-dogs-and-wolves/21032/> (2015).
37. Bradshaw, J., Dogs we understand; cats are mysterious, even though they are the most popular pet, *Washington Post*, <http://www.washingtonpost.com/national/health-science/dogs-we-understand-cats-are-mysterious-even-though-they-are-the-most-popular-pet/2013/10/14/2c59c6b0-26ca-11e3-ad0d-b7c8d2a594b9_story.html> (2013).
38. Empson, M., *Land and Labour: Marxism, Ecology and Human History* (Bookmarks, 2014), pp. 29–52.
39. Hill, J., Cats in ancient Egypt, *Ancient Egypt Online*, <http://www.ancientegyptonline.co.uk/cat.html> (2010).
40. Stockton, N., Scientists discover genes that helped turn fearsome wildcats into house cats, *Wired*, <http://www.wired.com/2014/11/genes-cat-domestication/> (2014).
41. All the burning questions you have about your cat's wild past, answered, *Huffington Post*, <http://www.huffingtonpost.com/2016/01/07/questions-about-cats-answered_n_8398800.html> (2016).
42. Olena, A., Understanding cats, *The Scientist*, <http://www.the-scientist.com/?articles.view/articleNo/37942/title/Understanding-Cats/> (2013).
43. Mueller, U. G. and Gerado, N., Fungus-farming insects: multiple origins and diverse evolutionary histories. *Proceedings National Academy Sciences USA* 99: 15247–9 (2002).
44. Meyer, R. S. and Purugganan, M. D., Evolution of crop species: genetics of domestication and diversification. *Nature Reviews. Genetics* 14: 840–52 (2013).
45. Ladizinsky, G., *Plant Evolution under Domestication* (Springer Science & Business Media, 2012), p. 190.
46. Ames, B. N., Profet, M. and Gold, L. S., Nature's chemicals and synthetic chemicals: comparative toxicology. *Proceedings National Academy Sciences USA* 87: 7782–6 (1990).
47. Washington University in St Louis, How rice twice became a crop and twice became a weed—and what it means for the future, *Science Newsline*, <http://www.sciencenewsline.com/articles/2013071717040014.html> (2013).
48. Brix, L., Humans have added new bones to the pig, *Science Nordic*, <http://sciencenordic.com/humans-have-added-new-bones-pig> (2012).

49. Gray, R., Did farming pigs change our sense of SMELL? Domestication of animals thousands of years ago may have driven evolution how we detect odours, *Daily Mail*, <http://www.dailymail.co.uk/sciencetech/article-3148389/Did-farming-pigs-change-sense-SMELL-Domestication-animals-thousands-years-ago-driven-evolution-detect-odours.html> (2015).
50. BGI Shenzen, First goat genome sets a good example for facilitating de novo assembly of large genomes, *Science Daily*, <http://www.sciencedaily.com/releases/2012/12/121223152629.htm> (2012).
51. Williams, S. C. P., Whence the domestic horse? *Science News*, <http://news.sciencemag.org/plants-animals/2012/05/whence-domestic-horse> (2012).
52. Diamond, J., The worst mistake in the history of the human race, *Discover Magazine*, <http://discovermagazine.com/1987/may/02-the-worst-mistake-in-the-history-of-the-human-race> (1999).
53. Harper, K. N. and Armelagos, G. J., Genomics, the origins of agriculture, and our changing microbe-scape: time to revisit some old tales and tell some new ones. *American Journal of Physical Anthropology* 57: 135–52 (2013).
54. Boyko, A. R., The domestic dog: man's best friend in the genomic era. *Genome Biology* 12: 216 (2011).
55. Bloom, P., The curious development of dog breeds, *My Magic Dog*, <http://mymagicdog.com/1438/the-curious-development-of-dog-breeds/> (2013).
56. Bodio, S. J., Darwin's other birds, *The Cornell Lab of Ornithology*, <http://www.allaboutbirds.org/Page.aspx?pid=1435> (2009).
57. McNamara, R., Charles Darwin and his voyage aboard H.M.S. Beagle, *About Education*, <http://history1800s.about.com/od/innovators/a/hmsbeagle.htm> (2016).
58. Cookson, C., Darwin's origin of the pigeon, *Financial Times*, <http://www.ft.com/cms/s/2/1399529a-7576-11e2-b8ad-00144feabdc0.html> (2013).
59. Pigeons and variation, *Christ's College, Cambridge*, <http://darwin200.christs.cam.ac.uk/node/88> (2015).
60. Natural selection, *Christ's College, Cambridge*, <http://darwin200.christs.cam.ac.uk/node/76> (2015).
61. Cyranoski, D., Genetics: pet project, *Nature News*, <http://www.nature.com/news/2010/100825/full/4661036a.html> (2010).
62. Callaway, E., 'I can haz genomes': cats claw their way into genetics, *Nature News*, <http://www.nature.com/news/i-can-haz-genomes-cats-claw-their-way-into-genetics-1.16708> (2015).
63. Rimbault, M. and Ostrander, E. A., So many doggone traits: mapping genetics of multiple phenotypes in the domestic dog. *Human Molecular Genetics* 21, R52–7 (2012).
64. Ledford, H., Dog DNA probed for clues to human psychiatric ills, *Nature News*, <http://www.nature.com/news/dog-dna-probed-for-clues-to-human-psychiatric-ills-1.19235> (2016).
65. Rietveld, M., The Hulk's incredible genome, *Genome News Network*, <http://www.genomenewsnetwork.org/articles/07_03/hulk.shtml> (2003).
66. Loewe, L., Genetic mutation. *Nature Education* 1: 113 (2008).

67. Markel, H., February 28: the day scientists discovered the double helix, *Scientific American*, <http://www.scientificamerican.com/article/february-28-the-day-scientists-discovered-double-helix/> (2013).
68. Sarasin, A. and Stary, A., Human cancer and DNA repair-deficient diseases. *Cancer Detection and Prevention* 21: 406–11 (1997).
69. Jones, S., Angelina Jolie's aunt Debbie Martin dies of breast cancer, *The Guardian*, <http://www.theguardian.com/film/2013/may/27/angelina-jolie-aunt-debbie-martin-dies-breast-cancer> (2013).
70. The genetics of cancer, *Cancer Net*, <http://www.cancer.net/navigating-cancer-care/cancer-basics/genetics/genetics-cancer> (2015).
71. BRCA1 and BRCA2: Cancer risk and genetic testing, *National Cancer Institute*, <http://www.cancer.gov/about-cancer/causes-prevention/genetics/brca-fact-sheet> (2015).
72. What is cystic fibrosis, *Cystic Fibrosis Trust*, <http://www.cysticfibrosis.org.uk/about-cf/what-is-cystic-fibrosis> (2016).
73. Montgomery, S., Natural selection, *Christ's College, Cambridge*, <http://darwin200.christs.cam.ac.uk/node/76> (2015).
74. Gregor Johann Mendel, *Complete Dictionary of Scientific Biography*, <http://www.encyclopedia.com/topic/Gregor_Johann_Mendel.aspx> (2008).
75. Watson, J. D., *DNA* (Arrow Books, 2004), p. 38.
76. Watson, J. D., *The Annotated and Illustrated Double Helix* (Simon & Schuster, 2012), p. 200.
77. Kandel, E. R., Genes, chromosomes, and the origins of modern biology, *Columbia University*, <http://www.columbia.edu/cu/alumni/Magazine/Legacies/Morgan/> (2016).
78. Benson, K. R., T. H. Morgan's resistance to the chromosome theory. *Nature Reviews. Genetics* 2: 469–74 (2001).
79. Carlson, E. A., H. J. Muller's contributions to mutation research. *Mutation Research* 752: 1–5 (2013).
80. Green, T., Hermann Muller: a genetics pioneer, *University of Texas News*, <http://news.utexas.edu/2010/01/19/hermann-muller-a-genetics-pioneer> (2010).
81. Wilhelm Conrad Röntgen: biographical, *The Nobel Foundation*, <http://www.nobelprize.org/nobel_prizes/physics/laureates/1901/rontgen-bio.html> (1901).
82. Bagley, M., Marie Curie: facts and biography, *Live Science*, <http://www.livescience.com/38907-marie-curie-facts-biography.html> (2013).
83. O'Carroll, E., Marie Curie: why her papers are still radioactive, *Christian Science Monitor*, <http://www.csmonitor.com/Technology/Horizons/2011/1107/Marie-Curie-Why-her-papers-are-still-radioactive> (2011).
84. Voosen, P., Hiroshima and Nagasaki cast long shadows over radiation science, *New York Times*, <http://www.nytimes.com/gwire/2011/04/11/11greenwire-hiroshima-and-nagasaki-cast-long-shadows-over-99849.html?pagewanted=all> (2011).
85. Mallo, M., Wellik, D. M. and Deschamps, J., Hox genes and regional patterning of the vertebrate body plan. *Developmental Biology* 344: 7–15 (2010).
86. Leyssen M. and Hassan, B. A., A fruitfly's guide to keeping the brain wired. *EMBO Reports* 8: 46–50 (2007).

87. Mobley, E., Activity of entire central nervous system captured on film for first time, *The Guardian*, <http://www.theguardian.com/science/2015/aug/11/activity-of-entire-central-nervous-system-captured-on-film-for-first-time> (2015).
88. Royer, N., The history of fancy mice, *American Fancy Rat and Mouse Association*, <http://www.afrma.org/historymse.htm> (2015).
89. Culliton, B. J., *The Monk in the Garden* by Robin Marantz Henig, *Genome News Network*, <http://www.genomenewsnetwork.org/articles/06_00/monk_excerpt.php> (2000).
90. Emani, C., The lab mouse story, *Dr. Sliderule's Archiac Science Ramblings*, <http://drsliderule.blogspot.co.uk/2016/01/the-lab-mouse-story.html> (2016).
91. Steensma, D. P., Kyle, R. A. and Shampo, M. A., Abbie Lathrop, the 'mouse woman of Granby': rodent fancier and accidental genetics pioneer. *Mayo Clinic Proceedings* 85: e83 (2010).
92. Rader, K., *Where the Wild Things Are Now: Domestication Reconsidered* (Oxford University Press, 2007), pp. 189–90.
93. Crow, J. F., C. C. Little, cancer and inbred mice. *Genetics* 161: 1357–61 (2002).
94. Gondo, Y., Trends in large-scale mouse mutagenesis: from genetics to functional genomics. *Nature Reviews. Genetics* 9: 803–10 (2008).
95. Arney, K., Interview: Prof Karen Steel—genes and deafness, *The Naked Scientists*, <http://www.thenakedscientists.com/HTML/interviews/interview/1000565/> (2014).
96. Quick statistics, *National Institute on Deafness and Other Communication Disorders*, <http://www.nidcd.nih.gov/health/statistics/pages/quick.aspx> (2015).
97. Steel, K., Mouse genetics for studying mechanisms of deafness and more: an interview with Karen Steel. Interview by Sarah Allan. *Disease Model Mechanisms* 4: 716–18 (2011).
98. O'Sullivan, G. J., O'Tuathaigh, C. M., Clifford, J. J., O'Meara, G. F., Croke, D. T. and Waddington, J. L., Potential and limitations of genetic manipulation in animals. *Drug Discovery Today Technology* 3: 173–80 (2006).
99. Mogensen, M. M., Rzadzinska, A. and Steel, K. P., The deaf mouse mutant whirler suggests a role for whirlin in actin filament dynamics and stereocilia development. *Cell Motility and the Cytoskeleton* 64: 496–508 (2007).
100. Friedman, J. M. and Halaas, J. L., Leptin and the regulation of body weight in mammals. *Nature* 395: 763–70 (1998).
101. Kroen, G. C., Food for sale everywhere fuels obesity epidemic, *Scientific American*, <http://www.scientificamerican.com/podcast/episode/food-for-sale-everywhere-fuels-obesity-epidemic/> (2015).

Chapter 2: Supersize My Mouse

102. Wolpert, L., Is cell science dangerous? *Journal of Medical Ethics* 33: 345–8 (2007).
103. Ball, P., Worried when science plays God? It's only natural, *The Guardian*, <http://www.theguardian.com/commentisfree/2015/feb/26/science-plays-god-three-parent-babies-sceptics> (2015).

104. Poulter, S., Britain to sprout 'Frankenfoods' after EU ruling: controversial crops could be grown from next year after being approved, *Daily Mail*, <http://www.dailymail.co.uk/news/article-2909128/Frankenfoods-grown-Britain-year-EU-ruling-controversial-crops.html> (2015).
105. Watson, J. D., *The Annotated and Illustrated Double Helix* (Simon & Schuster, 2012), p. 209.
106. Parrington, J., *The Deeper Genome* (Oxford University Press, 2015), pp. 33–8.
107. How the code was cracked, *Nobel Foundation*, <http://www.nobelprize.org/educational/medicine/gene-code/history.html> (2014).
108. Norman, J., Discovery of bacteriophages: viruses that infect bacteria, *History of Information*, <http://www.historyofinformation.com/expanded.php?id=4411> (2015).
109. The Nobel Prize in Physiology or Medicine 1978, *The Nobel Foundation*, <http://www.nobelprize.org/nobel_prizes/medicine/laureates/1978/> (1978).
110. Pray L., Restriction enzymes. *Nature Education* 1: 38 (2008).
111. Roberts, R. J., How restriction enzymes became the workhorses of molecular biology. *Proceedings National Academy Sciences USA* 102: 5905–8 (2005).
112. Birch, D., Hamilton Smith's second chance, *Baltimore Sun*, <http://articles.baltimoresun.com/1999-04-11/news/9904120283_1_hamilton-smith-scientist-nobel> (1999).
113. Cohen, S. N., DNA cloning: a personal view after 40 years. *Proceedings National Academy Sciences USA* 110: 15521–9 (2013).
114. Russo, E., The birth of biotechnology. *Nature* 421: 456–7 (2003).
115. Bacterial DNA: the role of plasmids, *Biotechnology Learning Hub*, <http://biotechlearn.org.nz/themes/bacteria_in_biotech/bacterial_dna_the_role_of_plasmids> (2014).
116. The banker and the biologist, *BBC News*, <http://news.bbc.co.uk/1/hi/magazine/7875331.stm> (2009).
117. Statistics and facts about the biotech industry, *Statista*, <http://www.statista.com/topics/1634/biotechnology-industry/> (2016).
118. Berg, P., Meetings that changed the world: Asilomar 1975: DNA modification secured. *Nature* 455: 290–1 (2008).
119. Brownlee, C., Biography of Rudolf Jaenisch. *Proceedings National Academy Sciences USA* 101: 13982–4 (2004).
120. Rudolf Jaenisch, *Science Watch*, <http://archive.sciencewatch.com/inter/aut/2009/09-mar/09marJaen/> (2009).
121. 1982: the transgenic mouse, *University of Washington*, <http://www.washington.edu/research/pathbreakers/1982b.html> (1996).
122. Stratton, K., Beyond luck, *Bellwether*, <http://www.vet.upenn.edu/docs/default-source/Research/brinster-story_bellwether.pdf?sfvrsn=0> (2012).
123. Jiang, T., Xing, B. and Rao J., Recent developments of biological reporter technology for detecting gene expression. *Biotechnology Genetic Engineering Revolution* 25: 41–75 (2008).
124. Gallagher, S., Seeing the knee in a new light: fluorescent probe tracks osteoarthritis development, *Tufts Now*, <http://now.tufts.edu/news-releases/seeing-knee-new-light-fluorescent-probe-tracks-osteoarthritis-development#sthash.eVnOQk7g.dpuf> (2015).

125. Szabala, B. M., Osipowski, P. and Malepszy, S. Transgenic crops: the present state and new ways of genetic modification. *Journal of Applied Genetics* 55: 287–94 (2014).
126. Rutherford, A. Why GM is the natural solution for future farming, *The Guardian*, <http://www.theguardian.com/science/2015/jan/31/gm-farming-natural-solution> (2015).
127. GM (genetic modification), *Soil Association*, <http://www.soilassociation.org/gm> (2015).
128. Gilbert, N., Case studies: a hard look at GM crops. *Nature* 497: 24–6 (2013).
129. GM genocide?, *The Economist*, <http://www.economist.com/blogs/feastandfamine/2014/03/gm-crops-indian-farmers-and-suicide> (2014).
130. Terminator gene halt a 'major U-turn', *BBC News*, <http://news.bbc.co.uk/1/hi/sci/tech/465222.stm> (1999).
131. Reinhardt, C. and Ganzel, W., The science of hybrids, *Wessel's Living History Farm*, <http://www.livinghistoryfarm.org/farminginthe30s/crops_03.html> (2003).
132. Are genetically modified plant foods safe to eat? *Green Facts*, <http://www.greenfacts.org/en/gmo/3-genetically-engineered-food/4-food-safety-labelling.htm> (2015).
133. GM food study was 'flawed', *BBC News*, <http://news.bbc.co.uk/1/hi/sci/tech/346651.stm> (1999).
134. Vidal, J. and Tran, M., Cut use of antibiotics in livestock, veterinary experts tell government, *The Guardian*, <http://www.theguardian.com/uk-news/2014/jul/07/reduce-antibiotics-farm-animals-resistant-bacteria> (2014).
135. McKie, R., After 30 years, is a GM food breakthrough finally here?, *The Guardian*, <http://www.theguardian.com/environment/2013/feb/02/genetic-modification-breakthrough-golden-rice> (2013).
136. Genes and human disease, *World Health Organization*, <http://www.who.int/genomics/public/geneticdiseases/en/index2.html> (2016).
137. Why use gene therapy for cystic fibrosis? *Oxford University Gene Medicine*, <http://www.genemedresearch.ox.ac.uk/genetherapy/cfgt.html> (2012).
138. Stem cell and gene therapy, *Immune Deficiency Foundation*, <http://primaryimmune.org/treatment-information/stem-cell-and-gene-therapy/> (2015).
139. Kay, M. A., Glorioso, J. C. and Naldini, L., Viral vectors for gene therapy: the art of turning infectious agents into vehicles of therapeutics. *Nature Medicine* 7: 33–40 (2001).
140. Life cycle of HIV, a retrovirus, *Sinauer Associates*, <http://www.sumanasinc.com/webcontent/animations/content/lifecyclehiv.html> (2002).
141. Why gene therapy caused leukemia in some 'boy in the bubble syndrome' patients, Science Daily, <http://www.sciencedaily.com/releases/2008/08/080807175438.htm> (2008).
142. Geddes, L., 'Bubble kid' success puts gene therapy back on track, *New Scientist*, <https://www.newscientist.com/article/mg22029413-200-bubble-kid-success-puts-gene-therapy-back-on-track/> (2013).
143. Genes and human disease, *World Health Organization*, <http://www.who.int/genomics/public/geneticdiseases/en/index2.html> (2016).

144. Prasad, A., Teratomas: the tumours that can transform into 'evil twins', *The Guardian*, <http://www.theguardian.com/commentisfree/2015/apr/27/teratoma-tumour-evil-twin-cancer> (2015).
145. Lewis, R. A., Stem cell legacy: Leroy Stevens, *The Scientist*, <http://www.the-scientist.com/?articles.view/articleNo/12717/title/A-Stem-Cell-Legacy--Leroy-Stevens/> (2000).
146. Lancaster, C., How teratomas became embryonic stem cells, *24th International Congress of History of Science, Technology and Medicine*, <http://www.ichstm2013.com/blog/2013/05/30/how-teratomas-became-embryonic-stem-cells/> (2013).
147. Prelle, K., Zink, N. and Wolf, E., Pluripotent stem cells: model of embryonic development, tool for gene targeting, and basis of cell therapy. *Anatomia, Histologia, Embryologia* 31: 169–86 (2002).
148. Krejci, L., Altmannova, V., Spirek, M. and Zhao, X., Homologous recombination and its regulation. *Nucleic Acids Research* 40: 5795–818 (2012).
149. Otto, S., Sexual reproduction and the evolution of sex. *Nature Education* 1: 182 (2008).
150. Laden, G., How long is a human generation? *Science Blogs*, <http://scienceblogs.com/gregladen/2011/03/01/how-long-is-a-generation/> (2011).
151. Todar, K., The growth of bacterial populations, *Online Textbook of Bacteriology*, <http://textbookofbacteriology.net/growth_3.html> (2012).
152. Moulton, G. E., Meiosis and sexual reproduction, *Infoplease*, <http://www.infoplease.com/cig/biology/meiosis-sexual-reproduction.html> (2015).
153. Jones, S., Angelina Jolie's aunt Debbie Martin dies of breast cancer, *The Guardian*, <http://www.theguardian.com/film/2013/may/27/angelina-jolie-aunt-debbie-martin-dies-breast-cancer> (2013).
154. Powell, S. N. and Kachnic, L. A. Roles of BRCA1 and BRCA2 in homologous recombination, DNA replication fidelity and the cellular response to ionizing radiation. *Oncogene* 22: 5784–91 (2003).
155. Welcsh, P. L. and King, M., BRCA1 and BRCA2 and the genetics of breast and ovarian cancer. *Human Molecular Genetics* 10: 705–13 (2001).
156. The Nobel Prize in Physiology or Medicine 2007, *Nobel Foundation*, <http://www.nobelprize.org/nobel_prizes/medicine/laureates/2007/> (2007).
157. Connor, S., The breakthrough of 'gene targeting', *The Independent*, <http://www.independent.co.uk/news/science/the-breakthrough-of-gene-targeting-394494.html> (2007).
158. Gumbel, A., Mario Capecchi: the man who changed our world, *The Independent*, <http://www.independent.co.uk/news/science/mario-capecchi-the-man-who-changed-our-world-396387.html> (2007).
159. Babinet, C. J., Transgenic mice: an irreplaceable tool for the study of mammalian development and biology. *American Society of Nephrology* 11: S88–S94 (2000).
160. Parrington, J. and Tunn, R., Ca(2+) signals, NAADP and two-pore channels: role in cellular differentiation. *Acta physiologica (Oxford)* 211: 285–96 (2014).
161. Berridge, M. J., Bootman, M. D. and Roderick, H. L., Calcium signalling: dynamics, homeostasis and remodelling. *Nature Reviews. Molecular and Cell Biology* 4: 517–29 (2003).

162. A case study of the effects of mutation: sickle cell anemia, *Understanding Evolution*, <http://evolution.berkeley.edu/evolibrary/article/mutations_06> (2016).
163. Kashir, J., Heindryckx, B., Jones, C., De Sutter, P., Parrington, J. and Coward K., Oocyte activation, phospholipase C zeta and human infertility. *Human Reproduction Update* 16: 690–703 (2010).

Chapter 3: Light as a Life Tool

164. Editors of Encyclopædia Britannica, Sun worship, *Encyclopædia Britannica*, <http://www.britannica.com/EBchecked/topic/573676/sun-worship> (2015).
165. Circadian rhythms fact sheet, *National Institutes of Health*, <http://www.nigms.nih.gov/Education/Pages/Factsheet_CircadianRhythms.aspx> (2015).
166. Sample, I., Jet lag and night shifts disrupt rhythm of hundreds of genes, study shows, *The Guardian*, <http://www.theguardian.com/science/2014/jan/20/jeg-lag-disrupts-genes-study> (2014).
167. Diverse eyes, *Understanding Evolution*, <http://evolution.berkeley.edu/evolibrary/article/0_0_0/eyes_02> (2005).
168. Rhodes, J., The beautiful flight paths of fireflies, *Smithsonian Magazine*, <http://www.smithsonianmag.com/arts-culture/beautiful-flight-paths-fireflies-180949432/?no-ist> (2014).
169. Bioluminescence, *National Geographic*, <http://education.nationalgeographic.com/education/encyclopedia/bioluminescence/?ar_a=1> (2016).
170. Robert Hooke, *Famous Scientists*, <http://www.famousscientists.org/robert-hooke/> (2016).
171. House, P., Robert Hooke and the discovery of the cell, *Science of Aging*, <http://www.science-of-aging.com/timelines/hooke-history-cell-discovery.php> (2009).
172. Hughes, E. and Pierson, R., The animalcules of Antoni Van Leeuwenhoek, *Journal of Obstetrics and Gynaecology* 35: 960 (2013).
173. Kelly, D., The first person who ever saw sperm cells collected them from his wife, *Gizmodo*, <http://throb.gizmodo.com/the-first-time-anyone-saw-sperm-1708170526> (2015).
174. Rosenhek, J., Sperm spotter, *Doctor's Review*, <http://www.doctorsreview.com/history/sperm-spotter/> (2008).
175. Freeman, L., A quick autopsy my love, then off to the ball: the eccentric behaviour of Dutch natural scientist Antoni van Leeuwenhoek and painter Johannes Vermeer, *Daily Mail*, <http://www.dailymail.co.uk/home/books/article-3052742/A-quick-autopsy-love-ball-eccentric-behaviour-Dutch-natural-scientist-Antoni-van-Leeuwenhoek.html> (2015).
176. What is electron microscopy? *University of Massachusetts*, <http://www.umassmed.edu/cemf/whatisem/> (2015).
177. Davis, N., What is cryo-electron microscopy, the Nobel prize-winning technique? The Guardian, <https://www.theguardian.com/science/2017/oct/04/what-is-cryo-electron-microscopy-the-chemistry-nobel-prize-winning-technique> (2017).

178. Parrington, J. and Coward, K., The spark of life. *Biologist (London)* 50: 5–10 (2003).
179. Grasa, P., Coward, K., Young, C. and Parrington, J., The pattern of localization of the putative oocyte activation factor, phospholipase Czeta, in uncapacitated, capacitated, and ionophore-treated human spermatozoa. *Human Reproduction* 23: 2513–22 (2008).
180. Heytens, E., Parrington, J., Coward, K., Young, C., Lambrecht, S., Yoon, S. Y., Fissore, R. A., Hamer, R., Deane, C. M., Ruas, M., Grasa, P., Soleimani, R., Cuvelier, C. A., Gerris, J., Dhont, M., Deforce, D., Leybaert, L. and De Sutter, P., Reduced amounts and abnormal forms of phospholipase C zeta (PLCzeta) in spermatozoa from infertile men. *Human Reproduction* 24: 2417–28 (2009).
181. Robinson, R., A close look at hearing repair, one protein at a time. *PLoS Biology* 11: e1001584 (2013).
182. Nordqvist, J., New mechanism of inner-ear repair discovered, *Medical News Today*, <http://www.medicalnewstoday.com/articles/261808.php> (2013).
183. Friday, L., Osamu Shimomura's serendipitous Nobel, *BU Today*, <http://www.bu.edu/today/2009/osamu-shimomura%E2%80%99s-serendipitous-nobel/> (2009).
184. Markoff, J., For witness to Nagasaki, a life focused on science, *New York Times*, <http://www.nytimes.com/2013/05/12/science/for-witness-to-nagasaki-a-life-focused-on-science.html?_r=0> (2013).
185. Kawaguichi, A., Nobel winner Shimomura returns to isle once again to seek fireflies, *Japan Times*, <http://www.japantimes.co.jp/news/2013/11/07/national/nobel-winner-shimomura-returns-to-isle-to-once-again-seek-sea-fireflies/#.Vqz1pLKLToN> (2013).
186. Shimomura, O., Johnson, F. H. and Saiga, Y., Extraction, purification and properties of aequorin, a bioluminescent protein from the luminous hydromedusan, Aequorea. *Journal of Cellular and Comparative Physiology* 59: 223–39 (1962).
187. Berridge, M. J., Lipp, P. and Bootman, M. D., The versatility and universality of calcium signalling. *Nature Reviews. Molecular Cell Biology* 1: 11–21 (2000).
188. Parrington, J., Davis, L. C., Galione, A. and Wessel, G., Flipping the switch: how a sperm activates the egg at fertilization. *Developmental Dynamics* 236: 2027–38 (2007).
189. Ito, J., Parrington, J. and Fissore, R. A., PLCζ and its role as a trigger of development in vertebrates. *Molecular Reproduction and Development* 78: 846–53 (2011).
190. Webb, S. E. and Miller, A. L., Calcium signalling during zebrafish embryonic development. *BioEssays* 22: 113–23 (2000).
191. The Nobel Prize in Chemistry 2008, *The Nobel Foundation*, <http://www.nobelprize.org/nobel_prizes/chemistry/laureates/2008/> (2008).
192. Davis, L. C., Morgan, A. J., Chen, J. L., Snead, C. M., Bloor-Young, D., Shenderov, E., Stanton-Humphreys, M. N., Conway, S. J., Churchill, G. C., Parrington, J., Cerundolo, V. and Galione A., NAADP activates two-pore channels on T cell cytolytic granules to stimulate exocytosis and killing. *Current Biology* 22: 2331–7 (2012).
193. Rose, S., Lynn Margulis obituary, *The Guardian*, <http://www.theguardian.com/science/2011/dec/11/lynn-margulis-obtiuary> (2011).

194. Caprette, D., The electron transport system of mitochondria, *Rice University*, <http://www.ruf.rice.edu/~bioslabs/studies/mitochondria/mitets.html> (2005).
195. Chial, H. and Craig, J., mtDNA and mitochondrial diseases. *Nature Education* 1: 217 (2008).
196. Piomboni, P., Focarelli, R., Stendardi, A., Ferramosca, A. and Zara, V., The role of mitochondria in energy production for human sperm motility. *International Journal of Andrology* 35: 109–24 (2012).
197. Ankel-Simons, F. and Cummins, J. M., Misconceptions about mitochondria and mammalian fertilization: implications for theories on human evolution. *Proceedings National Academy Sciences USA* 93: 13859–63 (1996).
198. Shitara, H., Kaneda, H., Sato, A., Iwasaki, K., Hayashi, J., Taya, C. and Yonekawa, H., Non-invasive visualization of sperm mitochondria behavior in transgenic mice with introduced green fluorescent protein (GFP). *FEBS Letters* 500: 7–11 (2001).
199. Sutovsky, P., Moreno, R. D., Ramalho-Santos, J., Dominko, T., Simerly, C. and Schatten, G., Ubiquitin tag for sperm mitochondria. *Nature* 402: 371–2 (1999).
200. Maher, B., Making new eggs in old mice, *Nature News*, <http://www.nature.com/news/2009/090411/full/news.2009.362.html#B2> (2009).
201. Richards, S., Ovarian stem cells in humans?, *The Scientist*, <http://www.the-scientist.com/?articles.view/articleNo/31793/title/Ovarian-Stem-Cells-in-Humans-/> (2012).
202. Couzin-Frankel, J., Feature: a controversial company offers a new way to make a baby, *Science News*, <http://news.sciencemag.org/biology/2015/11/feature-controversial-company-offers-new-way-make-baby> (2015).
203. Connor, S., Eggs unlimited: an extraordinary tale of scientific discovery, *The Independent*, <http://www.independent.co.uk/life-style/health-and-families/health-news/eggs-unlimited-an-extraordinary-tale-of-scientific-discovery-7624715.html> (2012).
204. Editorial, The human brain is the most complex structure in the universe. Let's do all we can to unravel its mysteries, *The Independent*, <http://www.independent.co.uk/voices/editorials/the-human-brain-is-the-most-complex-structure-in-the-universe-let-s-do-all-we-can-to-unravel-its-9233125.html> (2014).
205. Olson, S., How complex is a mouse brain? *Next Big Future*, <http://nextbigfuture.com/2012/05/how-complex-is-mouse-brain.html> (2012).
206. Brains of mice and humans function similarly when 'place learning', *KU Leuven*, <http://www.kuleuven.be/campus/english/news/2013/brains-of-mice-and-humans-function-similarly-when-place-learning> (2013).
207. Baker, M., Microscopy: bright light, better labels. *Nature* 478: 137–42 (2010).
208. Than, K., Brain cells colored to create 'brainbow', *Live Science*, <http://www.livescience.com/1977-brain-cells-colored-create-brainbow.html> (2007).
209. Jabr, F., Know your neurons: how to classify different types of neurons in the brain's forest, *Scientific American*, <http://blogs.scientificamerican.com/brainwaves/know-your-neurons-classifying-the-many-types-of-cells-in-the-neuron-forest/> (2012).

210. Richard Axel and Linda Buck awarded 2004 Nobel Prize in Physiology or Medicine, *HHMI News*, <http://www.hhmi.org/news/richard-axel-and-linda-buck-awarded-2004-nobel-prize-physiology-or-medicine> (2004).
211. A nose for science: Buck, '75, wins Nobel for decoding genetics of smell, *University of Washington*, <http://www.washington.edu/alumni/columns/dec04/briefings_buck.html> (2004).
212. Pieribone, V. and Gruber, D. F., *Aglow in the Dark: The Revolutionary Science of Biofluorescence* (Belknap Press of Harvard University Press, 2005), pp. 210–11.
213. Cherry, K., What is a neuron? *About Education*, <http://psychology.about.com/od/biopsychology/f/neuron01.htm> (2014).
214. Takeuchi, H. and Sakano, H., Neural map formation in the mouse olfactory system. *Cellular and Molecular Life Sciences* 71: 3049–57 (2014).
215. Cherry, K., What is an action potential? *About Education*, <http://psychology.about.com/od/aindex/g/actionpot.htm> (2014).
216. Depauw, F. A., Rogato, A., Ribera d'Alcala, M. and Falciatore, A. J., Exploring the molecular basis of responses to light in marine diatoms. *Journal of Experimental Botany* 63: 1575–91 (2012).
217. Crick, F. H. C., in The brain (*Scientific American*, <http://www.scientificamerican.com/magazine/sa/1979/09-01/> (1979), pp. 130–40.
218. Feilden, T., Switching on a light in the brain, *BBC News*, <http://www.bbc.co.uk/news/science-environment-20513292> (2012).
219. Colapinto, J., Lighting the brain, *New Yorker*, <http://www.newyorker.com/magazine/2015/05/18/lighting-the-brain> (2015).
220. Adams, A., Optogenetics earns Stanford professor Karl Deisseroth the Keio Prize in medicine, *Stanford News*, <http://news.stanford.edu/features/2014/optogenetics/> (2014).
221. Prigg, M., Researchers reveal neural switch that turns DREAMS on and off in seconds, *Daily Mail*, <http://www.dailymail.co.uk/sciencetech/article-3274586/Researchers-reveal-neural-switch-turns-DREAMS-seconds.html> (2015).
222. Aristotle, On memory and reminiscence, *Massachusetts Institute of Technology*, <http://classics.mit.edu/Aristotle/memory.html> (2009).
223. Mastin, L., The study of human memory, *The Human Memory*, <http://www.human-memory.net/intro_study.html> (2010).
224. Lømo T., The discovery of long-term potentiation. *Philosophical Transactions of the Royal Society of London. Series B Biological Sciences* 358: 617–20 (2003).
225. Callaway, E., Flashes of light show how memories are made, *Nature News*, <http://www.nature.com/news/flashes-of-light-show-how-memories-are-made-1.15330#/b2> (2014).
226. Agence France-Presse, Amnesia researchers use light to restore 'lost' memories in mice, *The Guardian*, <http://www.theguardian.com/science/2015/may/29/amnesia-researchers-use-light-to-restore-lost-memories-in-mice> (2015).
227. Shen, H., Activating happy memories cheers moody mice, *Nature News*, <http://www.nature.com/news/activating-happy-memories-cheers-moody-mice-1.17782> (2015).

228. Callaway, E., Be still my light-controlled heart, *Nature News*, <http://www.nature.com/news/2010/101111/full/news.2010.605.html> (2010).
229. Pathak, G. P., Vrana, J. D. and Tucker, C. L., Optogenetic control of cell function using engineered photoreceptors. *Biology of the Cell* 105: 59–72 (2013).
230. Costandi, M., Light switches on the brain, *The Guardian*, <http://www.theguardian.com/science/blog/2010/nov/17/light-switches-brain-optogenetics> (2010).
231. Gwynne, P., Genetically engineered protein responds remotely to red light, *Inside Science*, <http://www.insidescience.org/content/sending-light-through-skull-influence-brain-activity/1811> (2014).
232. Optogenetics shines with inner bioluminescence, *GEN News Highlights*, <http://www.genengnews.com/gen-news-highlights/optogenetics-shines-with-inner-bioluminescence/81251801/> (2015).
233. Wagner, D., Sound waves give San Diego neuroscientists control over brain cells, *KPBS*, <http://www.kpbs.org/news/2015/sep/28/sound-waves-give-san-diego-neuroscientists-control/> (2015).

Chapter 4: The Gene Scissors

234. Glauser, W. and Taylor, M., Hype in science: it's not just the media's fault, *Healthy Debate*, <http://healthydebate.ca/2015/03/topic/hype-in-science> (2015).
235. Mello, C., Phone interview by Parrington, J., 23 March (2015).
236. Mundasad, S., Row over human embryo gene editing, *BBC News*, <http://www.bbc.co.uk/news/health-32446954> (2015).
237. Gallagher, J., UK scientists edit DNA of human embryos, BBC News, <https://www.bbc.co.uk/news/health-41269200> (2017).
238. Cyranoski, D. and Ledford, H., Genome-edited baby claim provokes international outcry, Nature, <https://www.nature.com/articles/d41586-018-07545-0> (2018).
239. Dolgin, E., Stem cell rat race, *The Scientist*, <http://www.the-scientist.com/?articles.view/articleNo/27244/title/Stem-cell-rat-race/> (2009).
240. Telugu, B. P., Ezashi, T. and Roberts, R. M., The promise of stem cell research in pigs and other ungulate species. *Stem Cell Reviews* 6: 31–41 (2010).
241. Blair, K., Wray, J. and Smith, A., The liberation of embryonic stem cells. *PLoS Genetics* 7: e1002019 (2011).
242. Riordan, S. M., Heruth, D. P., Zhang, L. Q. and Ye, S. Q., Application of CRISPR/Cas9 for biomedical discoveries. *Cell and Bioscience* 5: 33 (2015).
243. Aida, T., Imahashi, R. and Tanaka, K., Translating human genetics into mouse: the impact of ultra-rapid in vivo genome editing. *Development Growth and Differentiation* 56: 34–45 (2014).
244. Maxmen, A., Easy DNA editing will remake the world. Buckle up, *Wired*, <http://www.wired.com/2015/07/crispr-dna-editing-2/> (2015).
245. Weeks, D. P., Spalding, M. H. and Yang, B., Use of designer nucleases for targeted gene and genome editing in plants. *Plant Biotechnology Journal* 14: 483–95 (2015).

246. Regalado, A., DuPont predicts CRISPR plants on dinner plates in five years, *MIT Technology Review*, <http://www.technologyreview.com/news/542311/dupont-predicts-crispr-plants-on-dinner-plates-in-five-years/> (2015).
247. George Church: the future without limit, *National Geographic*, <http://news.nationalgeographic.com/news/innovators/2014/06/140602-george-church-innovation-biology-science-genetics-de-extinction/> (2014).
248. Prakash, R., Zhang, Y., Feng, W. and Jasin, M., Homologous recombination and human health: the roles of BRCA1, BRCA2, and associated proteins. *Cold Spring Harbor Perspectives in Biology* 7: a016600 (2015).
249. Marx, V., Genome-editing tools storm ahead. *Nature Methods* 9: 1055–9 (2012).
250. Klug, A., The discovery of zinc fingers and their applications in gene regulation and genome manipulation. *Annual Reviews in Biochemistry* 79: 213–31 (2010).
251. Porteus, M. H. and Carroll, D., Gene targeting using zinc finger nucleases. *Nature Biotechnology* 23: 967–73 (2005).
252. Shekhar, C., Finger pointing: engineered zinc-finger proteins allow precise modification and regulation of genes. *Chemistry and Biology* 15: 1241–2 (2008).
253. Leong, I. U., Lai, D., Lan, C. C., Johnson, R., Love, D. R., Johnson, R. and Love, D. R., Targeted mutagenesis of zebrafish: use of zinc finger nucleases. *Birth Defects Research C Embryo Today* 93: 249–55 (2011).
254. Jagadeeswaran, P., Zinc fingers poke zebrafish, cause thrombosis! *Blood* 124: 9–10 (2014).
255. Joung, J. K. and Sander, J. D., TALENs: a widely applicable technology for targeted genome editing. *Nature Reviews. Molecular Cell Biology* 14: 49–55 (2013).
256. Ishino, Y., Shinagawa, H., Makino, K., Amemura, M. and Nakata, A., Nucleotide sequence of the iap gene, responsible for alkaline phosphatase isozyme conversion in Escherichia coli, and identification of the gene product. *Journal of Bacteriology* 169: 5429–33 (1987).
257. Jansen, R., Embden, J. D., Gaastra, W. and Schouls, L. M., Identification of genes that are associated with DNA repeats in prokaryotes. *Molecular Microbiology* 43: 1565–75 (2002).
258. Zimmer, C., Breakthrough DNA editor borne of bacteria, *Quanta Magazine*, <https://www.quantamagazine.org/20150206-crispr-dna-editor-bacteria/> (2015).
259. Pollack, A., Jennifer Doudna, a pioneer who helped simplify genome editing, *New York Times*, <http://www.nytimes.com/2015/05/12/science/jennifer-doudna-crispr-cas9-genetic-engineering.html?_r=0> (2015).
260. 2015 Genetics Prize: Jennifer Doudna, *Gruber Foundation*, <http://gruber.yale.edu/genetics/jennifer-doudna> (2015).
261. Stöppler, M. C., What is a 'flesh-eating' bacterial infection? *Medicine Net*, <http://www.medicinenet.com/flesh_eating_bacterial_infection/views.htm> (2015).
262. Connor, S., 'The more we looked into the mystery of CRISPR, the more interesting it seemed', *The Independent*, <http://www.independent.co.uk/news/science/the-more-we-looked-into-the-mystery-of-crispr-the-more-interesting-it-seemed-8925328.html> (2013).

263. Loria, K., The researchers behind 'the biggest biotech discovery of the century' found it by accident, *Tech Insider*, <http://www.techinsider.io/the-people-who-discovered-the-most-powerful-genetic-engineering-tool-we-know-found-it-by-accident-2015-6> (2015).
264. Rogers, A., A CRISPR cut, *Pomona College Magazine*, <http://magazine.pomona.edu/2015/spring/a-crispr-cut/> (2015).
265. Ledford, H., CRISPR, the disruptor, *Nature News*, <http://www.nature.com/news/crispr-the-disruptor-1.17673> (2015).
266. Shay, J. W. and Wright., W. E., Hayflick, his limit, and cellular ageing. *Nature Reviews. Molecular Cell Biology* 1: 72–6 (2000).
267. Zielinski, S., Henrietta Lacks' 'immortal' cells, *Smithsonian Magazine*, <http://www.smithsonianmag.com/science-nature/henrietta-lacks-immortal-cells-6421299/?no-ist> (2010).
268. Masters, J. R., Human cancer cell lines: fact and fantasy. *Nature Reviews. Molecular Cell Biology* 1: 233–6 (2000).
269. Skelin, M., Rupnik, M. and Cencic, A., Pancreatic beta cell lines and their applications in diabetes mellitus research. *ALTEX* 27: 105–13 (2010).
270. Trounson, A., A rapidly evolving revolution in stem cell biology and medicine. *Reproductive Biomedicine Online* 27: 756–64 (2013).
271. Agrawal, N., Dasaradhi, P. V., Mohmmed, A., Malhotra, P., Bhatnagar, R. K. and Mukherjee, S. K., RNA interference: biology, mechanism, and applications. *Microbiology and Molecular Biology Reviews* 67: 657–85 (2003).
272. Calcraft, P. J., Ruas, M., Pan, Z., Cheng, X., Arredouani, A., Hao, X., Tang, J., Rietdorf, K., Teboul, L., Chuang, K. T., Lin, P., Xiao, R., Wang, C., Zhu, Y., Lin, Y., Wyatt, C. N., Parrington, J., Ma, J., Evans, A. M., Galione, A. and Zhu, M. X., NAADP mobilizes calcium from acidic organelles through two-pore channels. *Nature* 459: 596–600 (2009).
273. Galione, A., Evans, A. M., Ma, J., Parrington, J., Arredouani, A., Cheng, X. and Zhu, M. X., The acid test: the discovery of two-pore channels (TPCs) as NAADP-gated endolysosomal Ca(2+) release channels. *Pflugers Archiv* 458: 869–76 (2009).
274. Grens, K., Feng Zhang: the Midas of methods, *The Scientist*, <http://www.the-scientist.com/?articles.view/articleNo/40582/title/Feng-Zhang--The-Midas-of-Methods/> (2014).
275. Maxmen, A., Easy DNA editing will remake the world. Buckle up, *Wired*, <http://www.wired.com/2015/07/crispr-dna-editing-2/> (2015).
276. Moore, J. D., The impact of CRISPR-Cas9 on target identification and validation. *Drug Discovery Today* 20: 450–7 (2015).
277. Shalem, O., Sanjana, N. E., Hartenian, E., Shi, X., Scott, D. A., Mikkelsen, T. S., Heckl, D., Ebert, B. L., Root, D. E., Doench, J. G. and Zhang F., Genome-scale CRISPR-Cas9 knockout screening in human cells. *Science* 343: 84–7 (2014).
278. Goldsmith, P., In vivo CRISPR-Cas9 screen sheds light on cancer metastasis and tumor evolution, *Broad Institute*, <https://www.broadinstitute.org/news/6607> (2015).

279. Platt, R. J., Chen, S., Zhou, Y., Yim, M. J., Swiech, L., Kempton, H. R., Dahlman, J. E., Parnas, O., Eisenhaure, T. M., Jovanovic, M., Graham, D. B., Jhunjhunwala, S., Heidenreich, M., Xavier, R. J., Langer, R., Anderson, D. G., Hacohen, N., Regev, A., Feng, G., Sharp, P. A. and Zhang F., CRISPR-Cas9 knockin mice for genome editing and cancer modeling. *Cell* 159: 440–55 (2014).
280. Snyder, B., New technique accelerates genome editing process, *Vanderbilt University*, <http://news.vanderbilt.edu/2014/08/new-technique-accelerates-genome-editing-process/> (2014).
281. Phillips, T. and Hoopes, L., Transcription factors and transcriptional control in eukaryotic cells. *Nature Education* 1: 119 (2008).
282. Saunders, T. L., Inducible transgenic mouse models. *Methods Molecular Biology* 693: 103–15 (2011).
283. Akst, J., Optogenetics meets CRISPR, *The Scientist*, <http://www.the-scientist.com/?articles.view/articleNo/43255/title/Optogenetics-Meets-CRISPR/> (2015).
284. Lavars, N., Scientists reduce blood sugar levels in mice by remote control, *Gizmag*, <http://www.gizmag.com/scientists-blood-sugar-level-mice-remote-control/35248/> (2014).
285. Fleischman, J., How Jennifer Doudna turned over the proverbial rock and found CRISPR, *American Society for Cell Biology*, <http://www.ascb.org/how-jennifer-doudna-turned-over-the-proverbial-rock-and-found-crispr/> (2015).
286. Burke, K. L., Interview with a gene editor, *American Scientist*, <http://www.americanscientist.org/issues/pub/interview-with-a-gene-editor> (2015).
287. Connor, S., Scientific split: the human genome breakthrough dividing former colleagues, *The Independent*, <http://www.independent.co.uk/news/science/scientific-split-the-human-genome-breakthrough-dividing-former-colleagues-9300456.html> (2014).
288. Regalado, A., Who owns the biggest biotech discovery of the century? *MIT Technology Review*, <http://www.technologyreview.com/featuredstory/532796/who-owns-the-biggest-biotech-discovery-of-the-century/> (2014).
289. Maxmen, A. M., Easy DNA editing will remake the world. Buckle up, *Wired*, <http://www.wired.com/2015/07/crispr-dna-editing-2/> (2015).
290. Sample, I., Pioneering scientists share £23m Breakthrough Prize pot at US awards, *The Guardian*, <http://www.theguardian.com/science/2014/nov/10/breakthrough-prize-scientists-23m-science-awards-2015> (2014).
291. Regalado, A., New CRISPR protein slices through genomes, patent problems, *MIT Technology Review*, <http://www.technologyreview.com/news/541681/new-crispr-protein-slices-through-genomes-patent-problems/> (2015).
292. Connor, S., Crispr: scientists' hopes to win Nobel Prize for gene-editing technique at risk over patent dispute, *The Independent*, <http://www.independent.co.uk/news/science/crispr-scientists-hopes-to-win-nobel-prize-for-gene-editing-technique-at-risk-over-patent-dispute-a6677436.html> (2015).
293. Lin, L., Eric Lander criticized for CRISPR article, *The Tech*, <http://tech.mit.edu/V135/N37/crispr.html> (2016).

294. Cohen, J., CRISPR patent fight revived, Science, <https://science.sciencemag.org/content/365/6448/15.2> (2019).
295. Knapton, S., DNA detectives win Nobel Prize for cancer cure breakthrough, *The Telegraph*, <http://www.telegraph.co.uk/news/science/science-news/11916833/DNA-detectives-win-Nobel-Prize-for-cancer-cure-breakthrough.html> (2015).
296. Katz, Y., Who owns molecular biology? *Boston Review*,<http://bostonreview.net/books-ideas/yarden-katz-who-owns-molecular-biology> (2015).
297. Regalado, A., CRISPR patent fight now a winner-take-all match, *MIT Technology Review*, <http://www.technologyreview.com/news/536736/crispr-patent-fight-now-a-winner-take-all-match/> (2015).
298. Zimmer, C., Breakthrough DNA editor borne of bacteria, *Quanta Magazine*, <https://www.quantamagazine.org/20150206-crispr-dna-editor-bacteria/> (2015).
299. Lanphier, E., Urnov, F., Haecker, S. E., Werner, M. A. and Smolenski, J. Don't edit the human germ line. *Nature* 519: 410–1 (2015).
300. Scrutinizing science: peer review, *Understanding Science: How Science Really Works*, <http://undsci.berkeley.edu/article/howscienceworks_16> (2015).
301. Cyranoski, D., Ethics of embryo editing divides scientists, *Nature News*, <http://www.nature.com/news/ethics-of-embryo-editing-divides-scientists-1.17131> (2015).
302. Cyranoski, D. and Reardon, S., Chinese scientists genetically modify human embryos, *Nature News*, <http://www.nature.com/news/chinese-scientists-genetically-modify-human-embryos-1.17378> (2015).
303. Sample, I., Scientists genetically modify human embryos in controversial world first, *The Guardian*, <http://www.theguardian.com/science/2015/apr/23/scientists-genetically-modify-human-embryos-in-controversial-world-first> (2015).
304. Harris, J., Phone interview by Parrington, J., 28 March (2015).
305. Ishii, T., E-mail interview by Parrington, J., 24 March (2015).
306. Ledford, H., CRISPR used to peer into human embryos' first days, Nature, <https://www.nature.com/news/crispr-used-to-peer-into-human-embryos-first-days-1.22646> (2017).
307. Begley, S. and Joseph, A., The CRISPR shocker: How genome-editing scientist He Jiankui rose from obscurity to stun the world, STAT, <https://www.statnews.com/2018/12/17/crispr-shocker-genome-editing-scientist-he-jiankui/> (2018).
308. Sample, I., The Guardian, Chinese scientist who edited babies' genes jailed for three years, <https://www.theguardian.com/world/2019/dec/30/gene-editing-chinese-scientist-he-jiankui-jailed-three-years> (2019).

Chapter 5: Next Year's Models

309. Cookson, C., Dr Harvey's extraordinary discovery, *Financial Times*, <http://www.ft.com/cms/s/2/3498ea54-2874-11e2-afd2-00144feabdc0.html> (2012).
310. UK Home Office, Research and testing using animals, <https://www.gov.uk/research-and-testing-using-animals> (2015).

311. Latham, S. R., *U.S. Law and Animal Experimentation: A Critical Primer* (Hastings Center Report, 2012).
312. Forty reasons why we need animals in research, *European Animal Research Association*, <http://eara.eu/campaign/forty-reasons-why-we-need-animals-in-research/> (2015).
313. Zhu, M. X., Evans, A. M., Ma, J., Parrington, J. and Galione, A. Two-pore channels for integrative Ca signaling. *Communicative and Integrative Biology* 3: 12–17 (2010).
314. Ralston, A., Operons and prokaryotic gene regulation. *Nature Education* 1: 216 (2008).
315. Nurse, P., The cell cycle and beyond: an interview with Paul Nurse. Interview by Jim Smith. *Disease Models and Mechanisms* 2: 113–5 (2009).
316. Editorial, Nematodes net Nobel. *Nature Cell Biology* 4: E244 (2002).
317. Vacaru, A. M., Unlu, G., Spitzner, M., Mione, M., Knapik, E. W. and Sadler, K. C., In vivo cell biology in zebrafish: providing insights into vertebrate development and disease. *Journal of Cell Science* 127: 485–95 (2014).
318. Parrington, J., *The Deeper Genome* (Oxford University Press, 2015), pp. 166–7.
319. Select Committee on Animals in Scientific Procedures Report, *House of Lords—UK Parliament*, <http://www.publications.parliament.uk/pa/ld200102/ldselect/ldanimal/150/15001.htm> (2002).
320. Ma, C., Animal models of disease, *Modern Drug Discovery*, <http://pubs.acs.org/subscribe/journals/mdd/v07/i06/pdf/604feature_ma.pdf> (2004).
321. Algar, J., What do we have in common with worms and flies? Our genomes, *Tech Times*, <http://www.techtimes.com/articles/14346/20140828/what-do-we-have-in-commong-with-worms-and-flies-our-genomes.htm> (2015)
322. Hove, J. R., In vivo biofluid dynamic imaging in the developing zebrafish. *Birth Defects Research C Embryo Today* 72: 277–89 (2004).
323. Kaustinen, K., A CRISPR approach, *DD News*, <http://www.ddn-news.com/news?newsarticle=9696> (2015).
324. Benowitz, S., A new role for zebrafish: larger scale gene function studies, *National Institutes of Health*, <http://www.nih.gov/news-events/news-releases/new-role-zebrafish-larger-scale-gene-function-studies> (2015).
325. Dunbar, R. I. and Shultz, S., Understanding primate brain evolution. *Philosophical Transactions of the Royal Society of London. Series B Biological Sciences* 362: 649–58 (2007).
326. Herzberg, N., Mice losing their allure as experimental subjects to study human disease, *The Guardian*, <http://www.theguardian.com/science/2015/mar/20/mice-clinical-trials-human-disease> (2015).
327. Parrington, J., Arnoult, C. and Fissore, R.A., The Eggstraordinary Story of How Life Begins, Molecular Reproduction and Development, 86: 4–19 (2018).
328. Knapton, S., Unhealthy lifestyle can knock 23 years off lifespan, *The Telegraph*, <http://www.telegraph.co.uk/news/health/news/11723443/Unhealthy-lifestyle-can-knock-23-years-off-lifespan.html> (2015).
329. Britain's obesity epidemic worse than feared, *The Telegraph*, <http://www.telegraph.co.uk/news/10566705/Britains-obesity-epidemic-worse-than-feared.html> (2014).

330. Boseley, S., Third of overweight teenagers think they are right size, study shows, *The Guardian*, <http://www.theguardian.com/society/2015/jul/09/overweight-teenagers-think-they-are-right-size-study> (2015).
331. Cooper, C., Obese men have just a '1 in 210' chance of attaining a healthy body weight, *The Independent*, <http://www.independent.co.uk/life-style/health-and-families/health-news/obese-men-have-just-a-1-in-210-chance-of-attaining-a-healthy-body-weight-10394887.html> (2015).
332. Tozzi, J., How Americans got so fat, in charts, *Bloomberg Business*, <http://www.bloomberg.com/news/articles/2016-01-07/how-americans-got-so-fat-in-charts> (2016).
333. Ashley E. A., Hershberger, R. E., Caleshu, C., Ellinor, P. T., Garcia, J. G., Herrington, D. M., Ho, C. Y., Johnson, J. A., Kittner, S. J., Macrae, C. A., Mudd-Martin, G., Rader, D. J., Roden, D. M., Scholes, D., Sellke, F. W., Towbin, J. A., Van Eyk, J., Worrall, B. B.; American Heart Association Advocacy Coordinating Committee, Genetics and cardiovascular disease: a policy statement from the American Heart Association. *Circulation* 126: 142–57 (2012).
334. Wessels, A. and Sedmera, D., Developmental anatomy of the heart: a tale of mice and man. *Physiological Genomics* 15: 165–76 (2003).
335. Experimenting on animals, *BBC Ethics*, <http://www.bbc.co.uk/ethics/animals/using/experiments_1.shtml> (2014).
336. Melina, R., Why do medical researchers use mice? *Live Science*, <http://www.livescience.com/32860-why-do-medical-researchers-use-mice.html> (2010).
337. Zaragoza, C., Gomez-Guerrero, C., Martin-Ventura, J. L., Blanco-Colio, L., Lavin, B., Mallavia, B., Tarin, C., Mas, S., Ortiz, A. and Egido, J., Animal models of cardiovascular diseases. *Journal of Biomedicine and Biotechnology* 2011: 497841 (2011).
338. Eissen, P., George Orwell and the politics of *Animal Farm*, *Paul Eissen*, <http://www.his.com/~phe/farm.html> (1997).
339. Genome-edited pigs created using innovative tech, *Feedstuffs Foodlink*, <http://feedstuffsfoodlink.com/story-genomeedited-pigs-created-using-innovative-tech-0-125733> (2015).
340. Swindle, M. M., Makin, A., Herron, A. J., Clubb, F. J., Jr and Frazier, K. S., Swine as models in biomedical research and toxicology testing. *Veterinary Pathology* 49: 344–56 (2012).
341. No pig in a poke, *The Economist*, <http://www.economist.com/news/science-and-technology/21674493-genome-engineering-may-help-make-porcine-organs-suitable-use-people-no-pig> (2015).
342. Zimmer, C., Editing of pig DNA may lead to more organs for people, *New York Times*, <http://www.nytimes.com/2015/10/20/science/editing-of-pig-dna-may-lead-to-more-organs-for-people.html?_r=0> (2015).
343. Limas, M., Can engineering the pig genome provide a safe new source of transplant organs? *Synbiobeta*, <http://synbiobeta.com/engineering-pig-genome-transplant-organs/> (2015).
344. Collins, N., Pig born using new GM approach, *The Telegraph*, <http://www.telegraph.co.uk/news/science/science-news/9995807/Pig-born-using-new-GM-approach.html> (2013).

345. Wang, Y., Du, Y., Shen, B., Zhou, X., Li, J., Liu, Y., Wang, J., Zhou, J., Hu, B., Kang, N., Gao, J., Yu, L., Huang, X. and Wei, H., Efficient generation of gene-modified pigs via injection of zygote with Cas9/sgRNA. *Scientific Reports* 5: 8256 (2015).
346. Betters, J. L. and Yu, L., NPC1L1 and cholesterol transport. *FEBS Letters* 584: 2740–7 (2010).
347. Gadea J., Garcia-Vazquez FA., Hachem A., Bassett A., Romero-Aguirregomezcorta J., Canovas S., Romar R., Parrington J. Generation of TPC2 knock out pig embryos by CRISPR/Cas technology, Reproduction in Domestic Animals, 53: 87–88 (2018).
348. Steenhuysen, J., Genome scientist Craig Venter in deal to make humanized pig organs, *Reuters*, <http://www.reuters.com/article/2014/05/06/health-transplants-pigs-idUSL2N0NR26320140506> (2014).
349. Reardon, S., Gene-editing record smashed in pigs, *Nature News*, <http://www.nature.com/news/gene-editing-record-smashed-in-pigs-1.18525> (2015).
350. Ogawa, T. and Bold, A. J., The heart as an endocrine organ. *Endocrine Connections* 3: R31–44 (2014).
351. Denner, J. and Tonjes, R. R., Infection barriers to successful xenotransplantation focusing on porcine endogenous retroviruses. *Clinical Microbiology Review* 25: 318–43 (2012).
352. Lovgren, S., HIV originated with monkeys, not chimps, study finds, *National Geographic*, <http://news.nationalgeographic.com/news/2003/06/0612_030612_hivvirusjump.html> (2003).
353. Kolata, G., When H.I.V. made its jump to people, *New York Times*, <http://www.nytimes.com/2002/01/29/science/when-hiv-made-its-jump-to-people.html?pagewanted=all> (2002).
354. Coghlan, A., Baboons with pig hearts pave way for human transplants, *New Scientist*, <https://www.newscientist.com/article/dn25508-baboons-with-pig-hearts-pave-way-for-human-transplants/> (2014).
355. Lewis, T., 'We all kind of marvel at how fast this took off', *Business Insider*, <http://www.businessinsider.com/how-crispr-is-revolutionizing-biology-2015-10?IR=T> (2015).
356. Servick, K., Gene-editing method revives hopes for transplanting pig organs into people, *Science News*, <http://news.sciencemag.org/biology/2015/10/gene-editing-method-revives-hopes-transplanting-pig-organs-people> (2015).
357. Servick, K., CRISPR slices virus genes out of pigs, but will it make organ transplants to humans safer? Science, <https://www.sciencemag.org/news/2017/08/crispr-slices-virus-genes-out-pigs-will-it-make-organ-transplants-humans-safer> (2017).
358. Parrington, J., The genetics of consciousness, *OUP Blog*, <http://blog.oup.com/2015/05/the-genetics-of-consciousness/> (2015).
359. Parrington, J., *The Deeper Genome* (Oxford University Press, 2015), pp. 181–94.
360. Duckworth, K., Mental illness facts and numbers, *National Alliance on Mental Illness*, <http://www2.nami.org/factsheets/mentalillness_factsheet.pdf> (2013).
361. Mental health facts and statistics, *Mind*, <http://www.mind.org.uk/information-support/types-of-mental-health-problems/statistics-and-facts-about-mental-health/how-common-are-mental-health-problems/> (2016).

362. Lobl, T., Is it time for the over-medicalisation of mental health to recede? *Recovery Wirral*, <http://www.recoverywirral.com/2013/03/is-it-time-for-the-over-medicalisation-of-mental-health-to-recede-report-in-the-independant-stop-medicalising-distress/> (2012).
363. Hicks, C., 'Dozens of mental disorders don't exist', *The Telegraph*, <http://www.telegraph.co.uk/news/health/10359105/Dozens-of-mental-disorders-dont-exist.html> (2013).
364. Mental health statistics, *Mental Health Foundation* <http://www.mentalhealth.org.uk/help-information/mental-health-statistics/> (2015).
365. Trafton, A. A., Turning point, *MIT Technology Review*, <http://www.technologyreview.com/article/533056/a-turning-point/> (2014).
366. Koshland, D. E. Sequences and consequences of the human genome. *Science* 246: 189 (1989).
367. Genome-wide association studies, *National Institutes of Health*, <https://www.genome.gov/20019523> (2015).
368. Neale, B. M. and Sklar, P., Genetic analysis of schizophrenia and bipolar disorder reveals polygenicity but also suggests new directions for molecular interrogation. *Current Opinion in Neurobiology* 30: 131–8 (2015).
369. Kavanagh, D. H., Tansey, K. E., O'Donovan, M. C. and Owen, M. J., Schizophrenia genetics: emerging themes for a complex disorder. *Molecular Psychiatry* 20: 72–6 (2015).
370. Psychiatric GWAS Consortium Coordinating Committee: Cichon, S., Craddock, N., Daly, M., Faraone, S. V., Gejman, P. V., Kelsoe, J., Lehner, T., Levinson, D. F., Moran, A., Sklar, P. and Sullivan, P. F., Genomewide association studies: history, rationale, and prospects for psychiatric disorders. *American Journal of Psychiatry* 166: 540–56 (2009).
371. Dougherty, E., From genes to brains: how advances in genomics are changing the study of neuroscience, *Brain Scan*, <http://mcgovern.mit.edu/news/newsletter/from-genes-to-brains-how-advances-in-genomics-are-changing-the-study-of-neuroscience/> (2014).
372. Teffer, K. and Semendeferi, K., Human prefrontal cortex: evolution, development, and pathology. *Progress in Brain Research* 195: 191–218 (2012).
373. Mitchell, J. F. and Leopold, D. A., The marmoset monkey as a model for visual neuroscience. *Neuroscience Research* 93: 20–46 (2015).
374. Manger, P. R., Cort, J., Ebrahim, N., Goodman, A., Henning, J., Karolia, M., Rodrigues, S. L. and Strkalj, G., Is 21st century neuroscience too focussed on the rat/mouse model of brain function and dysfunction? *Frontiers in Neuroanatomy* 2: 5 (2008).
375. Niu, Y., Shen, B., Cui, Y., Chen, Y., Wang, J., Wang, L., Kang, Y., Zhao, X., Si, W., Li, W., Xiang, A. P., Zhou, J., Guo, X., Bi, Y., Si, C., Hu, B., Dong, G., Wang, H., Zhou, Z., Li, T., Tan, T., Pu, X., Wang, F., Ji, S., Zhou, Q., Huang, X., Ji, W. and Sha, J., Generation of gene-modified cynomolgus monkey via Cas9/RNA-mediated gene targeting in one-cell embryos. *Cell* 156: 836–43 (2014).

376. Chen, Y., Zheng, Y., Kang, Y., Yang, W., Niu, Y., Guo, X., Tu, Z., Si, C., Wang, H., Xing, R., Pu, X., Yang, S. H., Li, S., Ji, W. and Li, X. J., Functional disruption of the dystrophin gene in rhesus monkey using CRISPR/Cas9. *Human Molecular Genetics* 24: 3764–74 (2015).
377. Sample, I., Genetically modified monkeys created with cut-and-paste DNA, *The Guardian*, <http://www.theguardian.com/science/2014/jan/30/genetically-modified-monkeys-cut-and-paste-dna-alzheimers-parkinsons> (2014).
378. Abbott, A., Biomedicine: the changing face of primate research, *Nature News*, <http://www.nature.com/news/biomedicine-the-changing-face-of-primate-research-1.14645> (2014).
379. Larson, C., Genome editing, *MIT Technology Review*, <http://www.technologyreview.com/featuredstory/526511/genome-editing/> (2014).
380. Zhang, S., China Is Genetically Engineering Monkeys With Brain Disorders, The Atlantic, <https://www.theatlantic.com/science/archive/2018/06/china-is-genetically-engineering-monkeys-with-brain-disorders/561866/> (2018).
381. Pennisi, E., 'Language gene' has a partner, *Science News*, <http://news.sciencemag.org/biology/2013/10/language-gene-has-partner> (2013).
382. Yong, E., Scientists 'humanise' Foxp2 gene in mice to probe origins of human language, *Not Rocket Science*, <http://scienceblogs.com/notrocketscience/2009/05/29/scientists-humanise-foxp2-gene-in-mice-to-probe-origins-of-h/> (2009).
383. Pennisi, E., Human speech gene can speed learning in mice, *Science News*, <http://news.sciencemag.org/biology/2014/09/human-speech-gene-can-speed-learning-mice> (2014).
384. Kirkey, S., Chinese scientists give monkeys human brain genes in 'morally risky' experiment, The Guardian, <https://www.theguardian.pe.ca/news/world/chinese-scientists-give-monkeys-human-brain-genes-in-morally-risky-experiment-301326/> (2019).
385. Regalado, A., Chinese scientists have put human genes in monkeys and yes – they may be smarter, MIT Technology Review, <https://www.technologyreview.com/s/613277/chinese-scientists-have-put-human-brain-genes-in-monkeysand-yes-they-may-be-smarter/> (2019).

Chapter 6: The Molecular Farm

386. A green and pleasant land, *Country Lovers*, <http://www.countrylovers.co.uk/places/histland.htm> (2011).
387. Slater, G., How the English people became landless, *Who Owns the World*, <http://homepage.ntlworld.com/janusg/landls.htm> (1913).
388. History of food, *Johns Hopkins Center for a Liveable Future*, <http://www.jhsph.edu/research/centers-and-institutes/teaching-the-food-system/curriculum/_pdf/History_of_Food-Background.pdf> (2008).
389. Levitt, T., US-style intensive farming isn't the solution to China's meat problem, *The Guardian*, <http://www.theguardian.com/environment/blog/2014/mar/03/us-intensive-farming-chinas-meat-problem> (2014).

390. Ledford, H., CRISPR conundrum: Strict European court ruling leaves food-testing labs without a plan, <https://www.nature.com/articles/d41586-019-02162-x> (2019).
391. Williams, Z., Twenty-five years of the gastropub: the revolution that saved British boozers, *The Guardian*, <http://www.theguardian.com/lifeandstyle/2016/jan/27/25-years-gastropub-revolution-saved-british-boozers-eagle-10-best> (2016).
392. Odland, S., Why are food prices so high? *Forbes*, <http://www.forbes.com/sites/steveodland/2012/03/15/why-are-food-prices-so-high/#2715e4857a0b6c0ef6634575> (2012).
393. Butler, P., Britain in nutrition recession as food prices rise and incomes shrink, *The Guardian*, <http://www.theguardian.com/society/2012/nov/18/breadline-britain-nutritional-recession-austerity> (2012).
394. Firger, J., Is cheap food to blame for the obesity epidemic? *CBS News*, <http://www.cbsnews.com/news/is-cheap-food-to-blame-for-the-obesity-epidemic/> (2014).
395. Briney, A., Green revolution, *About Education*, <http://geography.about.com/od/globalproblemsandissues/a/greenrevolution.htm> (2015).
396. Could gene editing help eradicate world hunger? *Futurism*, <http://futurism.com/could-gene-editing-help-eradicate-world-hunger/> (2015).
397. Rotman, D., why we will need genetically modified foods, *MIT Technology Review*, <http://www.technologyreview.com/featuredstory/522596/why-we-will-need-genetically-modified-foods/> (2013).
398. Garber, K., How global warming will hurt crops, *U.S. News*, <http://www.usnews.com/news/articles/2008/05/28/how-global-warming-will-hurt-crops> (2008).
399. Bawden, T. and Wright, O., Exclusive: the agricultural revolution—UK pushes Europe to embrace GM crops, *The Independent*, <http://www.independent.co.uk/news/uk/politics/exclusive-the-agricultural-revolution--uk-pushes-europe-to-embrace-gm-crops-8654595.html> (2013).
400. Why antibiotic resistance genes? *GMO Compass*, <http://www.gmo-compass.org/eng/safety/human_health/45.antibiotic_resistance_genes_transgenic_plants.html> (2015).
401. Bortesi, L. and Fischer, R., The CRISPR/Cas9 system for plant genome editing and beyond. *Biotechnology Advances* 33: 41–52 (2015).
402. Regalado, A., DuPont predicts CRISPR plants on dinner plates in five years, *MIT Technology Review*, <http://www.technologyreview.com/news/542311/dupont-predicts-crispr-plants-on-dinner-plates-in-five-years/> (2015).
403. Cyranoski, D., CRISPR tweak may help gene-edited crops bypass biosafety regulation, *Nature News*, <http://www.nature.com/news/crispr-tweak-may-help-gene-edited-crops-bypass-biosafety-regulation-1.18590> (2015).
404. Irish potato famine, *The History Place*, <http://www.historyplace.com/worldhistory/famine/begins.htm> (2000).
405. Donnelly, J., The Irish famine, *BBC History*, <http://www.bbc.co.uk/history/british/victorians/famine_01.shtml> (2011).

406. GM potato 'immune to blight', *The Telegraph*, <http://www.telegraph.co.uk/news/science/10643226/GM-potato-immune-to-blight.html> (2014).
407. Maralit, A., Banana extinction is in the horizon once more, *Food World News*, <http://www.foodworldnews.com/articles/44617/20151016/banana-cultivar-gros-michel-cavendish-panama-disease-tropical-race-4-tr4-banana-the-fate-of-the-fruit-that-changed-the-world-dan-koeppel-international-institute-of-tropical-agriculture.htm> (2015).
408. Talbot, D., Chinese researchers stop wheat disease with gene editing, *MIT Technology Review*, <http://www.technologyreview.com/news/529181/chinese-researchers-stop-wheat-disease-with-gene-editing/> (2015).
409. CRISPR cripples plant viruses. *Nature* 526: 8–9 (2015).
410. Zhang, D., Li, Z. and Li, J.-F., Genome editing: new antiviral weapon for plants. *Nature Plants* 1, 15146 (2015).
411. Baggaley, K., Restoring crop genes to wild form may make plants more resilient, *Science News*, <https://www.sciencenews.org/article/restoring-crop-genes-wild-form-may-make-plants-more-resilient> (2014).
412. Regalado, A., A potato made with gene editing, *MIT Technology Review*, <http://www.technologyreview.com/news/536756/a-potato-made-with-gene-editing/> (2015).
413. Ward, A., Progress in peanut allergy trials raises hopes, *Financial Times*, <http://www.ft.com/cms/s/0/4c4bedaa-18da-11e5-a130-2e7db721f996.html#axzz3pCw2Kd3G> (2015).
414. Novella, S., CRISPR and a hypoallergenic peanut, *Neurologica*, <http://theness.com/neurologicablog/index.php/crispr-and-a-hypoallergenic-peanut/> (2015).
415. Pollack, A., That fresh look, genetically buffed, *New York Times*, <http://www.nytimes.com/2012/07/13/business/growers-fret-over-a-new-apple-that-wont-turn-brown.html?pagewanted=1&_r=2&adxnnl=1&adxnnlx=1389618142-STd7jyKAZK9XNVtqjTIrSA> (2012).
416. Nagamangala Kanchiswamy, C., Sargent, D. J., Velasco, R., Maffei, M. E. and Malnoy, M., Looking forward to genetically edited fruit crops. *Trends in Biotechnology* 33: 62–4 (2015).
417. Pollack, A., By 'editing' plant genes, companies avoid regulation, *New York Times*, <http://www.nytimes.com/2015/01/02/business/energy-environment/a-gray-area-in-regulation-of-genetically-modified-crops.html> (2015).
418. Araki, M., Bioethicist calls for tighter regulation of non transgenic gene edited crops, *Genetic Literacy Project*, <http://www.geneticliteracyproject.org/2015/03/02/bioethicist-calls-for-tighter-regulation-of-non-transgenic-gene-edited-crops/> (2015).
419. Taylor, V., NOAA confirms 2010s were hottest decade on record, Mic, <https://www.mic.com/p/noaa-confirms-2010s-were-hottest-decade-on-record-20517672> (2020).
420. Nuccitelli, D., Global warming deniers are an endangered species, *The Guardian*, <http://www.theguardian.com/environment/climate-consensus-97-per-cent/2015/jul/22/global-warming-deniers-are-an-endangered-species> (2015).

421. Milman, O., James Hansen, father of climate change awareness, calls Paris talks 'a fraud', *The Guardian*, <http://www.theguardian.com/environment/2015/dec/12/james-hansen-climate-change-paris-talks-fraud> (2015).
422. Abraham, J., More evidence that global warming is intensifying extreme weather, *The Guardian*, <http://www.theguardian.com/environment/climate-consensus-97-per-cent/2015/jul/01/more-evidence-that-global-warming-is-intensifying-extreme-weather> (2015).
423. Lopez-Arredondo, D., Gonzalez-Morales, S. I., Bello-Bello, E., Alejo-Jacuinde, G. and Herrera, L., Engineering food crops to grow in harsh environments. *F1000Res* 4: 651 (2015).
424. Gowik, U. and Westhoff, P., The path from C3 to C4 photosynthesis. *Plant Physiologist* 155: 56–63 (2011).
425. Akst, J., Designer livestock, *The Scientist*, <http://www.the-scientist.com/?articles.view/articleNo/40081/title/Designer-Livestock/> (2014).
426. Ledford, H., Salmon approval heralds rethink of transgenic animals, *Nature News*, 527: 417–18 (2015).
427. Borrell, B., Why won't the government let you eat superfish? *Bloomberg Business*, <http://www.bloomberg.com/bw/articles/2014-05-22/aquadvantage-gm-salmon-are-slow-to-win-fda-approval> (2014).
428. Regalado, A., On the horns of the GMO dilemma, *MIT Technology Review*, <http://www.technologyreview.com/featuredstory/530416/on-the-horns-of-the-gmo-dilemma/> (2014).
429. Sanchez-Vizcaino, J. M., Mur, L., Gomez-Villamandos, J. C. and Carrasco, L., An update on the epidemiology and pathology of African swine fever. *Journal of Comparative Pathology* 152: 9–21 (2015).
430. Hunt, N., China reshapes global meat markets as swine fever rages, Reuters, <https://www.reuters.com/article/us-global-meat-asf/china-reshapes-global-meat-markets-as-swine-fever-rages-idUSKBN1XI0OT> (2019).
431. Devlin, H., Could these piglets become Britain's first commercially viable GM animals? *The Guardian*, <http://www.theguardian.com/science/2015/jun/23/could-these-piglets-become-britains-first-commercially-viable-gm-animals> (2015).
432. Zonca, C., New gene editing technology helps to beef up livestock nutrition, *ABC Rural*, <http://www.abc.net.au/news/2015-08-17/new-gene-editing-technology-helps-to-beef-up-livestock-nutrition/6703166> (2015).
433. Rogers, E., Drought-affected north Queensland farmers receive thousands of dollars through crowdfunding, *ABC Rural*, <http://www.abc.net.au/news/2015-12-03/drought-stricken-north-queensland-farmers-turn-to-crowdfunding/6998544> (2015).
434. Ortiz, E., First genetically edited cows arrive at UC Davis, *Center for Genetics and Society*, <http://www.geneticsandsociety.org/article.php?id=9062> (2015).
435. Cyranoski, D., Super-muscly pigs created by small genetic tweak, *Nature News*, <http://www.nature.com/news/super-muscly-pigs-created-by-small-genetic-tweak-1.17874> (2015).

Chapter 7: New Gene Therapy

436. Hogerzeil, H. V. and Mirza, Z., The world medicines situation, *World Health Organization*, <http://apps.who.int/medicinedocs/documents/s18772en/s18772en.pdf> (2011).
437. Burggren, W. W., Christoffels, V. M., Crossley, D. A., Enok, S., Farrell, A. P., Hedrick, M. S., Hicks, J. W., Jensen, B., Moorman, A. F., Mueller, C. A., Skovgaard, N., Taylor, E. W. and Wang, T., Comparative cardiovascular physiology: future trends, opportunities and challenges. *Acta Physiologica (Oxford)* 210: 257–76 (2014).
438. Sastry, A., Biggest obstacles to decent health care in the developing world are managerial, *Huffington Post*, <http://www.huffingtonpost.com/anjali-sastry/biggest-obstacles-to-dece_b_3659050.html> (2013).
439. Moore, A., Tracking down Martin Luther King, Jr's words on health care, *Huffington Post*, <http://www.huffingtonpost.com/amanda-moore/martin-luther-king-health-care_b_2506393.html> (2013).
440. Fact file on health inequities, *World Health Organization*, <http://www.who.int/sdhconference/background/news/facts/en/> (2015).
441. Bingham, J., Middle classes being robbed of eight years of active life, *The Telegraph*, <http://www.telegraph.co.uk/news/health/news/11854793/Middle-classes-being-robbed-of-eight-years-of-active-life.html> (2015).
442. Physical side effects, *American Cancer Society*, <http://www.cancer.org/treatment/treatmentsandsideeffects/physicalsideeffects/physical-side-effects-landing> (2015).
443. Devlin, H., Scientists find first drug that appears to slow Alzheimer's disease, *The Guardian*, <http://www.theguardian.com/science/2015/jul/22/scientists-find-first-drug-slow-alzheimers-disease> (2015).
444. McCarthy, M., Resistance to antibiotics is 'ticking time bomb': stark warning from Chief Medical Officer Dame Sally Davies, *The Independent*, <http://www.independent.co.uk/news/science/resistance-to-antibiotics-is-ticking-time-bomb--stark-warning-from-chief-medical-officer-dame-sally-davies-8528469.html> (2013).
445. Doudna, J. A. and Charpentier, E., Genome editing: the new frontier of genome engineering with CRISPR-Cas9. *Science* 346: 1258096 (2014).
446. Williams, S. C. and Deisseroth, K. Optogenetics. *Proceedings National Academy Sciences USA* 110: 16287 (2013).
447. Chial, H., Rare genetic disorders: learning about genetic disease through gene mapping, SNPs, and microarray data. *Nature Education* 1: 192 (2008).
448. Zhang, X., Exome sequencing greatly expedites the progressive research of Mendelian diseases. *Frontiers in Medicine* 8: 42–57 (2014).
449. Chong, J. X., Buckingham, K. J., Jhangiani, S. N., Boehm, C., Sobreira, N., Smith, J. D., Harrell, T. M., McMillin, M. J., Wiszniewski, W., Gambin, T., Coban Akdemir, Z. H., Doheny, K., Scott, A. F., Avramopoulos, D., Chakravarti, A., Hoover-Fong, J., Mathews, D., Witmer, P. D., Ling, H., Hetrick, K., Watkins, L., Patterson, K. E., Reinier, F., Blue, E., Muzny, D., Kircher, M., Bilguvar, K., López-Giráldez, F., Sutton, V. R., Tabor, H. K., Leal, S. M., Gunel, M., Mane, S., Gibbs, R. A., Boerwinkle, E., Hamosh, A.,

Shendure, J., Lupski, J. R., Lifton, R. P., Valle, D., Nickerson, D. A., Centers for Mendelian Genomics and Bamshad, M. J., The genetic basis of mendelian phenotypes: discoveries, challenges, and opportunities. *American Journal of Human Genetics* 97: 199–215 (2015).

450. Weatherall, D. J., Scope and limitations of gene therapy. *British Medical Bulletin* 51: 1–11 (1995).
451. Grens, K., CRISPR for cures? *The Scientist*, <http://www.the-scientist.com/?articles.view/articleNo/38561/title/CRISPR-for-Cures-/> (2013).
452. Trafton, A., Erasing a genetic mutation, *MIT News Office*, <http://newsoffice.mit.edu/2014/erasing-genetic-mutation> (2014).
453. Connor, S., Scientists 'edit' DNA to correct adult genes and cure diseases: new technique alters life-threatening mutations with pinpoint accuracy, *Belfast Telegraph*, <http://www.belfasttelegraph.co.uk/news/health/scientists-edit-dna-to-correct-adult-genes-and-cure-diseases-new-technique-alters-lifethreatening-mutations-with-pinpoint-accuracy-30205746.html> (2014).
454. Huntington's disease, *NHS Choices*, <http://www.nhs.uk/conditions/huntingtons-disease/pages/introduction.aspx> (2015).
455. Thomson, E. A., Huntington's disease gene is found, *MIT News*, <http://news.mit.edu/1993/huntington-0331> (1993).
456. Raven, C., Charlotte Raven: should I take my own life? *The Guardian*, <http://www.theguardian.com/society/2010/jan/16/charlotte-raven-should-i-take-my-own-life> (2010).
457. Armitage, H., Gene-editing method halts production of brain-destroying proteins, *Science News*, <http://news.sciencemag.org/biology/2015/10/gene-editing-method-halts-production-brain-destroying-proteins> (2015).
458. Taussig, N., Our beautiful sons could die before us, *The Guardian*, <http://www.theguardian.com/lifeandstyle/2014/aug/16/our-beautiful-sons-could-die-before-us> (2014).
459. Walsh, F., Gene editing treats disease in mice, BBC, <https://www.bbc.co.uk/news/health-35205954> (2016).
460. Cohen, J., Gene editing of dogs offers hope for treating human muscular dystrophy, Science, <https://www.sciencemag.org/news/2018/08/gene-editing-dogs-offers-hope-treating-human-muscular-dystrophy> (2018).
461. Genes and cancer, *American Cancer Society*, <http://www.cancer.org/acs/groups/cid/documents/webcontent/002550-pdf.pdf> (2014).
462. Washington University School of Medicine, DNA of 50 breast cancer patients decoded, *Science Newsline*, <http://www.sciencenewsline.com/articles/2011040313000012.html> (2011).
463. Jamieson, N. B., Chang, D. K. and Biankin, A. V., Cancer genetics and implications for clinical management. *Surgical Clinics of North America* 95: 919–34 (2015).
464. Walter and Eliza Hall Institute, New genome-editing technology to help treat blood cancers, *Science Daily*, <http://www.sciencedaily.com/releases/2015/03/150312202211.htm> (2015).

465. Fox, M., New gene-editing technique treats baby's leukemia, *NBC News*, <http://www.nbcnews.com/health/cancer/it-gone-gene-editing-technique-may-have-cured-babys-leukemia-n458786> (2015).
466. Sample, I., Baby girl is first in the world to be treated with 'designer immune cells', *The Guardian*, <http://www.theguardian.com/science/2015/nov/05/baby-girl-is-first-in-the-world-to-be-treated-with-designer-immune-cells> (2015).
467. Cyranoski, D., Chinese scientists to pioneer first human CRISPR trial, Scientific American, <https://www.scientificamerican.com/article/chinese-scientists-to-pioneer-first-human-crispr-trial/> (2016).
468. Gelernter, J., Genetics of complex traits in psychiatry. *Biological Psychiatry* 77: 36–42 (2015).
469. Kerner, B., Toward a deeper understanding of the genetics of bipolar disorder. *Frontiers in Psychiatry* 6: 105 (2015).
470. Harrison, P. J., Recent genetic findings in schizophrenia and their therapeutic relevance. *Journal of Psychopharmacology* 29: 85–96 (2015).
471. Di Forti, M., Iyegbe, C., Sallis, H., Kolliakou, A., Falcone, M. A., Paparelli, A., Sirianni, M., La Cascia, C., Stilo, S. A., Marques, T. R., Handley, R., Mondelli, V., Dazzan, P., Pariante, C., David, A. S., Morgan, C., Powell, J. and Murray, R. M., Confirmation that the AKT1 (rs2494732) genotype influences the risk of psychosis in cannabis users. *Biological Psychiatry* 72: 811–16 (2012).
472. Moore, T. H., Zammit, S., Lingford-Hughes, A., Barnes, T. R., Jones, P. B., Burke, M. and Lewis, G., Cannabis use and risk of psychotic or affective mental health outcomes: a systematic review. *Lancet* 370: 319–28 (2007).
473. Vincent, J., Gene editing could make pig organs suitable for human transplant one day, *The Verge*, <http://www.theverge.com/2015/10/14/9529493/pig-transplant-gene-editing> (2015).
474. Cox, D. B., Platt, R. J. and Zhang, F., Therapeutic genome editing: prospects and challenges. *Nature Medicine* 21: 121–31 (2015).
475. The Global HIV/AIDS Epidemic, HIV.gov, <https://www.hiv.gov/hiv-basics/overview/data-and-trends/global-statistics> (2020).
476. Cossins, D., How HIV destroys immune cells, *The Scientist*, <http://www.the-scientist.com/?articles.view/articleNo/38739/title/How-HIV-Destroys-Immune-Cells/> (2013).
477. HIV treatment, *Terrence Higgins Trust*, <http://www.tht.org.uk/myhiv/HIV-and-you/Your-treatment/HIV-treatment> (2015).
478. Roxby, P., 'Medical triumph' of prolonging HIV positive lives, *BBC News*, <http://www.bbc.co.uk/news/health-13794889> (2011).
479. Looney, D., Ma, A. and Johns, S., HIV therapy—the state of art. *Current Topics in Microbiology and Immunology* 389: 1–29 (2015).
480. Becker, Y., The molecular mechanism of human resistance to HIV-1 infection in persistently infected individuals: a review, hypothesis and implications. *Virus Genes* 31: 113–19 (2005).

481. Colen, B. D., A promising strategy against HIV, *Harvard Gazette*, <http://news.harvard.edu/gazette/story/2014/11/a-promising-strategy-against-hiv/> (2014).
482. Zaugg, J. and Wang, S., Chinese scientists use CRISPR tool on HIV patient for the first time, <https://edition.cnn.com/2019/09/13/health/china-crispr-hiv-intl-hnk/index.html> (2019).
483. Reardon, S., Gene edits to 'CRISPR babies' might have shortened their life expectancy, Nature <https://www.nature.com/articles/d41586-019-01739-w> (2019).
484. Grens, K., Genome editing cuts out HIV, *The Scientist*, <http://www.the-scientist.com/?articles.view/articleNo/40531/title/Genome-Editing-Cuts-Out-HIV/> (2014).
485. HIV/AIDS and hepatitis C (HCV), *Positive Help*, <http://www.positivehelpedinburgh.co.uk/hiv-hcv/> (2016).
486. Appleby, T. C., Perry, J. K., Murakami, E., Barauskas, O., Feng, J., Cho, A., Fox, D., Wetmore, D. R., McGrath, M. E., Ray, A. S., Sofia, M. J., Swaminathan, S. and Edwards, T. E., Viral replication: structural basis for RNA replication by the hepatitis C virus polymerase. *Science* 347: 771–5 (2015).
487. Keener, A. B., Combatting viruses with RNA-targeted CRISPR, *The Scientist*, <http://www.the-scientist.com/?articles.view/articleNo/42827/title/Combatting-Viruses-with-RNA-Targeted-CRISPR/> (2015).
488. Carroll, J., Better than RNAi? Emory team modifies CRISPR-Cas9 tech for viral infections, *Fierce Biotech Research*, <http://www.fiercebiotechresearch.com/story/better-rnai-emory-team-modifies-crispr-cas9-tech-viral-infections/2015-04-29> (2015).
489. Powers, J. H., Phoenix, J. A. and Zuckerman, D. M., Antibiotic uses and challenges: a comprehensive review from NRCWF, *Medscape*, <http://www.medscape.com/viewarticle/723457> (2010).
490. Asociación RUVID, Effects of antibiotics on gut flora analyzed, *Science Daily*, <http://www.sciencedaily.com/releases/2013/01/130109081145.htm> (2013).
491. North Carolina State University, Antibiotic 'smart bomb' can target specific strains of bacteria, *Science Daily*, <http://www.sciencedaily.com/releases/2014/01/140130110953.htm> (2014).
492. Rood, J., CRISPR chain reaction, *The Scientist*, <http://www.the-scientist.com/?articles.view/articleNo/42504/title/CRISPR-Chain-Reaction/> (2015).
493. Boyle, A., Gene method makes mutants more easily, and sparks concerns, *NBC News*, <http://www.nbcnews.com/science/science-news/gene-method-makes-mutants-more-easily-sparks-concerns-n326831> (2015).
494. Sample, I., Anti-malarial mosquitoes created using controversial genetic technology, *The Guardian*, <http://www.theguardian.com/science/2015/nov/23/anti-malarial-mosquitoes-created-using-controversial-genetic-technology> (2015).
495. Osborne, H., Mosquitoes genetically modified to be infertile in bid to reduce spread of malaria, *International Business Times*, <http://www.ibtimes.co.uk/mosquitoes-genetically-modified-be-infertile-bid-reduce-spread-malaria-1532178> (2015).

496. Scudellari, M., Self-destructing mosquitoes and sterilized rodents: the promise of gene drives, Nature, <https://www.nature.com/articles/d41586-019-02087-5> (2019).
497. Associated Press, Boom in gene-editing studies amid ethics debate over its use, *KRQE News 13*, <http://krqe.com/2015/10/18/boom-in-gene-editing-studies-amid-ethics-debate-over-its-use/> (2015).
498. Kamimura, K., Suda, T., Zhang, G. and Liu, D., Advances in gene delivery systems. *Pharmaceutical Medicine* 25: 293–306 (2011).
499. Fischer, A., Hacein-Bey-Abina, S. and Cavazzana-Calvo, M., Gene therapy of primary T cell immunodeficiencies. Gene 525: 170–3 (2013).
500. Robert Rosen, started research foundation after being diagnosed with rare blood cancer, dies, <https://www.chicagotribune.com/news/obituaries/ct-met-robert-rosen-obituary-20180124-story.html> (2018).
501. Begley, S., Advances in gene editing, and hype, underlie Editas move to go public, *STAT*, <http://www.statnews.com/2016/01/05/advances-gene-editing-editas/> (2016).
502. Chen, X. and Goncalves, M. A., Engineered Viruses as Genome Editing Devices. *Molecular Therapeutics* (2015).
503. LaFountaine, J. S., Fathe, K. and Smyth, H. D. Delivery and therapeutic applications of gene editing technologies ZFNs, TALENs, and CRISPR/Cas9. *International Journal of Pharmaceutics* 494: 180–94 (2015).
504. Cyranoski, D and Reardon, S., Chinese Scientists Genetically Modify Human Embryos, Scientific American,<https://www.scientificamerican.com/article/chinese-scientists-genetically-modify-human-embryos/> (2015).
505. Hollingsworth, J. and Yee, I., Chinese scientist who edited genes of twin babies is jailed for 3 years, CNN, <https://edition.cnn.com/2019/12/30/china/gene-scientist-china-intl-hnk/index.html> (2019).
506. Johnson, A. D., Richardson, E., Bachvarova, R. F. and Crother, B. I., Evolution of the germ line soma relationship in vertebrate embryos. *Reproduction* 141: 291–300 (2011).
507. Stanford, P. K., August Weismann's theory of the germ-plasm and the problem of unconceived alternatives. *History and Philosophy of the Life Sciences* 27: 163–99 (2005).
508. Sabour, D. and Scholer, H. R., Reprogramming and the mammalian germline: the Weismann barrier revisited. *Current Opinion in Cell Biology* 24: 716–23 (2012).
509. Chapman, K. M., Medrano, G. A., Jaichander, P., Chaudhary, J., Waits, A. E., Nobrega, M. A., Hotaling, J. M., Ober, C. and Hamra, F. K., Targeted germline modifications in rats using CRISPR/Cas9 and spermatogonial stem cells. *Cell Reports* 10: 1828–35 (2015).
510. Stouffs, K., Seneca, S. and Lissens, W., Genetic causes of male infertility. *Annales d'endocrinologie (Paris)* 75: 109–11 (2014).
511. Takehashi, M., Kanatsu-Shinohara, M. and Shinohara, T., Generation of genetically modified animals using spermatogonial stem cells. *Development Growth and Differentiation* 52: 303–10 (2010).

Chapter 8: Regenerating Life

512. Taub, R., Liver regeneration: from myth to mechanism. *Nature Reviews. Molecular Cell Biology* 5: 836–47 (2004).
513. About the liver, *Liver Directory*, <http://www.liverdirectory.com/the-liver/about-the-liver/> (2015).
514. Saraf, S. and Parihar, R., Sushruta: The first plastic surgeon in 600 B.C. *Internet Journal of Plastic Surgery* 4.2: 1–7 (2006).
515. Appelbaum, F. R., Hematopoietic-cell transplantation at 50. *New England Journal of Medicine* 357: 1472–5 (2007).
516. Office of the Director, 'Father of bone marrow transplantation', Dr E. Donnall Thomas, dies, *National Institutes of Health*, <https://www.nhlbi.nih.gov/about/directorscorner/messages/father-bone-marrow-transplantation-dr-e-donnall-thomas-dies> (2012).
517. Moore, K. A. and Lemischka, I. R., Stem cells and their niches. *Science* 311: 1880–5 (2006).
518. Wade, N., From one genome, many types of cells. But how? *New York Times*, <http://www.nytimes.com/2009/02/24/science/24chromatin.html?pagewanted=all&_r=> (2009).
519. Davidson, K. C., Mason, E. A. and Pera, M. F., The pluripotent state in mouse and human. *Development* 142: 3090–9 (2015).
520. Yu, J. and Thomson, J. A., Pluripotent stem cell lines. *Genes and Development* 22: 1987–97 (2008).
521. Lanner, F., Lineage specification in the early mouse embryo. *Experimental Cell Research* 321: 32–9 (2014).
522. Takashima, Y. and Suzuki, A., Regulation of organogenesis and stem cell properties by T-box transcription factors. *Cellular and Molecular Life Sciences* 70: 3929–45 (2013).
523. Mallo, M., Wellik, D. M. and Deschamps, J., Hox genes and regional patterning of the vertebrate body plan. *Journal of Developmental Biology* 344: 7–15 (2010).
524. Schroeder, I. S., Stem cells: are we ready for therapy? *Methods in Molecular Biology* 1213: 3–21 (2014).
525. Li, M. and Ikehara, S., Stem cell treatment for type 1 diabetes. *Frontiers in Cell and Developmental Biology* 2: 9 (2014).
526. What is type 1 diabetes? *Diabetes UK*, <https://www.diabetes.org.uk/Guide-to-diabetes/What-is-diabetes/What-is-Type-1-diabetes/> (2016).
527. Stem-cell breakthrough in treatment of diabetes, *Harvard Magazine*, <http://harvardmagazine.com/2014/10/melton-creates-beta-cells> (2014).
528. Weintraub, K., The Quest To Cure Diabetes: From Insulin To The Body's Own Cells, CommonHealth, <https://www.wbur.org/commonhealth/2019/06/27/future-innovation-diabetes-drugs> (2019).
529. Adelson, J. W. and Weinberg, J. K., The California stem cell initiative: persuasion, politics, and public science. *American Journal of Public Health* 100: 446–51 (2010).

530. Doherty, K., Regulation of stem cell research in Germany, *Euro Stem Cell*, <http://www.eurostemcell.org/regulations/regulation-stem-cell-research-germany> (2012).
531. Taylor, C. J., Bolton, E. M. and Bradley, J. A., Immunological considerations for embryonic and induced pluripotent stem cell banking. *Philosophical Transactions of the Royal Society of London. Series B Biological Sciences* 366: 2312–22 (2011).
532. McKie, R., Scientists clone adult sheep, *The Observer*, <http://www.theguardian.com/uk/1997/feb/23/robinmckie.theobserver> (1997).
533. Gurdon, J., Nuclear reprogramming in eggs. *Nature Medicine* 15: 1141–4 (2009).
534. Yoshimura, Y., Bioethical aspects of regenerative and reproductive medicine. *Human Cell* 19: 83–6 (2006).
535. Maffioletti, S. M., Gerli, M. F., Ragazzi, M., Dastidar, S., Benedetti, S., Loperfido, M., VandenDriessche, T., Chuah, M. K. and Tedesco, F. S., Efficient derivation and inducible differentiation of expandable skeletal myogenic cells from human ES and patient-specific iPS cells. *Nature Protocols* 10: 941–58 (2015).
536. Westphal, S. P., Cloned human embryos are stem cell breakthrough, *New Scientist*, <https://www.newscientist.com/article/dn4667-cloned-human-embryos-are-stem-cell-breakthrough/> (2004).
537. Baer, D., The amazing rise, fall, and rise again of Korea's 'king of cloning', *Business Insider*, <http://uk.businessinsider.com/the-amazing-rise-fall-and-rise-again-of-koreas-king-of-cloning-2015-9> (2015).
538. Mandavilli, A., Profile: Woo-Suk Hwang. *Nature Medicine* 11: 464 (2005).
539. Cyranoski, D., Human stem cells created by cloning. *Nature* 497: 295–6 (2013).
540. Baker, M., Stem cells made by cloning adult humans, *Nature News*, <http://www.nature.com/news/stem-cells-made-by-cloning-adult-humans-1.15107> (2014).
541. Mallet, K., GUMC researchers show adult human testes cells can become embryonic stem-like, capable of treating disease, *George University*, <http://explore.georgetown.edu/news/?ID=40657> (2009).
542. Takahashi, K. and Yamanaka, S., Induction of pluripotent stem cells from mouse embryonic and adult fibroblast cultures by defined factors. *Cell* 126: 663–76 (2006).
543. Gallagher, J., Gurdon and Yamanaka share Nobel Prize for stem cell work, *BBC News*, <http://www.bbc.co.uk/news/health-19869673> (2012).
544. Nemes, C., Varga, E., Polgar, Z., Klincumhom, N., Pirity M. K. and Dinnyes, A., Generation of mouse induced pluripotent stem cells by protein transduction. *Tissue Engineering Part C Methods* 20: 383–92 (2014).
545. Sample, I., Simple way to make stem cells in half an hour hailed as major discovery, *The Guardian*, <http://www.theguardian.com/science/2014/jan/29/make-stem-cells-major-discovery-acid-technique> (2014).
546. Rasko, J. and Power, C., What pushes scientists to lie? The disturbing but familiar story of Haruko Obokata, *The Guardian*, <http://www.theguardian.com/science/2015/feb/18/haruko-obokata-stap-cells-controversy-scientists-lie> (2015).

547. Curtis, M., Cracking the code of pancreatic development: beta cells from iPS cells, *Signals*, <http://www.signalsblog.ca/cracking-the-code-of-pancreatic-development-beta-cells-from-ips-cells/> (2015).
548. Kelland, K., Scientists create human liver from stem cells, *Reuters*, <http://uk.reuters.com/article/2013/07/04/us-liver-stemcells-idUSBRE9620Y120130704> (2013).
549. Gray, R., Tiny livers grown from stem cells could repair damaged organs, *The Telegraph*, <http://www.telegraph.co.uk/news/science/science-news/10157885/Tiny-livers-grown-from-stem-cells-could-repair-damaged-organs.html> (2013).
550. Willyard, C., The boom in mini stomachs, brains, breasts, kidneys and more. *Nature* 523: 520–2 (2015).
551. Baragona, S., Scientists create brain-like blobs in test tubes, *Voice of America*, <http://www.voanews.com/content/scientists-creat-brain-like-blobs-in-test-tubes/1738975.html> (2013).
552. Megraw, T. L., Sharkey, J. T. and Nowakowski, R. S., Cdk5rap2 exposes the centrosomal root of microcephaly syndromes. *Trends in Cell Biology* 21: 470–80 (2011).
553. Patterson, T., Human 'mini brains' grown in labs may help solve cancer, autism, Alzheimer's, *CNN*, <http://edition.cnn.com/2015/10/06/health/pioneers-brain-organoids/> (2015).
554. Thomson, H., First almost fully-formed human brain grown in lab, researchers claim, *The Guardian*, <http://www.theguardian.com/science/2015/aug/18/first-almost-fully-formed-human-brain-grown-in-lab-researchers-claim> (2015).
555. Williams, R., Mini brains model autism, *The Scientist*, <http://www.the-scientist.com/?articles.view/articleNo/43523/title/Mini-Brains-Model-Autism/> (2015).
556. Tenenbaum, D., Expert: editing stem cell genes will 'revolutionize' biomedical research, *University of Wisconsin-Madison*, <http://news.wisc.edu/23872> (2015).
557. Lewis, R., CRISPR Meets iPS: technologies converge to tackle sickle cell disease, *PLOS Blogs*, <http://blogs.plos.org/dnascience/2015/03/12/crispr-meets-ips-technologies-converge-tackle-sickle-cell-disease/> (2015).
558. Johns Hopkins Medicine, Custom blood cells engineered by researchers, *Science Daily*, <http://www.sciencedaily.com/releases/2015/03/150310123016.htm> (2015).
559. Repairing the genetic mutation caused by Duchenne muscular dystrophy, *Hema-Care*, <http://www.hemacare.com/blog/index.php/repairing-mutation-duchenne-muscular-dystrophy/> (2015).
560. Neuroscientists probe CRISPR transgenics and treatment paradigms, *Alzforum*, <http://www.alzforum.org/news/research-news/neuroscientists-probe-crispr-transgenics-and-treatment-paradigms> (2014).
561. Gene editing of human stem cells will 'revolutionize' biomedical research, *University of Wisconsin-Madison*, <http://www.med.wisc.edu/news-events/gene-editing-of-human-stem-cells-will-revolutionize-biomedical-research/46052> (2015).
562. Ledford, H., CRISPR, the disruptor, *Nature News*, <http://www.nature.com/news/crispr-the-disruptor-1.17673> (2015).

Chapter 9: Life as a Machine

563. Huelva Province: Rio Tinto Mines, *Andalucia.com*, <http://www.andalucia.com/province/huelva/riotinto/home.htm> (2015).
564. Dooley, T., In the time of the shootings, *Info Ayamonte*, <http://www.infoayamonte.com/index.php/the-snug/snug-articles> (2015).
565. Bluck, J., NASA scientists to drill for new, exotic life near acidic Spanish river, NASA, <http://www.nasa.gov/centers/ames/news/releases/2003/03_24AR_prt.htm> (2003).
566. Raddadi, N., Cherif, A., Daffonchio, D., Neifar, M. and Fava, F., Biotechnological applications of extremophiles, extremozymes and extremolytes. *Applied Microbiology and Biotechnology* 99: 7907–13 (2015).
567. Monash University, Antarctic life: highly diverse, unusually structured, *Science Daily*, <http://www.sciencedaily.com/releases/2015/06/150625091157.htm> (2015).
568. O'Callaghan, J., The deepest-ever sign of life on Earth? Evidence of organisms that lived 12 MILES under the crust 100 million years ago discovered, *Daily Mail*, <http://www.dailymail.co.uk/sciencetech/article-2810884/The-deepest-sign-life-Earth-Evidence-organisms-lived-12-MILES-crust-100-million-years-ago-discovered.html> (2014).
569. Hadhazy, A., Life might thrive 12 miles beneath Earth's surface, *Mother Nature Network*, <http://www.mnn.com/earth-matters/wilderness-resources/stories/life-might-thrive-12-miles-beneath-earths-surface> (2015).
570. Sjøgren, K., Live bacteria found deep below the seabed, *Science Nordic*, <http://sciencenordic.com/live-bacteria-found-deep-below-seabed> (2013).
571. Mullis, K., The Nobel Prize in Chemistry 1993, *Nobel Foundation*, <http://www.nobelprize.org/nobel_prizes/chemistry/laureates/1993/mullis-lecture.html> (1993).
572. Kary B. Mullis, *Encyclopaedia Britannica*, <http://www.britannica.com/biography/Kary-B-Mullis> (2015).
573. Farber, C., Interview Kary Mullis, *Spin*, <http://www.virusmyth.com/aids/hiv/cfmullis.htm> (1994).
574. Gonzalez, R. T., 10 famous geniuses and their drugs of choice, *Salon*, <http://www.salon.com/2013/08/16/10_famous_geniuses_who_used_drugs_and_were_better_off_for_it_partner/> (2013).
575. McClean, P., Polymerase chain reaction (or PCR), *Cloning and Molecular Analysis of Genes*, <https://www.ndsu.edu/pubweb/~mcclean/plsc431/cloning/clone9.htm> (1997).
576. Finnegan, G., Heat, salt, pressure, acidity: how 'extremophile' bacteria are yielding exotic enzymes, *Horizon*, <http://horizon-magazine.eu/article/heat-salt-pressure-acidity-how-extremophile-bacteria-are-yielding-exotic-enzymes_en.html> (2015).
577. Antibiotics, *NHS Choices*, <http://www.nhs.uk/conditions/Antibiotics-penicillins/pages/introduction.aspx> (2015).
578. Connor, S., Teixobactin discovery: scientists create first new antibiotic in 30 years—and say it could be the key to beating superbug resistance, *The Independent*, <http://

www.independent.co.uk/life-style/health-and-families/health-news/first-new-antibiotic-in-30-years-could-be-key-to-beating-superbug-resistance-9963585.html> (2015).

579. Sample, I., Craig Venter creates synthetic life form, *The Guardian*, <http://www.theguardian.com/science/2010/may/20/craig-venter-synthetic-life-form> (2010).
580. Giuliani, A., Licata, I., Modonesi, C. M. and Crosignani, P., What is artificial about life? *ScientificWorldJournal* 11: 670–3 (2011).
581. Coghlan, A., Craig Venter close to creating synthetic life, *New Scientist*, <https://www.newscientist.com/article/dn23266-craig-venter-close-to-creating-synthetic-life/> (2013).
582. Fikes, B. J., Life's core functions identified, *San Diego Union-Tribune*, <http://www.sandiegouniontribune.com/news/2015/aug/10/minimum-life-functions-palsson/> (2015).
583. Spector, D., How beer created civilization, *Business Insider*, <http://www.businessinsider.com/how-beer-led-to-the-domestication-of-grain-2013-12?IR=T> (2013).
584. Callaway, E., First synthetic yeast chromosome revealed, *Nature News*, <http://www.nature.com/news/first-synthetic-yeast-chromosome-revealed-1.14941> (2014).
585. Walsh, B., In the future we won't edit genomes – we'll just print new ones, MIT Technology Review, <https://www.technologyreview.com/s/610180/why-redesigning-the-humble-yeast-could-kick-off-the-next-industrial-revolution/> (2018).
586. Boh, S., Cooking up new ways to create food and medicines, *Straits Times*, <http://www.straitstimes.com/singapore/cooking-up-new-ways-to-create-food-and-medicines> (2015).
587. Ellis, T., Adie, T. and Baldwin, G. S., DNA assembly for synthetic biology: from parts to pathways and beyond. *Integrative Biology (Cambridge)* 3: 109–18 (2011).
588. Webb, S., Digging designer genomes, *Biotechniques* 59: 113– 17 (2015).
589. An institution for the do-it-yourself biologist, *DIY Bio*, <http://diybio.org/> (2016).
590. Dunne, C., Crazy bio-hacks: a mouse cloned from Elvis's DNA and a human-born dolphin, *Fast Company*, <http://www.fastcodesign.com/3020880/crazy-bio-hacks-a-mouse-cloned-from-elviss-dna-and-a-human-born-dolphin> (2013).
591. Biohacking: Democratization of Science or Just a Quirky Hobby? Labiotech, <https://www.labiotech.eu/features/biohacking-democratisation-science-hobby/> (2018).
592. Chamary, J. V., Welcome to gene club: underground genome editing, *BBC Focus Magazine*, <http://www.sciencefocus.com/feature/biohacking/welcome-gene-club-underground-genome-editing> (2016).
593. Krieger, L. M., Bay Area biologist's gene-editing kit lets do-it-yourselfers play God at the kitchen table, *San Jose Mercury News*, <http://www.mercurynews.com/science/ci_29372452/bay-area-biologists-gene-editing-kit-lets-do> (2016).
594. Biotechnology in the public interest, *BioBricks Foundation*, <http://biobricks.org/> (2016).
595. Brown, K.V., 'Stop Stabbing Yourself,' a Biohacker Tells His Daredevil Peers, Bloomeberg, <https://www.bloomberg.com/news/articles/2019-10-11/genetic-biohackers-are-finally-going-mainstream> (2019).

596. Ralston, A. and Shaw, K., Reading the genetic code. *Nature Education* 1: 120 (2008).
597. Samhita, L., Recoding life, *The Scientist*, <http://www.the-scientist.com/?articles.view/articleNo/38761/title/Recoding-Life/> (2014).
598. Zakaib, G. D., Genomes edited to free up codons, *Nature News*, <http://www.nature.com/news/2011/110714/full/news.2011.419.html> (2011).
599. Geddes, L., Reprogrammed bacterium speaks new language of life, *New Scientist*, <https://www.newscientist.com/article/mg22029402-800-reprogrammed-bacterium-speaks-new-language-of-life/> (2013).
600. Kwok, R., Chemical biology: DNA's new alphabet, *Nature News*, <http://www.nature.com/news/chemical-biology-dna-s-new-alphabet-1.11863> (2012).
601. Alexander, B., Synthetic life seeks work, *MIT Technology Review*, <http://www.technologyreview.com/news/540701/synthetic-life-seeks-work/> (2014).
602. Synthorx Inc., Synthorx advances its synthetic DNA technology to make its first full-length proteins incorporating novel amino acids, *PR Newswire*, <http://www.prnewswire.com/news-releases/synthorx-advances-its-synthetic-dna-technology-to-make-its-first-full-length-proteins-incorporating-novel-amino-acids-300130352.html> (2015).
603. Biello, D., New life made with custom safeguards, *Scientific American*, <http://www.scientificamerican.com/article/new-life-made-with-custom-safeguards/> (2015).
604. Gray, R., Do we really need DNA to form life? Breakthrough in synthetic enzymes could lead to the manufacture of organisms, *Daily Mail*, <http://www.dailymail.co.uk/sciencetech/article-2857172/Do-need-DNA-form-life-Breakthrough-synthetic-enzymes-lead-manufacture-organisms.html> (2014).
605. McLean, K., DNA robots: the medicinal revolution? *The Gist*, <http://the-gist.org/2015/09/dna-robots-the-medicinal revolution/> (2015).
606. Boyle, R., Bacteria can quickly swap genes with each other through a global network, *Popular Science*,<http://www.popsci.com/science/article/2011-11/bacteria-swap-gene-information-through-global-network> (2011).
607. Servick, K., Genome writing project aims to rally scientists around virus-proofing cells, Science, <https://www.sciencemag.org/news/2018/05/genome-writing-project-aims-rally-scientists-around-virus-proofing-cells> (2018).

Chapter 10: A Redesigned Planet?

608. Valencia, K., Creating Spiderman, *I, Science*, <http://www.isciencemag.co.uk/features/creating-spiderman/> (2012).
609. Konda, K., The origin of 'with great power comes great responsibility' and 7 other surprising parts of Spiderman's comic book history, *We Minored in Film*, <http://weminoredinfilm.com/2014/04/22/the-origin-of-with-great-power-comes-great-responsibility-7-other-surprising-parts-of-spider-mans-comic-book-history/> (2014).

610. Kreider, T., Our greatest political novelist? *New Yorker*, <http://www.newyorker.com/books/page-turner/our-greatest-political-novelist> (2013).
611. Findlay, A., Life after the Star Wars expanded universe: Kim Stanley Robinson's Mars Trilogy, *Reading at Recess*, <http://readingatrecess.com/2014/01/27/life-after-the-star-wars-expanded-universe-kim-stanley-robinson-mars-trilogy/> (2014).
612. Walter, N., Pigoons might fly, *The Guardian*, <http://www.theguardian.com/books/2003/may/10/bookerprize2003.bookerprize> (2003).
613. Findlay, A., Life after the Star Wars expanded universe: Margaret Atwood's Maddaddam Trilogy, *Reading at Recess*, <http://readingatrecess.com/2014/01/24/margaret-atwoods-maddaddam-trilogy/> (2014).
614. Yi, Y., Noh, M. J. and Lee, K. H., Current advances in retroviral gene therapy. *Current Gene Therapy* 11: 218–28 (2011).
615. Maggio, I., Holkers, M., Liu, J., Janssen, J. M., Chen, X. and Gonçalves, M. A., Adenoviral vector delivery of RNA-guided CRISPR/Cas9 nuclease complexes induces targeted mutagenesis in a diverse array of human cells. *Science Reports* 4: 5105 (2014).
616. Rizzuti, M., Nizzardo, M., Zanetta, C., Ramirez, A.and Corti, S., Therapeutic applications of the cell-penetrating HIV-1 Tat peptide. *Drug Discovery Today* 20: 76–85 (2015).
617. Kromhout, W. W., UCLA study shows cell-penetrating peptides for drug delivery act like a Swiss Army knife, *UCLA Newsroom*, <http://newsroom.ucla.edu/releases/ucla-engineering-study-shows-how-216290> (2011).
618. Gan, Y., Jing, Z., Stetler, R. A. and Cao, G., Gene delivery with viral vectors for cerebrovascular diseases. *Frontiers in Bioscience (Elite Edition)* 5: 188–203 (2013).
619. Shendure, J. and Akey, J. M., The origins, determinants, and consequences of human mutations. *Science* 349: 1478–83 (2015).
620. Breast cancer study: 50 women, 1700 genetic mutations, *Sci Tech Story*, <http://scitechstory.com/2011/04/05/breast-cancer-study-50-women-1700-genetic-mutations/> (2011).
621. Tyrrell, K. A., Navigating multiple myeloma with 'Google Maps' for the cancer genome, *University of Wisconsin*, <http://news.wisc.edu/23827> (2015).
622. Rahman, N., Realizing the promise of cancer predisposition genes. *Nature* 505: 302–8 (2014).
623. Sir Richard Doll: A life's research, *BBC News*, <http://news.bbc.co.uk/1/hi/health/3826939.stm> (2004).
624. Gayle, D., How some smokers stay healthy: genetic factors revealed, *The Guardian*, <http://www.theguardian.com/society/2015/sep/28/how-some-smokers-stay-healthy-genetic-factors-revealed> (2015).
625. Metabolic syndrome, *NHS Choices*, <http://www.nhs.uk/Conditions/metabolic-syndrome/Pages/Introduction.aspx> (2015).
626. Haggett, A., *Desperate Housewives, Neuroses and the Domestic Environment, 1945–1970* (Routledge, 2015), p. 109.

627. Pinto, R., Ashworth, M. and Jones, R., Schizophrenia in black Caribbeans living in the UK: an exploration of underlying causes of the high incidence rate. *British Journal of General Practice* 58: 429–34 (2008).
628. Edwards, S. L., Beesley, J., French, J. D. and Dunning, A. M., Beyond GWASs: illuminating the dark road from association to function. *American Journal of Human Genetics* 93: 779–97 (2013).
629. Singh, S., Kumar, A., Agarwal, S., Phadke, S. R. and Jaiswal, Y., Genetic insight of schizophrenia: past and future perspectives. *Gene* 535: 97–100 (2014).
630. Ledford, H., First robust genetic links to depression emerge, *Nature News*, <http://www.nature.com/news/first-robust-genetic-links-to-depression-emerge-1.17979> (2015).
631. CONVERGE consortium, Sparse whole-genome sequencing identifies two loci for major depressive disorder. *Nature* 523: 588–91 (2015).
632. Keener, A. B., Genetic variants linked to depression, *The Scientist*, <http://www.the-scientist.com/?articles.view/articleNo/43557/title/Genetic-Variants-Linked-to-Depression/> (2015).
633. Harmsen, M. G., Hermens, R. P., Prins, J. B., Hoogerbrugge, N. and de Hullu, J. A., How medical choices influence quality of life of women carrying a BRCA mutation. *Critical Reviews in Oncology/Hematology* 96: 555–68 (2015).
634. Wittersheim, M., Buttner, R. and Markiefka, B., Genotype/phenotype correlations in patients with hereditary breast cancer. *Breast Care (Basel)* 10: 22–6 (2015).
635. Webb, J., Switching on happy memories 'perks up' stressed mice, *BBC News*, <http://www.bbc.co.uk/news/science-environment-33169548> (2015).
636. Sutherland, S., Revolutionary neuroscience technique slated for human clinical trials, *Scientific American*, <http://www.scientificamerican.com/article/revolutionary-neuroscience-technique-slated-for-human-clinical-trials/> (2016).
637. Campbell, D., Recession causes surge in mental health problems, *The Guardian*, <http://www.theguardian.com/society/2010/apr/01/recession-surge-mental-health-problems> (2010).
638. O'Neill, B., Five things that *Brave New World* got terrifyingly right, *The Telegraph*, <http://blogs.telegraph.co.uk/news/brendanoneill2/100247159/five-things-that-brave-new-world-got-terrifyingly-right/> (2013).
639. Cooper, D. K., Ekser, B. and Tector, A. J., A brief history of clinical xenotransplantation. *International Journal of Surgery* 23: 205–10 (2015).
640. Palomo, A. B., Lucas, M., Dilley, R. J., McLenachan, S., Chen, F. K., Requena, J., Sal, M. F., Lucas, A., Alvarez, I., Jaraquemada, D. and Edel, M. J., The power and the promise of cell reprogramming: personalized autologous body organ and cell transplantation. *Journal of Clinical Medicine* 3: 373–87 (2014).
641. Dennett, D., Where am I?, *New Banner*, <http://www.newbanner.com/SecHumSCM/WhereAmI.html> (1978).
642. Kempermann, G., Song, H. and Gage, F. H., Neurogenesis in the adult hippocampus. *Cold Spring Harbor Perspectives in Medicine* 5: a018812 (2015).
643. Newborn neurons help us adapt to environment, *Business Standard*, <http://www.business-standard.com/article/news-ians/newborn-neurons-help-us-adapt-to-environment-115022300534_1.html> (2015).

644. 2015 Alzheimer's disease facts and figures, *Alzheimer's Association*, <http://alz.org/facts/overview.asp?utm_source=gdn&utm_medium=display&utm_content=topics&utm_campaign=ff-gg&s_src=ff-gg&gclid=Cj0KEQiArJe1BRDe_uz1uu-QjvYBEiQACUj60hgp09jdxoFm1zRohjNDbihofMMk2YMEZveNpX9LjxYaArK98P8HAQ> (2015).
645. Mullin, E., Are there different types of Alzheimer's disease? *Forbes*, <http://www.forbes.com/sites/emilymullin/2015/09/30/are-there-different-types-of-alzheimers-disease/> (2015).
646. Cell Transplantation Center of Excellence for Aging and Brain Repair, Stem cells found to play restorative role when affecting brain signaling process, *Science Newsline*, <http://www.sciencenewsline.com/summary/2014060519050019.html> (2014).
647. Atlasi, Y., Looijenga, L. and Fodde, R., Cancer stem cells, pluripotency, and cellular heterogeneity: a WNTer perspective. *Current Topics in Developmental Biology* 107: 373–404 (2014).
648. Agence France-Presse, Stem cells grow beating heart, *Discovery News*, <http://news.discovery.com/tech/biotechnology/stem-cells-grow-beating-heart-130814.htm> (2013).
649. Cyranoski, D., Stem cells: Egg engineers, *Nature News*, <http://www.nature.com/news/stem-cells-egg-engineers-1.13582> (2013).
650. Sample, I., Scientists use skin cells to create artificial sperm and eggs, *The Guardian*, <http://www.theguardian.com/society/2014/dec/24/science-skin-cells-create-artificial-sperm-eggs> (2014).
651. Top 10 old guys who fathered kids, *Shark Guys*, <http://www.thesharkguys.com/lists/top-10-old-guys-who-fathered-kids/> (2010).
652. 'Limit' to lab egg and sperm use, *BBC News*, <http://news.bbc.co.uk/1/hi/health/7346535.stm> (2008).
653. Sample, I., GM embryos: time for ethics debate, say scientists, *The Guardian*, <http://www.theguardian.com/science/2015/sep/01/editing-embryo-dna-genome-major-research-funders-ethics-debate> (2015).
654. Reuters, Genetically modified human embryos should be allowed, expert group says, *The Guardian*, <http://www.theguardian.com/science/2015/sep/10/genetically-modified-human-embryos-should-be-allowed-expert-group-says> (2015).
655. Vogel, G., Scientists use gene-editing technology to knock out genes in human embryos for first time, Science, <https://www.sciencemag.org/news/2017/09/scientists-use-gene-editing-technology-knock-out-genes-human-embryos-first-time> (2017).
656. Researcher Who Led Team That Genetically Edited Babies Sentenced to Prison in China Time, <https://time.com/5756656/he-jiankui-sentenced-gene-editing-babies/> (2019).
657. Stern, H. J., Preimplantation genetic diagnosis: prenatal testing for embryos finally achieving its potential. *Journal of Clinical Medicine* 3: 280–309 (2014).

658. Rochman, B., Family with a risk of cancer tries to change its destiny, *Wall Street Journal*, <http://www.wsj.com/articles/SB10001424052702304703804579379211430859016> (2014).
659. Regalado, A., Engineering the perfect baby, *MIT Technology Review*, <http://www.technologyreview.com/featuredstory/535661/engineering-the-perfect-baby/> (2015).
660. What is intra-cytoplasmic sperm injection (ICSI) and how does it work?, *Human Fertilisation and Embryology Authority*, <http://www.hfea.gov.uk/ICSI.html> (2015).
661. What causes male infertility? *Stanford University*, <https://web.stanford.edu/class/siw198q/websites/reprotech/New%20Ways%20of%20Making%20Babies/causemal.htm> (2015).
662. Lander, E. S., Brave new genome, *New England Journal of Medicine* 373: 5–8 (2015).
663. Knapton, S., Humans will be 'irrevocably altered' by genetic editing, warn scientists ahead of summit, *The Telegraph*, <http://www.telegraph.co.uk/news/science/science-news/12025316/Humans-will-be-irrevocably-altered-by-genetic-editing-warn-scientists-ahead-of-summit.html> (2015).
664. Plucker, J., The Cyril Burt affair, *Human Intelligence*, <http://www.intelltheory.com/burtaffair.shtml> (2013).
665. Sailer, S., Nature vs. nurture: two pairs of identical twins interchanged at birth, *Unz Review: An Alternative Media Selection*, <http://www.unz.com/isteve/nature-vs-nurture-two-pairs-of-identical-twins-switched-at-birth/> (2015).
666. Kamin, L. J., *The Science and Politics of I.Q.* (Routledge, 1974), p. 50.
667. Callaway, E., 'Smart genes' prove elusive, *Nature News*, <http://www.nature.com/news/smart-genes-prove-elusive-1.15858> (2014).
668. Sample, I., New study claims to find genetic link between creativity and mental illness, *The Guardian*, <http://www.theguardian.com/science/2015/jun/08/new-study-claims-to-find-genetic-link-between-creativity-and mental-illness> (2015).
669. Connor, S., Scientists find that schizophrenia and bipolar disorder are linked to creativity, *The Independent*, <http://www.independent.co.uk/news/science/scientists-find-that-schizophrenia-and-bipolar-disorder-are-linked-to-creativity-10305708.html> (2015).
670. Biography of Wolfgang Amadeus Mozart, *Wolfgang Amadeus*, <http://www.wolfgang-amadeus.at/en/biography_of_Mozart.php> (2016).
671. Behrman, S., Mozart: musical beauty in an age of revolution, *Socialist Worker*, <http://socialistworker.co.uk/art/7930/Mozart%3A+musical+beauty+in+an+age+of+revolution> (2006).
672. Wilde, R., The location of Mozart's grave, *About Education*, <http://europeanhistory.about.com/od/famouspeople/a/dyk11.htm> (2015).
673. Schultz, O. and Rivard, L., Case study in genetic testing for sports ability, *Genetics Generation*, <http://www.nature.com/scitable/forums/genetics-generation/case-study-in-genetic-testing-for-sports-107403644> (2013).
674. Scott, M. and Kelso, P., One club wants to use a gene-test to spot the new Ronaldo: is this football's future? *The Guardian*, <http://www.theguardian.com/football/2008/apr/26/genetics> (2008).

675. Prince-Wright, J., Cristiano Ronaldo's extra ankle bone helps him score stunners…seriously? *NBC Sports*, <http://soccer.nbcsports.com/2014/05/27/cristiano-ronaldos-extra-ankle-bone-helps-him-score-stunners-seriously/> (2014).
676. Fenn, A., Cristiano Ronaldo: Real Madrid star's journey to the Ballon d'Or, *BBC Sport*, <http://www.bbc.co.uk/sport/0/football/25719657> (2014).
677. Edgley, R., The sports science behind Lionel Messi's amazing dribbling ability, *Bleacher Report*, <http://bleacherreport.com/articles/2375473-the-sports-science-behind-lionel-messis-amazing-dribbling-ability> (2015).
678. Chase, C., How Einstein saw the world, *Creative by Nature*, <https://creativesystems thinking.wordpress.com/2014/02/16/how-einstein-saw-the-world/> (2014).
679. Lloyd, R., Charles Darwin: strange and little-known facts, *Live Science*, <http://www.livescience.com/3307-charles-darwin-strange-facts.html> (2009).
680. Jane Gray, *Darwin Correspondence Project*, <https://www.darwinproject.ac.uk/jane-gray> (2015).
681. All aboard the *Beagle! Christ's College, Cambridge*, <http://darwin200.christs.cam.ac.uk/node/19> (2015).
682. Einstein at the patent office, *Swiss Federal Institute of Intellectual Property*, <https://www.ige.ch/en/about-us/einstein/einstein-at-the-patent-office.html> (2011).
683. Human Fertilisation and Embryology Authority, <http://www.hfea.gov.uk/> (2015).
684. Connor, S., IVF embryos to be genetically manipulated as scientists investigate repeated miscarriages, The Independent, <https://www.independent.co.uk/news/science/ivf-embryos-to-be-genetically-manipulated-as-scientists-investigate-repeated-miscarriages-10506064.html> (2015).
685. National Institutes of Health Guidelines on Human Stem Cell Research, *National Institutes of Health*, <http://stemcells.nih.gov/policy/pages/2009guidelines.aspx> (2009).
686. Reardon, S., NIH reiterates ban on editing human embryo DNA, *Nature News*, <http://www.nature.com/news/nih-reiterates-ban-on-editing-human-embryo-dna-1.17452> (2015).
687. Drainie, B., Oryx and Crake, *Quill and Quire*, <http://www.quillandquire.com/review/oryx-and-crake/> (2003).
688. Hylton, W. S., How ready are we for bioterrorism? *New York Times*, <http://www.nytimes.com/2011/10/30/magazine/how-ready-are-we-for-bioterrorism.html?_r=0> (2011).
689. Flight, C., Silent weapon: smallpox and biological warfare, *BBC History*, <http://www.bbc.co.uk/history/worldwars/coldwar/pox_weapon_01.shtml> (2011).
690. Harding, A., The 9 deadliest viruses on Earth, *Live Science*, <http://www.livescience.com/48386-deadliest-viruses-on-earth.html> (2014).
691. HIV transmission, *Centers for Disease Control and Prevention*, <http://www.cdc.gov/hiv/basics/transmission.html> (2015).
692. Ebola virus disease, *World Health Organization*, <http://www.who.int/mediacentre/factsheets/fs103/en/> (2015).

693. Maron, D. F., Weaponized Ebola: is it really a bioterror threat? *Scientific American*, <http://www.scientificamerican.com/article/weaponized-ebola-is-it-really-a-bioterror-threat/> (2014).
694. Fyffe, S., U.S. needs a new approach for governance of risky research, Stanford scholars say, *Stanford University*, <http://news.stanford.edu/news/2015/december/biosecurity-research-risks-121715.html> (2015).
695. Stewart, S., Evaluating Ebola as a biological weapon, *Stratfor*, <https://www.stratfor.com/weekly/evaluating-ebola-biological-weapon> (2014).
696. Mills, N., The anthrax scare: not a germ of truth, *The Guardian*, <http://www.theguardian.com/commentisfree/cifamerica/2011/sep/15/anthrax-iraq> (2011).
697. Pigoon, *Technovelgy*, <http://www.technovelgy.com/ct/content.asp?Bnum=1177> (2015).
698. Smith, R. H., Margaret Atwood: life without certainty, *Be Thinking*, <http://www.bethinking.org/culture/margaret-atwood-life-without-certainty> (2012).
699. Francis, A., Need a lung? Humanized pig organs for transplant will be available in the near future, *Tech Times*, <http://www.techtimes.com/articles/6644/20140508/need-a-lung-humanized-pig-organs-for-transplant-will-be-available-in-the-near-future.htm> (2014).
700. Devlin, H., First human-pig 'chimera' created in milestone study, The Guardian, <https://www.theguardian.com/science/2017/jan/26/first-human-pig-chimera-created-in-milestone-study> (2017).
701. Hayasaki, E., Better Living Through Crispr: Growing Human Organs in Pigs, Wired, <https://www.wired.com/story/belmonte-crispr-human-animal-hybrid-organs/> (2019).
702. Home Office, UK statistics, *Speaking of Research*, <http://speakingofresearch.com/facts/uk-statistics/> (2013).
703. Izpisua Belmonte, J. C., Callaway, E. M., Caddick, S. J., Churchland, P., Feng, G., Homanics, G. E., Lee, K. F., Leopold, D. A., Miller, C. T., Mitchell, J. F., Mitalipov, S., Moutri, A. R., Movshon, J. A., Okano, H., Reynolds, J. H., Ringach, D., Sejnowski, T. J., Silva, A. C., Strick, P. L., Wu, J. and Zhang, F., Brains, genes, and primates. *Neuron* 86: 617–31 (2015).
704. Chen, J., Cao, F., Liu, L., Wang, L. and Chen, X., Genetic studies of schizophrenia: an update. *Neuroscience Bulletin* 31: 87–98 (2015).
705. Gray, J., Walking wounded: our often barbaric struggle to cure mental illness, *New Statesman*, <http://www.newstatesman.com/culture/books/2015/04/walking-wounded-our-often-barbaric-struggle-cure-mental-illness> (2015).
706. Boly, M., Seth, A. K., Wilke, M., Ingmundson, P., Baars, B., Laureys, S., Edelman, D. B. and Tsuchiya, N., Consciousness in humans and non-human animals: recent advances and future directions. *Frontiers in Psychology* 4: 625 (2013).
707. Graham, S. A. and Fisher, S. E., Understanding language from a genomic perspective. *Annual Review of Genetics* 49: 131–60 (2015).
708. French, C. A. and Fisher, S. E., What can mice tell us about Foxp2 function? *Current Opinion in Neurobiology* 28: 72–9 (2014).

709. Samuel, S., Vox, Scientists added human brain genes to monkeys. Yes, it's as scary as it sounds. <https://www.vox.com/future-perfect/2019/4/12/18306867/china-genetics-monkey-brain-intelligence> (2019).
710. Yang, C., Ontogeny and phylogeny of language. *Proceedings National Academy Sciences USA* 110: 6324–7 (2013).
711. Konopka, G. and Roberts, T. F., Animal models of speech and vocal communication deficits associated with psychiatric disorders. *Biological Psychiatry* 79: 53–61 (2016).
712. Malynn, D., Film review: *Rise of the Planet of the Apes*, *Bio News*, <http://www.bionews.org.uk/page_104605.asp> (2011).
713. Ekshtut, S., Tuber or not tuber, *Russian Life*, <http://www.russianlife.com/pdf/potatoes.pdf> (2000).
714. Houllier, F., Biotechnology: bring more rigour to GM research. *Nature* 491: 327 (2012).
715. Ayers, C., Extinctathon: would you like to play? *Play-Extinctathon*, <play-extinctathon.tumblr.com/> (2015).
716. Owen, C., Animal welfare, *Issues Today*, <http://www.independence.co.uk/pdfs/26_animalwelfare_ch1.pdf> (2009).
717. Boseley, S. and Davidson, H., Global obesity rise puts UN goals on diet-related diseases 'beyond reach', *The Guardian*, <http://www.theguardian.com/society/2015/oct/09/obesitys-global-spread-un-goals-diet-related-diseases-fail> (2015).
718. Reardon, S., Dramatic rise seen in antibiotic use, *Nature News*, <http://www.nature.com/news/dramatic-rise-seen-in-antibiotic-use-1.18383> (2015).
719. Middlehurst, C., Can China kick its animal antibiotic habit? The Guardian, <https://www.theguardian.com/environment/2018/jun/19/can-china-kick-its-animal-antibiotic-habit> (2018).
720. Local and regional food systems, *GRACE Communications Foundation*, <http://www.sustainabletable.org/254/local-regional-food-systems> (2015).
721. Cernansky, R., The rise of Africa's super vegetables, *Nature News*, <http://www.nature.com/news/the-rise-of-africa-s-super-vegetables-1.17712> (2015).
722. McKie, R., £1,000 for a micro-pig: Chinese lab sells genetically modified pets, *The Observer*, <http://www.theguardian.com/world/2015/oct/03/micropig-animal-rights-genetics-china-pets-outrage> (2015).
723. Cyranoski, D., Gene-edited 'micropigs' to be sold as pets at Chinese institute, *Nature News*, <http://www.nature.com/news/gene-edited-micropigs-to-be-sold-as-pets-at-chinese-institute-1.18448> (2015).
724. Phillips, R., Couple loves cloned best friend, *CNN*, <http://edition.cnn.com/2009/LIVING/02/06/cloned.puppy/index.html?iref=topnews> (2009).
725. Baer, D., This Korean lab has nearly perfected dog cloning, and that's just the start, *Tech Insider*, <http://www.techinsider.io/how-woosuk-hwangs-sooam-biotech-mastered-cloning-2015-8> (2015).
726. Derbyshire, D., How genetics can create the next superstar racehorse, *The Observer*, <http://www.theguardian.com/science/2014/jun/22/horse-breeding-genetics-thoroughbreds-racing-dna> (2014).
727. Bland, A. A leap into the unknown: cloned eventing horse Tamarillo is groomed for success, *The Independent*, <http://www.independent.co.uk/news/science/a-leap-

into-the-unknown-cloned-eventing-horse-tamarillo-is-groomed-for-success-8827747.html> (2013).

728. Green, R., Red Rum, *BBC Liverpool*, <http://www.bbc.co.uk/liverpool/localhistory/journey/stars/red_rum/profile.shtml> (2014).

729. Roman, J., Woolly mammoth remains discovered in Siberia set to be cloned, *Tech Times*, <http://www.techtimes.com/articles/94121/20151012/woolly-mammoth-remains-discovered-in-siberia-set-to-be-cloned.htm> (2015).

730. Pinkstone, J., Woolly mammoths are one step closer to being brought back from the dead: Scientists restart a 28,000-year-old cell from the extinct creature and implant it inside a MOUSE, Daily Mail, <https://www.dailymail.co.uk/sciencetech/article-6798787/Mammoth-moves-frozen-cells-come-life-just.html> (2019).

731. Wu, B., Bringing extinct animals back to life no longer just part of the movies, *Science Times*, <http://www.sciencetimes.com/articles/4932/20150327/bringing-extinct-animals-back-life-longer-part-movies.htm> (2015).

732. Church, G., George Church: de-extinction is a good idea, *Scientific American*, <http://www.scientificamerican.com/article/george-church-de-extinction-is-a-good-idea/> (2013).

733. Hotz, R. L., Bone yields dinosaur DNA, scientists believe: paleontology: experts call it a historic first. But skeptics say it might be from bacterial decay instead, *Los Angeles Times*, <http://articles.latimes.com/1994-11-18/news/mn-64303_1_ancient-dna-scquence> (1994).

734. Switek, B., Scrappy fossils yield possible dinosaur blood cells, *National Geographic*, <http://phenomena.nationalgeographic.com/2015/06/16/scrappy-fossils-yield-possible-dinosaur-blood-cells/> (2015).

735. Castro, J., Archaeopteryx: the transitional fossil, *Live Science*, <http://www.livescience.com/24745-archaeopteryx.html> (2015).

736. Harris-Lovett, S., 'Jurassic World' paleontologist wants to turn a chicken into a dinosaur, *Los Angeles Times*, <http://www.latimes.com/science/sciencenow/la-sci-sn-horner-dinosaurs-20150612-story.html> (2015).

737. Hogenboom, M., Chicken grows face of a dinosaur, *BBC Earth*, <http://www.bbc.co.uk/earth/story/20150512-bird-grows-face-of-dinosaur> (2015).

738. Geggel, L., When will we see an actual dino-chicken? *Discovery News*, <http://news.discovery.com/animals/dinosaurs/when-will-we-see-a-dino-chicken-15052.htm> (2015).

739. Lachniel, M., An analysis of *Blade Runner*, *Blade Runner Insight*, <http://www.br-insight.com/an-analysis-of-blade-runner> (1998).

740. Goldenberg, S., Western Antarctic ice sheet collapse has already begun, scientists warn, *The Guardian*, <http://www.theguardian.com/environment/2014/may/12/western-antarctic-ice-sheet-collapse-has-already-begun-scientists-warn> (2014).

741. Quick facts on ice sheets, *National Snow and Ice Data Center*, <https://nsidc.org/cryosphere/quickfacts/icesheets.html> (2015).

742. Kemper, A. and Martin, R., New York, London and Mumbai: major cities face risk from sea-level rises, *The Guardian*, <http://www.theguardian.com/sustainable-business/blog/major-cities-sea-level-rises> (2013).

743. Mortimer, C., COP21: James Hansen, the father of climate change awareness, claims Paris agreement is a 'fraud', *The Independent*, <http://www.independent.co.uk/environment/cop21-father-of-climate-change-awareness-james-hansen-denounces-paris-agreement-as-a-fraud-a6771171.html> (2015).
744. Kunzig, R., Will Earth's ocean boil away? *National Geographic*, <http://news.nationalgeographic.com/news/2013/13/130729-runaway-greenhouse-global-warming-venus-ocean-climate-science/> (2015).
745. McGrath, M., Global warming increases 'food shocks' threat, *BBC News*, <http://www.bbc.co.uk/news/science-environment-33910552> (2015).
746. Carrington, D., World's climate about to enter 'uncharted territory' as it passes 1C of warming, *The Guardian*, <http://www.theguardian.com/environment/2015/nov/09/worlds-climate-about-to-enter-uncharted-territory-as-it-passes-1c-of-warming> (2015).
747. Irfan, U., The UN Climate Action Summit was a disappointment, Vox, <https://www.vox.com/2019/9/24/20880416/un-climate-action-summit-2019-greta-thunberg-trump-china-india> (2019).
748. Beil, L., Getting creative to cut methane from cows, *Science News*, <https://www.sciencenews.org/article/getting-creative-cut-methane-cows?mode=pick&context=166> (2015).
749. Youris.com, The case for low methane-emitting cattle, *Science Daily*, <http://www.sciencedaily.com/releases/2014/01/140110131013.htm> (2014).
750. Doré, L., Bill Gates says that capitalism cannot save us from climate change, *The Independent*, <http://i100.independent.co.uk/article/bill-gates-says-that-capitalism-cannot-save-us-from-climate-change--b1xNpbL8O_x> (2015).
751. Vale, P., Bill Gates dismisses free market's ability to counter climate change because the private sector is 'inept', *Huffington Post*, <http://www.huffingtonpost.co.uk/2015/11/02/bill-gates-climate-change-private-sector-inept_n_8452166.html> (2015).
752. Walsh, F., Call for $2bn global antibiotic research fund, *BBC News*, <http://www.bbc.co.uk/news/health-32701896> (2015).
753. Amelinckx, A., California passes the country's strongest regulations for antibiotic use in livestock, *Modern Farmer*, <http://modernfarmer.com/2015/10/california-antibiotic-livestock-regulations/> (2015).
754. CVM Updates, FDA Releases Annual Summary Report on Antimicrobials Sold or Distributed in 2018 for Use in Food-Producing Animals, U.S. Food and Drug Administration, <https://www.fda.gov/animal-veterinary/cvm-updates/fda-releases-annual-summary-report-antimicrobials-sold-or-distributed-2018-use-food-producing> (2019).
755. Wade, N., Gene drives offer new hope against diseases and crop pests, *New York Times*, <http://www.nytimes.com/2015/12/22/science/gene-drives-offer-new-hope-against-diseases-and-crop-pests.html> (2015).
756. McMullan, T., Hacking the brain: how technology is curing mental illness, *Alphr*, <http://www.alphr.com/science/1000875/hacking-the-brain-how-technology-is-curing-mental-illness> (2015).

INDEX OF NAMES

INDEX OF SUBJECTS

(Page numbers in italics refer to figures.)

THE DEEPER GENOME

Why there is more to the human genome than meets the eye

John Parrington

978-0-19-968874-6 | Paperback | £9.99

"*The Deeper Genome*...provides an elegant, accessible account of the profound and unexpected complexities of the human genome, and shows how many ideas developed in the 20th century are being overturned."

Clare Ainsworth, *New Scientist*

Things have changed since the early heady days of the Human Genome Project. But the emerging picture is if anything far more exciting. We are learning more about ourselves, and about the genetic aspects of many diseases. In its complexity, flexibility, and ability to respond to environmental cues, the human genome is proving to be far more subtle than we ever imagined.

ONE PLUS ONE EQUALS ONE

Symbiosis and the evolution of complex life

John Archibald

978-0-19-875812-9 Paperback £9.99

"*One Plus One Equals One* is an eloquent account, at times verging on the poetic. With serious scholarship, it illuminates a rare scientific endeavour." - **Nancy A. Moran,** *Nature*

It is natural to look at biotechnology in the 21st century with a mix of wonder and fear. But biotechnology is not as 'unnatural' as one might think. All living organisms use the same molecular processes to replicate their genetic material and the same basic code to 'read' their genes. Here, John Archibald shows how evolution has been 'plugging-and-playing' with the subcellular components of life from the very beginning, and continues to do so today. For evidence, we need look no further than the inner workings of our own cells. Molecular biology has allowed us to gaze back more than three billion years, revealing the microbial mergers and acquisitions that underpin the development of complex life.